D0151617

CONSTRUCTION ESTIMATING: An Accounting and Productivity Approach

CONSTRUCTION ESTIMATING: An Accounting and Productivity Approach

Second Edition

James J. Adrian, Ph.D., P.E., C.P.A.
Bradley University
Adrian and Associates, President

Published by
STIPES PUBLISHING COMPANY
10–12 Chester Street
Champaign, Illinois 61820

ISBN 0-87563-439-7

Copyright ©1982, 1993
James J. Adrian

Contents

Preface

There is no business skill of more importance to the construction industry than estimating. Accurate estimates are necessary to the project owner, to the design team, and to the construction team. The project owner judges the feasibility of a project on the basis of overall cost estimates. The design team prepares detailed estimates in order to determine the scope and cost of each proposed design element. The construction contractor stays in business because of accurate project estimates. They are fundamental to his existence. A good contractor estimate is the basis on which the contractor wins or loses a contract for the project, and clearly sets out the potential profit most likely to result from efficient and timely performance.

When one thinks of how to prepare a construction estimate, it is usual to view it as the task of determining quantities of work to be performed. This determination of quantities of work is commonly referred to as the taking-off of quantities. However, the preparation of an accurate estimate is more than that quantitative determination. It also entails determination of the productivity of the resources used to perform the work and determination of the cost of the resources. It follows that both accounting and the productivity analysis are disciplines that provide the alternative means essential to the accurate preparation of a comprehensive estimate.

This book has as a major objective to aid the reader in both the determination of work quantities and the determination of the cost of performing the work. Since the determination of the cost of the work is dependent on informed use of accounting and productivity principles and techniques, these topics are emphasized in each chapter. Hence the title of the book: *Construction Estimating: An Accounting and Productivity Approach.*

Chapter 1 introduces the reader to construction estimating and gives an overview of the process. Then, given the book's emphasis on relating estimating to productivity and accounting principles, Chapters 2 and 3 present these two topics. The objective of these two chapters is to give the estimator sufficient knowledge for the practical application of productivity and accounting principles to the preparation of an estimate. For that reason, no prior reader knowledge is assumed. The author carefully presents the two topics step-by-step.

While the emphasis of this book is on the preparation of a general contractor's estimate for the construction project itself, the need is also recognized for estimates prior to the construction phase. Therefore, Chapter 4 addresses the problem of the owner's feasibility estimate. Several designer estimating techniques are presented in Chapter 5. The remaining chapters focus on the preparation of the general contractor's estimate for a project.

The author believes that it is important that the estimator view the contractor construction estimate as part of other project management functions, which include project planning and scheduling, and project control. In addition, the estimator should view the complete estimate in orderly steps that include quantity take-off, costing of the work, determination of overhead costs, and the determination of an appropriate profit. Given the need to view contractor estimating in this comprehensive manner, Chapters 6 through 9 focus carefully on these topics.

Chapters 10 through 14 are the ones generally included in all estimating books. In succession they address the work common to most general contractors: sitework, concrete, masonry, carpentry, and structural steel. Each chapter sets out the principles and techniques relevant to taking-off quantities of work. In addition and uniquely, each chapter targets the factors relevant to pricing the work. Each also illustrates the use of the productivity technique that enables the estimator to properly analyze the optimal means for performing the work. This enables the estimator to select the lowest cost method as well as to determine costs by means of that method.

Chapter 15 addresses various miscellaneous types of construction work and the estimating and screening of subcontractor estimates. Thus the author helps the reader-user realize that a general contractor's estimate must also include estimates from the other contractors—the specialty contractors.

Chapters 16 and 17 concern job overhead, company overhead, and the percentage of profit to be included in the construction contractor's bid/estimate. Inadequate overhead determination is one of, if not the greatest, cause of inaccurate contractor estimates.

The reader-user will find the book unique in its combination of detailed "nuts and bolt" take-off techniques and of business theory relevant to the pricing function. The book clearly illustrates how the estimator's knowledge of accounting and of productivity analysis offer him the alternative construction methods that are essential to the estimating task.

The book should prove useful to the practitioner as well as to the construction student. For contractor practitioners the book will be timely since it complements one recently published by the American Institute of Certified Public Accountants. Called *Audit Guide for Construction Contractors*, it places increased emphasis on the contractor's implementation of the sound estimating principles that ensure compliance with audit requirements for acceptable financial statements. Moreover, increasing construction project complexities and uncertainties draw attention to the need for the accurate construction estimates advocated here in *Construction Estimating*.

For efficient use as a textbook, the author has completed each chapter with exercises that enable the student/reader to test his knowledge of the principles covered in that chapter. Finally, Chapter 18 reviews current use of computer-assisted applications of standard work codes for large estimating jobs.

Appendix A contains interest tables used in the construction industry. Appendix B lists standard codes used in computer-assisted estimating for large construction projects.

It is the author's hope that ultimately the book will aid in understanding and solving some of the difficulties associated historically with projects that exceed budget constraints. Given the practical means for controlling the cost estimate/budget through sound accounting and productivity techniques, the project owner, designer, and contractor should all benefit from accurate construction estimating.

The author is indebted to many practitioners for the knowledge accumulated from them that has helped in the preparation of this book. But most of all, he is grateful to his family for once again providing him an environment in which he can serve others through his work.

J.A.A.

CONSTRUCTION ESTIMATING: An
Accounting and Productivity Approach

1

Construction Estimating: Its Importance in the Construction Industry

*Accurate Estimating: Fundamental to Success   *
*Inaccurate Estimating: A Serious Industry Problem   *
*Estimating: More Art than Science   *   Kinds of Estimates in
the Construction Process   *   Types of Construction and
Types of Contractors   *   Relationship between Estimating,
Accounting, and Productivity*

ACCURATE ESTIMATING: FUNDAMENTAL TO SUCCESS

Estimating is a fundamental part of the construction industry. The success or failure of a project is in great part dependent on the accuracy of several major estimates, including the project feasibility estimate, all the preconstruction estimates including conceptual design estimates, the system estimates, and finally on the detailed contractor estimates. In effect, accurate estimates optimize good contracting.

Often the quality of the project design, along with the ability to start construction and complete it on schedule, is dependent on the accuracy of cost estimates made throughout the design phase of a project. Similarly, the subsequent accuracy of a contractor-prepared detailed estimate dictates the financial success of the contractor. Simply put, estimating is the nuts and bolts of the construction industry. The preparation of accurate estimates leads to the success or failure of the project—the financial profit (or loss) of the project owner, and of the project owner's team, which includes the contractors engaged.

The importance of construction estimates is highlighted by the fact that each individual entity involved in the construction process must make significant financial commitments that depend in large part on the accuracy of a relevant estimate. A project owner is dependent first upon those who prepare an accurate feasibility estimate, then upon the timely competence of the chosen design team which prepares the initial and costly construction drawings. Similarly, the project owner's further financial commitment to the preparation of more detailed and also costly architect/engineer prepared contract documents, which include working drawings and specifications, is dependent on the preparation and accuracy of the earlier designer estimates. The project or second-phase design drawings may prove to be as much as ten percent of the project's total cost. Thus, if design phase estimates are grossly inaccurate, the result could be a project that becomes unfeasible at the bidding stage with the result that either the project owner or the owner's design team have to absorb those relatively high costs expended in design document preparation.

Like the project owner, the contractor is very dependent on the competent preparation of estimates. Unlike many other kinds of manufacturers, who determine product cost by actual calculation from cost data collected as the product is built and subsequently priced, the construction contractor is commonly called upon to give the owner client a price before production costs can be actually known. (Note: The contractor is considered part of the manufacturing industry.) Compare the construction contractor to the car manufacturer who produces an automobile and measures the actual cost of producing it as it is produced. Based on this measurement of actual cost and the manufacturer's desired profit margin, the car manufacturer then establishes a sales price for the automobile. The need for accurate estimates is not as important to the manufacturer as is the manufacturer's ability to determine the market for its product.

Unlike the automobile manufacturer, the "forecast" nature of the construction industry is such that the project owner's engagement of the contractor is usually based on the contractor's "bid price" to perform future work with yet-to-be purchased materials. While it is true that occasionally a project owner may engage a contractor and pay the contractor for actual costs incurred plus an agreed "fixed" percent profit, as the costs are incurred, this is rare. (Note: this is termed a time-and-material or unit-cost contract.) More often than not, because of construction differences, the project owner must, in effect, force the contractor to commit to a price before the determination of actual costs. Thus the construction estimate is of first importance to the contractor.

INACCURATE ESTIMATING: A SERIOUS INDUSTRY PROBLEM

Notwithstanding the real-time importance of the contractor estimate, the fact remains that an estimate is by definition, a forecast of future events.

Given the numerous component parts and costs of a project, be it a project as small as a residential unit or as large as a skyscraper, it is not realistic to expect a construction estimate to be one hundred percent accurate. Because of the uncertainties of future events, construction estimating will never be totally scientific. Instead the preparation of accurate construction estimates, be it a project owner feasibility estimate, a designer estimate, or a contractor estimate, is partly a science and partly an art.

While one cannot expect a construction estimate to be one hundred percent accurate, the fact is that seriously inaccurate estimates have often plagued the construction industry. This has been especially true in recent years. In part, the increased inaccuracy of estimates in recent years has likely resulted because of widely fluctuating material costs that add to the traditional difficulties associated with predicting labor productivity.

The question of how accurate a construction estimate should be to ensure project success and financial success of the owner and contractors is relative. Obviously one can expect the degree of accuracy to vary with the amount of effort and time taken to prepare an estimate and the degree of detail in contract document preparation. As evidence of the inaccuracy of typical construction estimates, Figure 1.1 illustrates the success/failure range of different types of estimates as determined by a study of sample estimates and final costs studied by the author. The illustration compares the dollar value of the estimate relative to the actual project cost as determined from actual cost records. The ranges of accuracy/inaccuracy represent averages from the analysis of numerous estimates. Naturally a specific project owner, architect, or contractor estimate may be significantly more accurate than the ranges show, depending on the superior skills of an individual estimator and the degree and kind of uncertainties that characterize the project and are shrewdly accounted for.

While a given estimate may in fact be very accurate, it is fair to say that the degree of estimating accuracy is in general poor in the construction industry. Due in part to the ever increasing inflationary trend in manufacturing costs, estimates have, more often than not, been low relative to actual

	Range in Accuracy	Approximate Manhours
Unit cost per square foot estimate	40%	1
Trade estimate	30%	2
Factor estimate	20%	5
Parameter estimate	15%	10
Contractor-detailed estimate	7%	150

Figure 1.1 *Expected Accuracy of Estimates Assuming a $2,000,000 Project*

project costs. The result has been lengthy project delays, serious cost overruns, and on an occasion, financial failure for the contractor. It is a rare occasion when a project is found to have been built for its estimated cost.

ESTIMATING: MORE ART THAN SCIENCE

Estimating is not a process unique to the construction industry. To the contrary, estimating is practiced by the housewife, the farmer, the business manager, the military planner, and the reader. Estimating can and is practiced by many individuals. As such anyone can prepare an estimate to include a cost estimate of a construction project. The real challenge is not the preparation of an estimate, it is the preparation of a reliable estimate.

Many problems, including the construction of a building, are subject to unknown forces. Thus, the accuracy of an estimate of these unknown forces is proportional to the span of time between the estimate and the forecast event. Three broad states or conditions exist, which affect the accuracy of any estimate:

1. Events having certainty.
2. Events recognizing risk.
3. Events having uncertainty.

Certainty Factors: The simplest state of nature in any event is **certainty**. However, this simple state seldom exists, especially when considering events surrounding a construction project.

If certainty does exist, and it may, the condition enables the most reliable estimate. For example, a cost estimate would be easiest if it were based on fixed production rates and labor costs. In preparing some estimates, experience teaches us that a practicable set of assumptions often has a high degree of occurrence—that is, is in a state of certainty. This expectation is fully warranted in some cases; for example, when labor costs, production rates, and cost indexes are conspicuously stable in a given construction market—especially in the short run for short-term projects. In preparing an estimate to include a construction estimate, we often make these assumptions regarding certainty in order to expedite a workable means of analysis. The event of certainty assumes that each action undertaken results in the same outcome and has a probability of one.

Risk Factors: Situations involving the state of **risk** are appropriate whenever the estimator can obtain good estimates of the probability of future conditions. In order to determine these probabilities, the estimator likely has collected data regarding past events, and analyzed the results. Risk is defined as a condition in which each of several outcomes is assigned a probability whose sum equals one. However, many uncontrollable factors in the future, by their very nature, can at best only be anticipated. For example, the probabilities of different working conditions, including weather patterns, can only be predicted (not guaranteed) by experts. Sometimes these probabilities are entirely subjective, because actual measurement is either impossible or undesirable because it is too complex or too time consuming.

Uncertainty Factors: Unlike the disciplines of engineering and physical laws, which depend to a great extent on well-defined physical cause and effect relationships, the construction of a project is in great part dependent on the free-will actions of people. Thus, the estimator does not have the certain opportunity to command and control the circumstances that follow his estimating work. This state of **uncertainty** may better characterize the preparation of most construction estimates than one of certainty. This condition exists whenever there is uncertainty regarding the probabilities or relative values that describe a specific set of states. For these states, we have little bits and pieces of information so poorly understood that we are unable to assign any probability ranking. This is the qualification for the condition of uncertainty. It is assumed that the outcomes are identified in some detail. This is the situation often facing the estimator preparing a cost estimate of a construction project.

Given conditions of risk and uncertainty, when actual events or costs are compared to estimated events or costs, a difference is usually found. In some circumstances actual costs may never be known. Nonetheless, it is interesting to speculate on the deviation that would be revealed if they were known. This deviation would be the sum of the following:

1. Risks and uncertainty components associated with the future.
2. Mistakes in calculations.
3. Errors of belief.

Risks and uncertainty by their very nature will likely be major reasons for the deviation of an estimate from the actual cost. For example, the fact that an unexpected and unusual number of rain days occur during the construction of a project will cause a construction estimate to be in error.

Mistakes may result from imprecision and technical errors in computation or from blunders of various sorts. For example, the construction estimator may either make an error in extending math calculations, or may erroneously add a column of numbers twice. These types of mistakes may go unnoticed, or on occasion may even compensate each other. The solution to "mistakes" in preparing a construction estimate is uncompromisingly accurate arithmetic and strict attention to the orderly methods that result in faultless computation.

Errors of belief result through ignorance or inadvertence. Examples include the failure to recognize quantity-price breaks, the omission of cost items, or the overlooking of a scheduled contractual increase in direct labor cost. These types of errors create estimating deviation from actual costs that can be prevented by well-thought-out policies and practices in the estimating process.

Despite a variety of possible ways to make an estimate, resources are always restricted because time, money, and estimating staff are limited in fact. Given the uncertainties of a cost estimate, one reaches a point where the objectives must be satisfied with whatever time, money, and staff are available. The ideal would be to have the estimate coincide with the reconciled actual cost. But no procedure, or mathematical technique, or policy employed by an estimator is without its flaws or is able to guarantee perfect estimates. Although flaws in estimates may be obvious, these procedures and techniques are used for the simple reason that they are the best means available at the time. This is the environment that surrounds the construction cost estimate.

KINDS OF ESTIMATES IN THE CONSTRUCTION PROCESS

As noted previously, a construction project is characterized by the need for and preparation of several kinds of cost estimates. The degree of detail of these successive estimates and even the number that are prepared varies in part according to the type and size of the project. For example, the differing type and role of a feasibility estimate depends on whether a project is a publicly or privately funded project.

Figure 1.2 illustrates a typical series of events that occur from the inception of a project through completion of construction. The sequence of the different types of cost estimates commonly in the construction process are also illustrated in Figure 1.2.

A project starts with an owner's conception or definition of a project. It may be one of several types and may be started for varying purposes. The project may be funded by public money or by private money. Independent of and regardless of the type of project being considered, the project owner usually performs some type of feasibility estimate at this first stage. The objective of the feasibility estimate is to determine whether or not the project should be built, whatever the financing. The thrust of the project takes one of three different forms depending on the type of project.

Investment Project: For this type of project, the feasibility estimate is characterized by particular attention to project cost appreciation and tax benefits to include depreciation and favorable capital gain tax treatment.

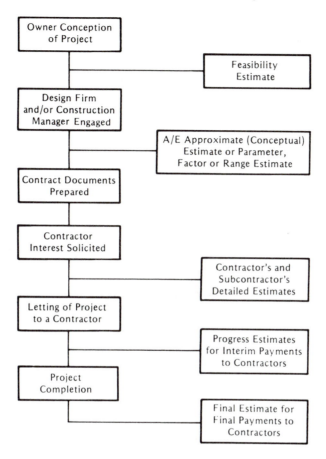

Figure 1.2 *Phases in Estimates in the Construction Process*

Project Serving a Business Function: For this type of project, the feasibility estimate focuses on documentation of financial benefits of the project; for example, the forecasted improved sales volume from a proposed new business or from expected decreases in production costs occasioned by a more efficient plant for a company.

Public Construction Project: For this type of project, the feasibility estimate focuses on a quantification of the needs of the taxpayer in regard to the use of the project.

Typically the feasibility estimate is performed by the project owner, or by an accountant hired specifically to prepare the feasibility estimate. The following lists some of the typical preparers of feasibility estimates for various types of projects.

TYPE OF PROJECT	PREPARER OF FEASIBILITY ESTIMATE
INVESTMENT PROJECT	Project owner's accountant, or a consultant, or a construction manager
BUSINESS-SERVING PROJECT	The project owner's own engineering staff
PUBLIC PROJECT	Public agency with jurisdiction over project

In addition to serving the objective of determining whether a project should be built, the feasibility estimate also serves the purpose of aiding the project owner to obtain funding for a project. This funding may come from a lending institution, from the raising of equity capital, or from the selling of bonds. Whoever the providers of the project funds, they cannot be expected to offer funds without a cost justification study—that is, for sound reasons supported by a feasibility estimate.

Assuming a project is judged feasible and funding is obtained, the next step in the process is the engagement of a design firm by the project owner, as illustrated in Figure 1.2. The architect- or engineer-based design firm starts with the development of programming, preliminary, and schematic drawings. The purpose of these drawings are to define project spacial requirements and the quality and type of construction.

During the preparation of these early phase estimates, the project design team will prepare approximate estimates of project cost. Also called parameter, factor, or range estimates, they will often be no more than a unit cost per square foot—or a functional budgetary estimate. These types of estimates are at best, "ballpark" estimates. Unfortunately, the accuracy of such an estimate is often not sufficiently supportive at this important phase of a project. In order to properly satisfy the needs of the project owner and to adequately develop the project to the project owner's cost and quality objectives, the designer should be able to conceptualize the "budget" cost of the project as a function of these initial drawings.

Usually the project design team goes through several iterations when preparing these early phase designs. Each iteration is accompanied by a cost estimate.

Because of the importance of cost estimates during the preparation of contract documents, and because some designers lack sufficient staff or experience for accurate estimating, the project owner (or the project designer) occasionally has to engage a construction consultant to prepare a cost estimate in the design phase of the project, or to verify the accuracy of the estimate prepared by the project designer. The construction consultant might be a contractor, a construction management firm, or an independent construction estimating consulting firm. The construction consultant might utilize a conceptual estimating technique in its attempt to prepare an accurate estimate. This is also illustrated in Figure 1.2.

Assuming project contract documents are completed and a budget is satisfactorily established, the project owner next invites contractors to bid on the new project. This invitation may be open to any contractor willing to bid the project, or may be a qualified or invitational-only type of contractor solicitation. Open invitations are characteristic of public construction projects, whereas qualified or invitational solicitations are characteristic of private construction projects.

Independent of and regardless of the type of contractor solicitation, the designer-prepared contract documents become the contractor's means of preparing a cost estimate and bid for the project. The major contract documents include the following 6 kinds:

1. Project drawings.   4. Special conditions.
2. Specifications.   5. Agreements.
3. General conditions.   6. Addendums.

Usually the project drawings and specifications play the most significant role in the contractor's preparation of a cost estimate for a project. The **drawings** indicate the quantity of work to be performed, whereas the **specifications** for the most part indicate the quality of the work that is to be performed.

It is at this point in time that perhaps the best known construction cost estimate is prepared—the detailed contractor estimate. See Figure 1.2 also.

Most building construction projects are built by one or more prime contractors and several subcontractors. The general contractor often subcontracts a significant portion of its work to other contractors; that is, the project subcontractors. This is especially true in regard to the specialty trades, including the electrical, mechanical, and plumbing work.

Usually the project's general contractor will request bids from three different firms before awarding a contract to a specific subcontractor. This is especially true for public construction projects, where in fact there often is a legal requirement to obtain three competitive bids for each contract to be awarded.

The competitive bidding process—coupled with the fact that a project's prime or general contractor subcontracts specialty work to several subcontractors—results in the preparation of several separate combinations of bids by the same contractors and subcontractors competing for a single project. This process is illustrated in Figure 1.3. Some experts suggest that this process is seriously inefficient.

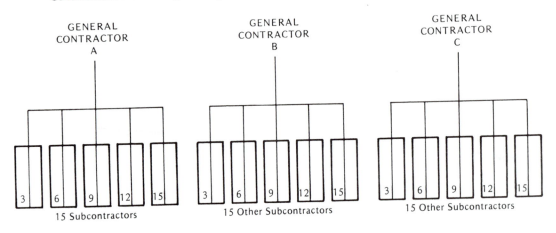

Figure 1.3   *Redundant Bidding Possible on a Large Project*

Numerous firms spend expensive manhours preparing a cost estimate and subsequently are not awarded any work. For example, if as few as three general contractors compete for a project and each in turn seeks bids from as few as five subcontractors, as many as forty-eight contractors might prepare an estimate and yet only six can be awarded the work. See Figure 1.3 also. (Note: which is a small number relative to what would likely be awarded.)

Obviously the 42 unsuccessful bidders (as shown in Figure 1.3) must pass along the costs of their unsuccessful bids to those project owner's with whom they are successful. In effect, all project owners bear the costs associated with the preparation of successful and unsuccessful bids. This, in effect, is the "insurance" that the project owner pays in order to receive low bids in the competitive bidding process.

Assuming the project owner receives competitive bids and judges one better or best qualified to perform the work than the others, that contractor is selected by the project owner who then awards the contract. This process is referred to as the **letting**.

The detailed contractor estimate is not the only cost estimate prepared during the construction of a project. Typically several different types of cost estimates will be needed during the construction phase of the project. Perhaps the most frequent construction cost estimate prepared after the detailed contractor estimate, which awards the contract, is an estimate of the cost of work related to contract change orders.

Seldom is there a construction project that does not result in one or more change order. A change order is essentially a change in the quantity or type of work to be performed relative to what was stated in the initial agreed upon contract documents. Often a change order occurs because of a change of owner preference or because of an error or omission in the contract documents.

Each time there is a change order, regardless of the reason, there is a need for the preparation of a cost estimate of the work to be performed. The estimate serves as the basis for determining the compensation to be paid by the project owner to the contractor. The occurrence of such estimates is illustrated in Figure 1.2.

During the construction of the project, either the project owner, or the architect/engineer design agent, or the contractor will prepare several construction-phase interim estimates. These are sometimes referred to as **progress estimates** and serve to establish the contractor's percentage of completion at a particular phase of the project. The calculation of the percentage of completion in turn is used to determine interim contractor billings to the project owner for the contractor's income statement. This is also illustrated in Figure 1.2.

At or near the completion of the construction on a project, it may be necessary to make a determination of the final quantities of work put in place. This is especially true if some of the quantities were initially uncertain and not set out in the contract documents. This type of estimate is often referred to as the **final estimate** because the contractor now determines the final quantities to be billed for final payment by the project owner. The occurrence of the final estimate is also illustrated in Figure 1.2.

In addition to the estimates listed in Figure 1.2, other cost estimates may be prepared during either the design or construction phase of a project. However, these "other" estimates would be the exception rather than the rule. The estimates in Figure 1.2 may also occur at a point in time other than that shown, depending on the type and scope of project and project owner.

TYPES OF CONSTRUCTION AND TYPES OF CONTRACTORS

When one talks about the preparation of a cost estimate for a construction project, one must distinguish and specify the type of project being considered and also distinguish and specify the type of contractor preparing the estimate. Construction projects and firms vary significantly, depending on the size of project and the work to be performed. While the principles of construction cost estimating discussed in this book are for the most part applicable to any type of project or contractor, the primary emphasis of subsequent chapters address the preparation of an estimate for a general contractor for building construction rather than for heavy or industrial construction.

Notwithstanding the "building construction" focus of this book on the general contractor estimate, it is useful to compare the characteristics of different types of construction projects and different types of contractors. Understanding the chief characteristics of other kinds of projects and contractors will help the estimator detail the preparation of those items specific to building construction projects.

Classification of Projects

Construction projects are often classified according to one of the following three functional descriptions:

1. Building Construction
2. Heavy & Highway (Engineering) Construction
3. Industrial Construction

Building Construction: This category describes buildings that are erected for habitational, institutional, educational, light construction, commercial, social, and recreational purposes. Both single and multi-family residential units are included in the **building construction** category. Also included are office buildings, corporate headquarter buildings, schools, commercial structures, and so forth. Building construction is generally regarded as the mainstay of the construction industry and usually contributes about one-half of the total annual dollar volume of construction. Most building construction projects are financed by private capital and are designed by private architectural or architectural/engineering firms.

Heavy and Highway Construction: Also commonly referred to as engineering construction, this category of construction includes structures that involve materials such as earth, steel, and concrete. Included in the types of projects are bridges, roadways, sewage and treatment plants, utility projects, and railroad construction.

Heavy and highway or engineering construction is primarily planned and designed by professional engineers. The contractor constructing these types of projects is very dependent on the use of high cost earth-moving equipment to perform this work. Most such projects are publicly financed.

Industrial Construction: The erection of these projects is associated with the manufacturing or processing of commercial products or industrial services. Included in these types of projects are steel mills, petroleum refineries, and electric or nuclear power generating stations. These structures are highly technical in nature and are frequently built by specialized contracting firms that do both the design and construction of the project.

Types of Contractors

There are numerous ways to classify contractors—by type of work performed, by size, and by method of contracting. Within these classifications contractors can be further identified as to organizational, legal, and financial structures.

The most commonly used method to classify contractors is by type of work performed. The following four classifications are often made as to type of contractor:

1. Residential contractor
2. General building contractor
3. Specialty contractor
4. Heavy and highway contractor

Residential Contractors: Construction work by this contractor is characterized by severe cycles in work volume. The **residential contractor** is routinely impacted by implementation of the policies of federal, state, and local governments. Primarily due to governmental changes in fiscal and monetary policies, new residential housing starts may be greater than two million units for the construction industry one year while they may be as low as one million the following year. Wide variations in housing starts from one year to the next and from one season to the next often create unique management problems for the residential contractor. The ability to balance workload and to provide for an adequate cash flow through slack periods can make the difference between a prosperous contractor and a bankrupt contractor.

Residential contractors have the advantage of being able to operate on a low volume because they normally have relatively low overhead—as opposed to the building or heavy and highway contractor. Whereas the latter must compete against large contractors with relatively sound management practices, the residential contractor competes against much smaller contractors who normally have less sophisticated management skills. This is not to say that the residential contractor does not have competition; the opposite is in fact true.

However, the similar size of average competitors and their "wholly-owned" personal management skills often allow residential contractors to work from their own homes as one-man operations. Thus, residential contractors are numerous—often leading to excessive competition, low profit margins, and considerable potential for bankruptcy. However, it is not all negative. Residential contractors are flexible, have the advantages that go with smallness and specialization and, given sound estimating methods, have great potential for both solvency and quality.

Most of the residential contractor's projects are of relatively short duration. As a result, on the one hand, the residential contractor ordinarily experiences fewer accounting problems in the area of revenue recognition than either building or heavy and highway contractors do. On the other hand, the residential contractor's strong dependence on the economy, leads to wide swings in volume and results in a high failure rate for the residential contractor.

Of all types of contractors, the residential contractor is often the one with the weakest estimating and accounting systems, and the least accurate accounting records. This lack of estimating-accounting sophistication often contributes significantly to the residential contractor's high failure rate.

The residential contractor may build projects on a contract basis or on a speculative basis. When the contractor builds on a contract basis, he commits himself to a fixed-dollar value contract with an owner who usually defines his own project. The contractor forfeits his ownership of the project as he receives payments.

An alternative to this arrangement is the **speculative builder**. The so-called "spec" residential contractor is actually the owner of the project while it is being built. The contractor may in fact be the owner *after* the project is built if he decides to rent it or has the unfortunate fate of not being able to find a renter or buyer. The advantages of the speculative builder are that he is free to choose his own site and plans, his estimate of cost is not tied to the success of the project in that he "sells" the building after the true cost is determined, and it may be easier to build in this manner because he does not have to compete for a contract. Most importantly, when the market is receptive, his profit margin may be relatively high. Whereas the contract builder may have to settle for about a 3% profit margin, the spec builder may be able to attain a 10% profit margin if he "moves" his projects after they are built. This higher profit margin for the spec builder stems from profits associated with the site on which the project is built, increased productivity due to the repetitive type of units which the contractor builds, and a more constant and controlled workload.

However, all these business advantages favoring the speculative builder are based on the assumption that he can immediately rent or sell the project once it is built. The spec builder constructs projects on his own or borrowed money and not with an outside owner's as in the case of the contract builder. As such, the spec builder ties up his money or, in the case of a loan, continues to pay interest during construction and beyond construction if a buyer is not found. The spec builder subjects himself to a high degree of risk because he continues to have a cash outflow and no cash inflow if the cost of money increases and a potential buyer or renter cannot be secured. The speculative contractor attempts to compensate for this risk with a higher profit margin than the contract builder.

General Building Contractors: These contractors handle a wide range of construction projects. These include most industrial and commercial types of buildings. Because of the larger project scale of the buildings handled by **general contractor**, individual contracts are usually let for a significantly larger dollar value than those let by a residential contractor. Examples of projects that the general building contractor builds include school buildings, office buildings, hospitals, retail stores, apartment buildings, and manufacturing plants.

More often than not, the general building contractor contracts with a project owner to take full responsibility for delivery of the construction project. While the general building contractor has total responsiblity for the time and cost of the project, he will likely subcontract with several specialty contractors to perform a significant portion of the work. When the general building contractor is engaged for total responsibility of a project, he is referred to as the general contractor or **GC** for the project. Because the general contractor has a contract with the project owner, he is also referred to as a prime contractor.

Because of the large scale nature of most projects that a general building contractor undertakes and agrees to construct, and because of the long-range business commitments of the owner, the general building contractor often has to secure various bonds from a surety company. In effect, he purchases insurance payable to those to whom he guarantees his service. Often the ability to carry out operations as a general building contractor is dependent upon the contractor's ability to secure bonds, including a bid bond and performance bond.

If the contractor is awarded the contract and then does not sign the contract or cannot provide a performance bond, the bid bond assures that the project owner will be reimbursed. This reimbursement, subject to the maximum bid bond penalty, is for the difference between the contractor's bid and the bid of the next lowest responsible bidder. The responsibility to pay this difference is the contractor's or the surety company providing the bond.

The performance bond further protects the project owner against the contractor's failure to perform the contract in accordance with its terms. If the contractor does not perform, the surety company that has provided the performance bond to the contractor can become liable to the project owner to complete the project according to the contract terms.

As a prerequisite for selling bonds to the contractor, the surety company will often require that contractor to submit financial data including audited financial statements. The financial data greatly influences the surety company's willingness to write a bond for the contractor.

The ability to generate profits as a general building contractor is very dependent on the ability to estimate and control the costs of many different types of construction work, building components, and resources including several labor crafts and numerous types of material. This emphasizes the need of the general building contractor to implement an effective estimating cost system, as well as sound project planning, scheduling, and control procedures.

Specialty Contractors: Firms in this category include those engaged in electrical, plumbing, mechanical, and other specialized types of construction. These firms are usually subcontractors and are highly dependent on labor skills and labor productivity.

In the past, the specialty contractor has typically worked as a subcontractor to the residential, general building, or less frequently, the heavy and highway contractor. However, with the recent popularity of the construction management (**CM**) process, and design-build contracts, it is becoming common for the specialty contractor to have a direct (prime) contract with the project owner for the specialty aspect of the construction project.

The specialty contractor's dependence on external parties is often dictated by the contractual relationship with the project owner (whether as a subcontractor or as a prime contractor). This relationship also affects the specialty contractor's liability and billing and collection procedures.

Heavy and Highway Contractors: The characteristics of contractors in this part of the construction industry are usually quite unlike any of the others. The **heavy and highway contractor** constructs projects that involve large earthwork and a heavy use of concrete and structural steel materials. Typical projects include roadways, bridges, dams, and occasionally large industrial projects. Usually such projects are of substantial duration.

Because of the types of projects the heavy and highway contractor builds, and the materials that he puts in place, the contractor is usually very dependent on the use of heavy-duty construction equipment. The profitability of the heavy and highway contractor is often related to his ability to manage the equipment.

The heavy and highway contractor tends to perform a larger annual volume of work than the residential or building contractor. Necessarily, the type of projects he performs are typically large dollar projects. Because of the dependence on considerable fixed assets for financing these types of large projects, it is relatively difficult for even an experienced contractor to go into business as a heavy and highway contractor. As a result, there are considerably fewer heavy and highway contractors than residential or building contractors.

Because of the large dollar value of typical heavy and highway construction projects, a considerable number of the individual projects are now joint ventures. In addition, the monetary value of projects and the heavy and highway contractor's dependence on equipment and equipment financing result in a strong dependence on external parties—including lending institutions, equipment finance companies, and surety companies. Because a high percentage of the work performed by the heavy and highway contractor is public construction, involving federal, state, and municipal projects, the heavy and highway contractor is also dependent on public agencies and sub-ject to the numerous laws and regulations governing public construction.

The project management skills needed by the heavy and highway con-tractor vary somewhat from those of either residential or building contrac-tors. For example, the quantity takeoff process for heavy and highway con-tractors is less complex than that for the building contractor. However, the pricing of the work is no less important. The heavy and highway contractor tends to be less dependent on labor crafts than the building contractor. Therefore the heavy and highway contractor subcontracts a smaller percent-age of work than the residential or building contractor.

INTERDEPENDENCE OF RELATIONSHIP BETWEEN ESTIMATING, ACCOUNTING, AND PRODUCTIVITY

The preparation of an accurate construction estimate requires several skills regardless of whether it is the project owner's feasibility estimate, the design-er estimate, or the detailed contractor estimate. While numerous hours and individual steps are required to prepare any estimate, essentially the con-struction estimate can be thought of in the following sequence:

1. Determination of the work to be performed.
2. Determination of the productivity of resources to be used for the work.
3. Determination of the cost of materials to be fabricated, and the cost of resources to be used for the work.

In regard to Step 2, "productivity" can be thought of as the amount of work put in place per unit of resource or labor.

Knowledge or practice of any one or two of the above estimating steps without the third will prove to be ineffective and more than likely will result in an inaccurate estimate. The preparation of a construction estimate, the knowledge of productivity, and the knowledge of accounting (that is, anal-ysis of costs) are interdependent. The relationship between estimating, pro-ductivity, and accounting is illustrated in Figure 1.4.

The relationship between estimating, productivity, and cost accounting will be emphasized and illustrated throughout the remaining chapters of this book. As evidence of this emphasis, Chapter 2 addresses itself totally to the topic of construction productivity, and Chapter 3 reviews the topic of cost accounting. Both these chapters are presented to enable the reader to better appreciate the emphasis on these two topics in all the following: Chapters 4 through 18.

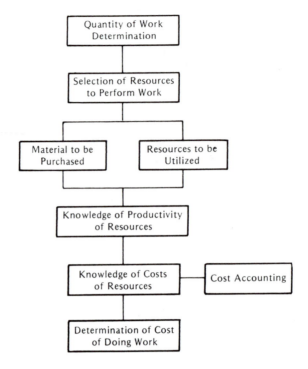

Figure 1.4 *Relationship Between Estimating, Accounting, and Productivity*

EXERCISE 1.1

Estimating is a fundamental part of the construction process. However it has been suggested that the typical process by which a project owner awards contracts to construction contractors results in excessive estimating time and cost. To illustrate estimating time and cost that may be part of a construction project, let us assume that a project owner puts out a projected $1,000,000 project for bid to ten potential general contractors. Let us assume that each of these ten firms submits a bid. However, none of the potential general contractors can perform all of the projected construction work. Instead let us assume that each general contractor plans to subcontract one-half of the million dollars worth of work to specialty contractors. Let us further assume that this subcontracted work is to be awarded through ten separate specialty contracts. In other words, the project will be built by a general contractor and ten specialty contractors. As a means of seeking lowest cost subcontractors, each general contractor solicits three subcontractor bids for each of the ten subcontracts. Let us assume that each general contractor solicits those bids from a separate group of subcontractors.

Using a general guideline that each one hundred thousand dollars of construction work requires twenty hours of time for quantity take-off, pricing, and other steps necessary to preparing an estimate, and assuming that each estimating manhour equates to a twenty-dollar cost, calculate the total contractor estimating time and cost that typifies the way project owners commonly secure construction contractors.

EXERCISE 1.2

Estimating is by definition a forecast. While it is not realistic to believe that an estimate of cost for a construction project will exactly equal the actual cost, ideally the range of variation should be minimal. The need to minimize the difference between the actual cost for a project and the estimated cost for a construction contractor's estimate is highlighted by the fact that approximately 2.5% is the average net profit margin for building contractors. In other words, after recognition of all costs, the contractor only realizes approximately 2.5 cents of profit on a dollar of construction project revenue. This profit margin is based on nationally published averages for all building contractors.

Let us assume that a contractor has kept records of the firm's estimated cost for performing 10 typical projects, has also collected on-site cost records, and has carefully determined the actual cost for constructing the projects. The data is tabulated as follows:

PROJECT	CONTRACT AMOUNT	ESTIMATED COST	ACTUAL COST
101	$1,400,000	$1,344,000	$1,300,000
102	800,000	784,000	800,000
103	2,100,000	2,037,000	2,050,000
104	950,000	902,500	900,000
105	1,850,000	1,784,500	1,800,000
106	280,000	266,000	255,000
107	900,000	873,000	850,000
108	800,000	784,000	812,000
109	1,600,000	1,536,000	1,562,000
110	600,000	588,000	599,000

Answer the following questions regarding the ten projects above.

a. How many of the projects resulted in a profit?
b. What was the individual and average net profit margin percentage for the ten projects? (*Figure the contract amount minus actual cost divided by the contract amount multiplied by 100.*)
c. What was the individual project and average project planned net profit margin percentage for the ten projects? (*Figure the contract amount minus estimated cost divided by the contract amount multiplied by 100.*)
d. For the individual projects and the average project, what was the range

of variation of the actual cost of a project from the estimated? (*Figure the estimated cost minus the actual cost divided by the estimated cost and multiplied by 100.*)

e. How many projects had an estimated cost minus actual cost variation greater than the estimated profit for the project?

f. What conclusion can one reach from the answer to Question **e**?

EXERCISE 1.3

One of the primary purposes of a construction estimate is to establish a guaranteed price that a project owner agrees to pay the construction contractor and that the contractor in turn has a right to charge for building a project. One might argue that there is no need for a contractor to prepare a detailed contractor estimate if the project owner does not solicit a lump sum guaranteed dollar amount contract, but instead advertises for a time-and-material contract or a cost-plus percentage contract. Is this true? If it is not, list specific purposes of the contractor's detailed estimate over and above establishing a guaranteed dollar contract amount.

2

Productivity Records: Ways to Determine Standards for Construction Estimating

*Productivity Differently Defined by Those Differently Impacted * Productivity and Construction Estimating * Decline of Productivity and Lack of Productivity Standards * Productivity, Estimating, and the Environment * Productivity vs. Unit Cost Standards * Accounting vs. Scientific Based Productivity Standards * Developing Scientific Based Productivity Standards * Method Productivity Delay Model * The Production Function: Modeling the Construction Process to Establish a Productivity Standard * Comparing Alternative Construction Methods*

PRODUCTIVITY DIFFERENTLY DEFINED BY THOSE DIFFERENTLY IMPACTED

The term productivity has different meanings to different people. Many individuals automatically think only of labor and/or labor unions when the term "productivity" is mentioned. Others associate capital expenditures with the term. Each of these interpretations is only partly correct.

The U.S. Department of Commerce defines productivity *as dollars of output per manhour of labor input.* Using this definition, the Department of Commerce reports on U.S. productivity annually. In order to make relative comparisons from one period to the next, these federal statisticians adjust the numerator (that is, dollars of output) for the consumer price index.

When one looks at the definition of productivity given in Figure 2.1, the implication is that there is only one way to increase productivity—through

PRODUCTIVITY = $'s of Output per Manhour of Labor Input
(Numerator Adjusted for Inflation).

Figure 2.1 *U.S. Department of Commerce Definition of Productivity, with the Dollars Adjusted Annually*

more labor effort. This is not true. There are many ways to increase productivity. For example, the following recommendations represent 8 explicit ways to increase productivity.

1. Substitution of equipment for labor force.
2. Use of more efficient equipment and tools.
3. Use of better materials.
4. Improved production management.
5. Control of an adverse environment.
6. Harder working efforts.
7. Improved training of labor force.
8. Lessened governmental regulations.

The list above does not exhaust the means available to increase productivity. Moreover, it should now be clear to the reader that there are ways to increase productivity other than by expecting labor to "work harder." For example, providing a worker with better tools may actually result in that worker being able to expend less effort while he increases his productivity. In fact, the use of the better tool may enable the worker to put out more work with the same or less work effort. Clearly, productivity is more than a labor force issue, and that is why the author of this text prefers to define productivity as the dollars of output per dollars of input. But, while this "re-definition" substitutes dollars of input for manhours so that productivity is not a labor-only topic; the fact is, it is not a widely accepted definition. The definition standardized by the Department of Commerce remains the accepted definition. Therefore the reader should infer that the "Figure 2.1" productivity definition is meant wherever the term "productivity" is used in this chapter.

PRODUCTIVITY AND CONSTRUCTION ESTIMATING

Chapter 1 discussed the relationship between estimating, productivity, and cost accounting. In effect, an accurate evaluation of productivity is the intermediate step between establishing the work to be performed for a construction project and establishing the cost of performing that work.

It is true that some estimators establish the cost of doing specified work by merely multiplying the quantity of work to be performed by the unit cost per unit of work. That sort of calculation yields a total cost for performing the work. However the fact remains that implicit in the use of a unit cost per unit of work is the consideration of both the productivity of the producing resource or resources and the unit cost per unit of the producing resource or resources.

Three elements are vital to an accurate estimate: (1) determination of the quantity of work; (2) identification of the productivity needed to perform the and the most difficult to estimate. Project drawings are usually rather specific about the quantity of work to be performed. Similarly, the cost of rework; and (3) calculation of the unit cost of the resources to be used for the work. The second, productivity, is the element most subject to uncertainty sources over a relatively short time period are ordinarily stable. Nevertheless, given the wide variation in the productivity of the resources that are part of the construction production process, the forecasting or estimating of productivity is undoubtedly the leading risk factor in a construction estimate.

Productivity of construction producing resources to include labor and equipment is dependent on numerous factors, including uncontrollable weather, worker morale, and management supervision. These few are not the only factors. It is the estimator's ability to identify the many factors that impact productivity that in great part dictates the accuracy of a construction estimate. Clearly an understanding of productivity—including its forecasting and measuring—and the contractor's ability to improve productivity both affect the preparation of a construction estimate.

Knowledge of alternative production systems and their related productivities also affect the preparation of a construction estimate and the subsequent profit (or loss) the contractor obtains from performing the work. For example, a contractor can utilize several different means to place concrete. The contractor may pump the concrete, or use a crane with an attached bucket or conveyor system, or some combination of these. The contractor's choice of the most economical means of placing the concrete and the ability to prepare an accurate cost estimate are both dependent on the ability to know the productivity of the alternative means and methods for placing concrete.

The recognition of alternative means and methods for performing a specific type of construction work emphasizes the need for a strong relationship and communication channel between estimating and the actual production process. Unless the estimator identifies the means and methods that field personnel, such as a superintendent, can and will actually utilize to perform work, the estimator may assume an erroneous means and method and thus incorrectly forecast the productivity and related cost. An estimator cannot accurately prepare a cost estimate for work unless he understands the means and methods utilized in the field for that work.

DECLINE OF PRODUCTIVITY AND LACK
OF PRODUCTIVITY STANDARDS

Perhaps the Number One and Number Two problems of the construction industry are its declining rate of productivity increase and its lack of productivity standards. In fact these two problems—the decline in rate of increase in productivity and the lack of productivity standards—are interrelated. In a study of industry productivity, a federal agency concluded that an industry's ability to increase productivity is dependent on the degree that it can set productivity standards.

The lack of productivity standards and the declining rate of productivity increase in the construction industry are both highlighted when one looks at statistics published by governmental agencies or collected from industry studies. First of all, there is the general matter of productivity in the United States. Over the last ten years, industrial productivity has only been increasing at a rate of about 2.7% annually. If one compares this rate of increase with those of some other countries, it is evident that several others have been increasing their productivity at a rate in excess of 5% annually (all figures adjusted for inflation each year). Based on these numbers, it appears that the United States has an overall productivity problem. In fact, a very simple definition of inflation is when costs increase faster than productivity increases.

A second and more detailed measure of productivity highlights the problem of the construction industry in regard to productivity. The U.S. Department of Commerce reports on the productivity of individual industries. An excerpt of this data is illustrated in Figure 2.2. Whereas total U.S. productivity has been increasing at a 2.7% rate annually, the productivity of the construction industry has been increasing at a rate of less than one percent a year. In fact over the time period studied (1970 to 1980), the construction industry rates as one of, if not the worst, industry in regard to increasing productivity.

Industry	Percent of Productivity Increase
Agriculture	3.64%
Construction	0.80%
Government	1.64%
Manufacturing	2.60%
Mining	3.17%
Public Utilities	5.40%
Transportation	4.60%

Figure 2.2   *Annual Productivity Increase by Industry for 1970 thru 1980*

A third-level analysis of productivity statistics is more relevant to the construction estimating function. This analysis measures the productivity of the construction method itself. If one analyzes the typical construction method, one finds that the construction method includes about 45% non-productive time. Naturally every industry has non-productive time in its production process. But the nature of the construction process, including the fact that it occurs in a variable environment, is typically a unique production process for each project, and is necessarily decentralized in regard to the physical location of each building project relative to the contractor's management office location—these are all facts that result in a higher percentage of non-productive time than most industries. This in part can be expected. However, the expenditure of nearly half of all work time for non-productive time for a typical construction method is too much for efficiency and reasonable profit.

Using a management technique that the author has developed (the Method Productivity Delay Model or MPDM), a great deal of construction job-site data has been collected in order to analyze some causes of non-productive time. Non-productive time is therefore separated by source into three broad categories. See Figure 2.3. It is important to note that approximately a third of all non-productive time can be traced to industry-related factors, another third to labor-related factors, and another third to management inefficiencies. A more detailed breakout of some of the reasons causes for non-productive time is illustrated in Figure 2.4.

Productive	55%
Non-Productive	45%
Total Time	100%

Sources of Non-Productive Time

Labor	1/3
Management	1/3
Industry	1/3

Figure 2.3 *Identified Sources of Non-Productive Time*

This high percentage of non-productive time in the typical construction method routinely impacts construction estimating in several ways. For one, the fact that actual productivity can vary significantly—from twenty-five percent to seventy-five percent—causes the contractor serious difficulty when he tries to estimate productivity and the related costs. Similarly these possible negative variations can and do create a high degree of contractor financial risk each and every time a project is estimated and performed. On a positive note, it is the degree of successful elimination of non-productive

INDUSTRY-RELATED FACTORS	LABOR-RELATED FACTORS	MANAGEMENT-RELATED FACTORS
Uniqueness of many projects	High percentage labor cost	Poor cost systems and control
Locations at which projects are built	Variability of labor productivity	Poor project planning
Adverse weather and climate seasonality	Supply-demand characteristics of industry	Poor method planning for measuring, predicting and productivity
Dependence on the economy	Little potential for labor learning	
Small size of firms	Risk of worker accidents	
Lack of R and D	Union work rules	
Restrictive building codes	Low motivation for worker	
Government labor and environmental laws		

Figure 2.4 *Detailed Causes of Reasons for Low Productivity in the Construction Industry*

time in the typical construction method that can and does provide the contractor the potential to outperform competitors and also to better his own estimated productivity and costs. The fact remains—the cost of performing work is strongly related to the productivity attained in performing the work.

PRODUCTIVITY, ESTIMATING, AND THE ENVIRONMENT

The strong relationship between productivity, construction estimating, and overall project planning is highlighted when one recognizes the dependence of productivity on the environment. This is especially true in geographic locations affected by a varying climate. Consider the midwestern United States. Given its typical cold winters that routinely result in difficult working conditions, productivity for various labor crafts is usually only a fraction of what it is during the summer months.

As an example, let us assume that mason productivity in January in Chicago is only 60% of what can be expected in June. It follows that the labor cost component of the construction project estimate will normally be higher in January than if the same work is performed in June. This calculation of a

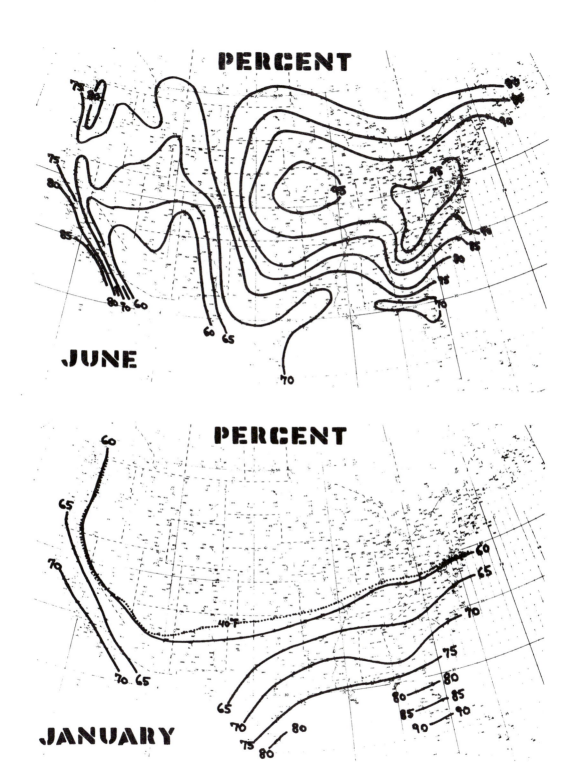

Figure 2.5 *Effect of Seasonal Temperatures on Construction Productivity* (Grimm & Wagner)

percent is illustrated in Figure 2.5. In effect, the cost of doing the work is dependent of the labor productivity which in turn is dependent on environmental factors.

Labor productivity is not limited to the influence of environmental factors. Figure 2.6 lists some additional factors that variously affect productivity. In effect, the estimator must be able to predict productivity's dependence on these and other factors, each and every time a construction estimate is prepared.

The dependence of estimating and productivity on the environment also draws attention to the relationship of estimating to overall project planning.

Worker Morale	Job-site Location
Job-site Management	Availability of
Skill of Workers	Other Work
Time of the Year	Material Flow to Project
Day of the Week	Crew Size
Time of the Day	Working Conditions

Figure 2.6   *Factors that Affect Labor Productivity*

The building of a construction project entails performing a series of construction activities, many of which are interrelated. (Note: These activities are sometimes called work items). Consider the small project that is illustrated in Figure 2.7. This type of network representing the relationship between a sequence of project activities is commonly referred to as a Critical

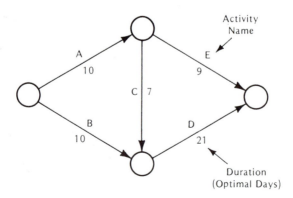

Figure 2.7   *Example CPM Diagram for a Small Project*

Path Method (**CPM**) diagram. The project activities and their relationships to one another are shown by the arrows. Duration for each activity or item is represented by the number of workdays below each arrow. Ordinarily an activity cannot start until the preceding one is completed. When simultaneous "starts" are possible, arrows go forward on two or more paths. Note the A and B paths from encircled 1 and the E and D paths to encircled 4.

The activity durations (in workdays) shown along the arrows representing the activities are the optimal estimated minimum times for the activities, given optimal working conditions. A workday is assumed to be the current 8-hour day. The contractor for the project assumes that Activities A and D are not affected by the weather. Therefore the productivity of these activities can be considered constant.

Let us further assume that the contractor believes that Activities B, C, and E are affected by the weather. Based on historical data, the contractor therefore assigns the activities productivity efficiency factors that express the expected productivity of the activities in any one month as a fraction of what he considers ideal productivity. Ideal productivity is assigned a productivity efficiency factor of 1.00. The assigned productivity efficiency factors for Activities B, C, and E are shown in Figure 2.8.

Factor 0.40	Factor 0.50	Factor 0.60	Factor 0.70	Factor 0.80	Factor 0.90	Factor 1.00
October November December	January September	February	March	August	April July	May June

Figure 2.8 *Productivity Efficiency Factors for Work Items by Months when Performed in Midwestern United States*

The contractor desires to start the project on the morning of September 15, 19X0 (a Wednesday). He decides he is not going to work on weekends or holidays, and is going to avoid working more than 8 hours a day. Thus, his policy is to work an 8-hour 5-day week, except when a holiday occurs on a weekday.

Let us assume the contractor wants to schedule the project on an earliest start-time basis. He wants to fit the schedule to a calendar to determine dates for ordering equipment, hiring labor, and procuring materials for the various project activities. A sample calendar for the months September through December 19**X0**, is shown in Figure 2.9 (19**X0** denotes start-up year; 19**Y0** would denote project years following the start-up year).

Activity A, in Figure 2.7, is one of the initial activities of the project. It has an earliest start time (EST) of September 15, 19**X0**. Given optimal conditions, Activity A has an estimated duration of 10 days. As previously stated,

		SEPTEMBER							OCTOBER				
S	M	T	W	T	F	S	S	M	T	W	T	F	S
				1	2	3	4					1	2
5	6	7	8	9	10	11	3	4	5	6	7	8	9
12	13	14	15	16	17	18	10	11	12	13	14	15	16
19	20	21	22	23	24	25	17	18	19	20	21	22	23
26	27	28	29	30			24 31 25	26	27	28	29	30	

		NOVEMBER							DECEMBER				
S	M	T	W	T	F	S	S	M	T	W	T	F	S
	1	2	3	4	5	6				1	2	3	4
7	8	9	10	11	12	13	5	6	7	8	9	10	11
14	15	16	17	18	19	20	12	13	14	15	16	17	18
21	22	23	24	25	26	27	19	20	21	22	23	24	25
28	29	30					26	27	28	29	30	31	

Figure 2.9 *Calendar for Example Problem in Year 19X0*

Activity A has productivity efficiency factors of one for all months of the year at the project location. Therefore, the 10 required workdays will take 10 actual days, regardless of the time of the year in which the activity is performed. Since the contractor is not going to work on weekends (September 18, 19, 25, and 26), the earliest finish time (EFT) of Activity A is day's end on September 28, 19**X0**.

However, Activity B can also start at the beginning of the project. Therefore, its EST is on the morning of September 15, 19**X0**. Activity B requires 10 optimal workdays. There are 12 working days available in September. From an inspection of Figure 2.8, it may be observed that Activity B has a productivity efficiency factor of 0.50 in September. Therefore, only 6 actual optimal workdays (12 × 0.5) are finished in September, owing to the lower productivity. It is observed that the productivity efficiency factor is 0.40 for Activity B in October. There are 4 actual optimal workdays left to be done. Owing to the productivity efficiency factor for October, these 4 optimal duration days will actually take 10 workdays (4/0.4). Therefore, the EFT of Activity B is day's end on October 14, 19**X0**.

Activity C cannot start until Activity A is finished. Therefore, the EST of Activity C is the morning of September 29, 19**X0**. (a Wednesday). Activity C has an estimated duration of seven optimal workdays. There are two available working days left in September. The productivity efficiency factor is 0.50 in September for Activity C. Therefore, only one optimal workday is completed in September, leaving seven days for October. Because Activity C's productivity efficiency factor is 0.40 in October, it takes 15 actual days. (6/0.4) to finish the six optimal workdays. Therefore, the EFT of Activity C is day's end on October 21, 19**X0** (a Thursday). Of course, this assumes that the contractor does not work overtime or weekends.

Activity D cannot start until Activities B and C are finished. Activity B can be finished at the end of October 14, 19**X0**. Activity C can be finished at the end of October 21, 19**X0**. Therefore, the EST of Activity D is the morning of October 22, 19**X0** (a Friday). Activity D takes 21 optimal workdays. Activity D has a productivity efficiency factor of 1.00 for all months at the project location. There are six workdays available for Activity D in October. Therefore, 15 workdays are required in November. Remembering that the contractor does not work weekends, we find that the EFT of Activity D is day's end of November 19, 19**X0**.

Activity E also cannot start until activity A is finished. Therefore, the EST of Activity E is the morning of September 29, 19**X0** (a Wednesday). Activity E has an estimated duration of nine optimal workdays. The productivity efficiency factor for Activity E is 0.50 for September. There are two actual workdays left in September. However, owing to the productivity efficiency factor of 0.50, only one optimal workday is actually completed during these two days. There are 21 available workdays in October. The EFT of Activity E is, therefore, day's end of October 28, 19**X0**. Thus, the contractor can finish the project at day's end of November 19, 19**X0**. This corresponds to the EFT of Activity D.

In summary, both the productivity of a specific activity and the productivities of the other activities that make up a project affect the cost of any one activity. Consider Activity C in the project illustrated in Figure 2.7. The cost of doing the work for Activity C is dependent on when it is performed. If it is performed in October, November, or December, its cost will be 2.5 times its cost if performed in May or June.

Let us also consider Activity D. Without recognition of the lessened productivity efficiency factors for prior activities, one might assume that Activity D would be performed in an earlier time period than its October 22 to November 19 time period. In effect, the productivity, and cost of Activity D is dependent on the time at which it is performed which in turn is dependent on its technological relationship to other activities. Clearly this is a strong relationship between productivity, estimating, overall project planning, and the working weather-conditioned environment.

PRODUCTIVITY VS. UNIT COST STANDARDS

An important component of any estimating system is the collection of past project data and the structuring of the data for use in estimating future projects. In regard to structuring past project data for use in estimating direct labor cost for future projects, the contractor can work up either productivity data or unit cost data (or both). For example, in establishing data on the direct labor cost for carpenters who place wall forms, the contractor might establish a historical productivity data file of 7.0 manhours/100 square feet of contact area or, as an alternative, a unit cost of $1.05/square foot of contact area. It is common to find that a contractor utilizes either productivity or unit cost as the historical data base.

The question can be asked as to which base—productivity data or unit cost data—is preferable. Obviously either of these data bases can be converted from one to the other so long as the labor wage rate of the worker is recognized. However, the fact that a given unit of cost may consist of the output of several workers and include workers of varying crafts and wage rates, it may be difficult to convert unit costs to a manhour productivity record.

A strong case can be made for structuring and using direct labor productivity data instead of direct labor unit costs. This is based on the fact that historical productivity data is not as sensitive to change as a function of time as unit cost data is. Over the last ten years (1970-1980), direct labor productivity has been relatively constant. Figure 2.2 indicates that productivity in the construction industry has averaged an increase of less than one percent annually over the time period from 1970-1980. During this same time period, construction wage rates have increased by as much as 15% in a single year. The result has been that in a given year a historical direct labor unit cost may become outdated by as much as that same 15% in a single year. Given the fact that historical data is only as good as it is accurate at the time it is being used for forecasting future costs, direct labor productivity data has proven in recent years to be more advantageous for estimating future costs than have direct labor unit costs.

Given historical direct labor productivity data for a work process, the estimator can easily convert the productivity data to a total direct labor cost and a direct labor unit cost for the work process. For example, let us assume the following productivity data has been structured for the placing of wall forms for concrete per 100 square feet of contact area (sfca); thus:

Carpenters: 7.0 hours   **Laborers:** 3.0 hours

Let us further assume that the current carpenter wage rate is $14.00 per hour and the laborer wage rate is $12.00 per hour. The calculation for the total direct labor cost per 100 square feet of contact area can be calculated as follows:

Carpenters: (7.0 hours) ($14.00/hour) = **$ 98.00**
Laborers: (3.0 hours) ($12.00/hour)   = **36.00**
Total Direct Labor Cost/100 sfca   = **$134.00**

Similarly the direct labor unit cost per square foot of contact area can be calculated as follows:

$$\textbf{Unit Cost} \ = \ \frac{\text{Total Direct Labor Cost}}{\text{Work Quantity}} \ = \ \frac{\$134.00}{100 \text{ sfca}} \ = \ \textbf{\$1.34/sfca}$$

Assuming the carpenter and laborer productivities to be relatively constant, it becomes easy to modify the calculated total direct labor cost and/or unit cost as the carpenter or laborer wage rates change as a function of time.

ACCOUNTING VERSUS SCIENTIFIC PRODUCTIVITY STANDARDS

Productivity work standards provide the basis of the estimator forecasting the duration and direct labor cost of a proposed project. There are essentially two alternative means of developing work standards. A productivity work standard can be established by using historical accounting data or by analyzing a work process and then developing a scientific standard. Such a scientific standard is commonly referred to as an "engineering" standard.

More often than not, the construction firm and the estimator have utilized accounting based standards. The process of developing an accounting standard is relatively simple. The estimator merely collects labor timecard data regarding labor hours expended in a given time period and measures the work put in place during that same time period. For example, let us assume that a contractor determines via the collection of labor timecards that 400 carpenter hours are expended in a given week placing forms for concrete walls on a given construction project. During this same period, inspection of the job site indicates that the contractor has placed 5,000 square feet of contact area of forms. By dividing the labor hours expended by the work performed during the same time period, a productivity standard of 8 manhours per 100 square feet of contact area is determined. This process is illustrated in Figure 2.10.

The theory behind the structuring and use of historical accounting based data for estimating future projects is based on mathematics, the science of

---

Carpenter Hours
  Expended in the Work Week   = 400 Manhours

Work Performed
  in the Time Period   = 5,000 square feet of
                                      contact area (sfca)

$$\text{Productivity} = \frac{\text{Hours}}{\text{Work}} = \frac{400 \text{ hours}}{(50) \ 100 \text{ sfca}} = \frac{8 \text{ hours}}{100 \text{ sfca}}$$

Figure 2.10 *Calculation of a Productivity Standard*

numbers. In particular, with the mathematical discipline referred to as statistics, it can be proved that the more data one collects regarding the occurrence of an event (such as the productivity of a work process), the more reliable the cumulative data is in regard to predicting a future event. For example, if a contractor has collected productivity data regarding the placement of masonry block from 10 of his own past projects, it follows that the data is more reliable regarding the prediction of future masonry productivity data than if the contractor has only accumulated data from the performance of three past projects. Obviously this statement is only true if in fact the observed events or work processes are similar to future events or work processes.

The critic of labor productivity work standards structured from past accounting data draws attention to four potential weaknesses underlying these standards. First, it might be argued that a construction contractor never builds the exact construction work process under the exact working conditions more than once. For example, in performing outdoor masonry work, make-up of a labor crew may change, a planned wall may have more or fewer blockouts or lintels than walls previously built, the effectiveness of the supervising foreman may vary, or the previous environment of mild temperatures and low rainfall conditions may never be duplicated. While this difficulty with accounting standards is somewhat valid, it can also be argued that the cumulation of a significant amount of data will "average" these conditions and can in fact be indicative of a future event.

The difficulty of an accounting standard being or not being indicative of a future event is related to the detailedness used to define the cost object or work item that is being measured. The more minute or detailed the cost object or work item, the fewer the number of external events that can affect the event. For example, if the technique for all types of concrete placement on a construction project is defined as the work item for developing a productivity standard, the standard may be too broad or gross on the one hand to be indicative of the expected productivity for a specific type of concrete placement such as for a suspended slab. On the other hand, if a contractor narrows—and focuses—the definition of the work item productivity standard to placement of concrete for a slab, and maybe even limits the standard to specific elevation of a slab such as a suspended slab, that standard will likely be more reliable in estimating the productivity on future work of this type. In other words, the degree of detailedness in defining productivity standards dictates to some degree their reliability in predicting future events.

A second common criticism against standards based on past accounting data and focusing on one of the four potential weaknesses is that the data and resulting standards may not be current. The very time that it takes to collect, process, and structure job-site accounting data may sometimes result in that data being outdated before it is available for estimating future work. While the timeliness of accounting data does present a problem for the construction firm in regard to its usefulness for measuring and controlling a job in progress, the lack of timeliness characteristic may be overemphasized in regard to using the data for future estimating. The previous section of this chapter contrasted the reliability of productivity data to unit cost data especially as regards the time factor. It was noted that, unfortunately, construction productivity has been relatively unchanged over the last decade—thus, it is constant data. In effect, as the industry now operates, labor productivity data does not become substantially outdated with time, so the potential of a timeliness problem with past data is lessened for construction estimating.

A third potential difficulty in using past project accounting data for developing estimating standards relates to the difficulty of obtaining accurate accounting data at the job site. Standards based on accounting data are only as good as the accuracy of the data itself.

The construction process is relatively unique in that the production process is decentralized from the company's headquarters. While the construction firm may be headquartered in one city, it may construct projects within a large geographic area including several cities and towns. The result is that the job-site accounting data may have to be collected by "production" personnel and subsequently forwarded to central headquarters for processing. Like most production personnel—including foremen and superintendents—job site personnel tend to discount the need to collect accurate accounting related data at the job site. They tend to be doers and makers or more "production" conscious, and as a result tend to discount the data collection function.

The natural difficulties encountered when production personnel have to collect accurate job site data can be lessened by development of a reliable and mandatory job-site cost system that includes data collection procedures. Cost systems and the importance of and a design for collecting accurate job-site data is further discussed in Chapter 8.

Perhaps the least cited, but potentially most difficult aspect of productivity standards based on historical data is the fourth—the fact that productivity inefficiencies have become the "norm" or accepted standard. Earlier in this chapter we indicated that as much as 45% of total time for a construction work process is non-productive. Further it was noted that a significant portion of this non-productive time can be eliminated by improved management.

Because accounting based standards in effect use averages of non-productive and productive time to establish the standard, they do not indicate the potential a contractor might achieve in terms of productivity. Instead, they represent what has been achieved, including historical productivity inefficiencies. In effect, inefficiencies inevitably become the standard for future estimating.

Admittedly, if one cannot expect to reduce or eliminate inefficiencies, one should use a standard that includes the inefficiency when estimating the result of a future event. If this is not done, the estimator will be guilty of preparing a low-cost estimate with subsequent possible detrimental financial results.

Notwithstanding the need to eliminate inefficiencies and non-productive time, it is a fact that both will occur and must be reckoned in a realistic estimate. But it can also be argued that the mere fact that the estimator or contractor does not know the attainable productivity standard (that is, a standard based on correction of "non-productive" time) results in never seeking and never attaining the attainable. In effect, it can be further argued that not knowing what standard is attainable often results in not accomplishing that productivity standard.

Given the fact that there is evidence that the above is true—that a firm's (or industry's) ability to increase productivity depends to some degree on the ability to set an attainable standard—unrealistically low standard setting might be the most valid criticism of accounting-based standards for construction productivity. Nevertheless, too much "realism" may be one of the significant reasons for the low productivity that is characteristic of the construction industry.

An alternative to an accounting-based productivity standard is a scientific- or engineering-based standard. Such a standard is developed by "modeling" the work process and analyzing the potential output as a function of the work input of the resources that contribute to the work process. There are several different "models" or management techniques available to the estimator to develop a scientific-based productivity work standard. A few of these techniques judged applicable to the construction process are presented by the author later in this chapter.

In addition to viewing a standard as either accounting-based or scientific-based, cost or productivity standards can also be classified in one of the following three types:

1. Basic standards are unchanging standards. They provide the basis for comparing actual costs or production through the years by means of the same criterion. Such accounting reports effectively spotlight whatever trends develop. However, trends can lose their significance when a too short time elapses between changes in product and method.
2. Perfection or ideal or maximum efficiency or theoretical standards reflect all the industrial engineer's dream of a "factory heaven." Perfection standards specify the absolute minimum costs for the maximum productivity possible under the best conceivable operating conditions, using existing specifications and equipment. Ideal standards, like other standards, are used whenever management feels that ideal standards will provide psychologically attractive [productive] goals.
3. Currently attainable standards set forth the costs or productivity that should be incurred under efficient operating conditions. They are difficult but possible to achieve. Attainable standards are looser than

ideal standards because they include allowances for ordinary machine breakdowns and lost time. However, attainable standards are usually set high enough for people to consider that the achievement of such a standard performance is a satisfying accomplishment. Moreover, variances from that standard are more likely to be seen as slightly unfavorable than favorable. But favorable variances may be attained by a little more than expected efficiency of performance.

Recalling our earlier discussion of accounting-based construction productivity standards helps us identify this type of a standard as a "basic" standard. Similarly a scientific-based standard (for example, those types discussed in the following section) might be viewed as a "perfection or ideal or maximum efficiency" standard. Realistically what is described as a "currently attainable" standard might be what a contractor might reach for in performing construction work. Unfortunately, the construction industry's inattention to perfection or ideal or maximum efficiency standards has pretty well resulted in achievement of the low productivity that characterize its basic standards rather than attainable productivity standards.

DEVELOPING SCIENTIFIC-BASED PRODUCTIVITY STANDARDS

Previously we cited the weaknesses likely when accounting data is used to develop standards for estimating future productivity or cost. Given these weaknesses, an alternative standard, one built on a scientific analysis of the work process, is proposed.

Several models or management techniques are available to analyze a work process. Included in these techniques are motion analysis, work sampling, process charts, and the like. Unfortunately some models or techniques do not fit the environment of the construction process, which is complex. Numerous labor inputs, materials, equipment and a variable environment make it difficult to model the typical construction work process. Thus several models or techniques used elsewhere are not applicable to the construction industry.

While no one model or management technique is optimal or all encompassing for each and every construction work process, some are better than others. In the view of the author, three models best serve the construction estimator in developing a scientific-based productivity standard. They are time study, the method productivity delay model, and the production function.

Time Study

To give an example of how an estimator might establish a productivity or work standard, let us consider the analysis of a construction process of carpenters placing sheets of plywood for forming of a concrete slab. Assume that the work process breaks down into the following tasks:

1. Carpenters walking to get plywood.
2. Carpenters picking up plywood.
3. Carpenters taking plywood to placement location.
4. Carpenters placing and nailing plywood in place.

The observer of such a work process breaks that process into specific tasks because, by analyzing the total work process as a function of its successive tasks, it may well be possible to determine ways to reduce or eliminate some of the time required for specific tasks, and thus determine a more productive work process. For example, the time required for the carpenters to walk to and pick up plywood may be decreased by a more judicious storage location of the plywood.

Once it is determined what tasks are to be analyzed, the times for each are recorded by observing several cycles of the overall work process. The individual task times are summed to yield a cycle time. These cycle times are used to calculate a select, a normal, and a standard time.

Let us assume that the observer of the placement of the plywood collects the task and cycle times illustrated in Figure 2.11. The sum of the time for the 9 cycles is 1,410 seconds or 23.5 minutes. The select time for the work process is calculated as follows:

$$\textbf{Select Time} = \frac{\text{total time}}{\text{\#cycles}} \text{ or } \frac{1,410 \text{ seconds}}{9 \text{ cycles}} = 156.67 \text{ seconds (2.61 minutes)}$$

This select time is just an average of the cycle times. In order to calculate a normal time a rating factor must be chosen. Rating factors are generally set

Activity:
1. Walking to get plywood   3. Taking plywood to its location
2. Picking plywood up   4. Nailing plywood down

	CYCLES								
	1	2	3	4	5	6	7	8	9
Activity 1	30	38	22	26	28	34	26	22	29
Activity 2	12	20	22	22	12	14	16	18	20
Activity 3	48	54	56	58	42	44	48	54	87
Activity 4	54	62	48	54	64	48	48	34	96
Total time (seconds)	*144*	*174*	*148*	*160*	*146*	*140*	*138*	*128*	*232*
Total time (minutes)	2.4	2.9	2.5	2.7	2.4	2.3	2.3	2.1	3.8

Figure 2.11 *Time Study Data on Placing Plywood Used in a Concrete Slab*

between .8 and 1.2 numerically. If the observer feels work was slower than usual, a factor greater than 1 should be chosen. If it is judged that the work is faster than normal, a factor of less than 1 should be used. Let us assume a rating factor of 1.1. The normal time is then calculated as follows:

Normal time = select time (rating factor)
= 156.67 seconds (1.1)
= 172.34 seconds (2.87 minutes)

To determine the standard time for a work process an allowance factor should be recognized for personal needs. This factor allows for a reasonable amount of personal time for conversation, smokes, drinks, and other human needs. The factor is generally assumed to be between 15 and 20 percent. Let us assume an allowance factor of 15 percent. Using this allowance factor, the standard time for the work process is calculated as follows:

Standard time = normal time + normal time (allowance factor)
= 172.34 seconds + 172.34 seconds (.15)
= 172.34 seconds + 25.85 seconds
= 198.19 seconds (3.30 minutes)

This standard time tells us that one piece of plywood should be laid down every 3 minutes and 18 seconds. Given data on the size of the plywood sheet placed and the size of the carpenter crew, the standard time of 198.19 seconds per sheet of plywood can be converted to amount of square feet of forming per carpenter manhour.

Let us assume that each plywood panel is 32 square feet of area and the crew consists of four carpenters. A square feet per manhour standard can be determined as follows:

$$\text{Standard productivity} = \frac{3600 \text{ sec./hr.}}{198.19 \text{ sec./panel}} \cdot \frac{32 \text{ sq. ft./panel}}{4 \text{ carpenter hours/hour}}$$

Standard productivity = 145.32 square feet/carpenter hour

or

Standard productivity = 0.688 carpenter hours/100 square feet

The estimator can then proceed to use this calculated productivity to estimate the cost of similar work for a future project.

The estimator may also use the collected times in Figure 2.11 to determine an improved work process and thus establish an increased productivity standard. For example, analysis of the data in Figure 2.11 indicates that the work task identified as "carpenter walking to get plywood" averages 28.33 seconds for each cycle or approximately 18.1% of the total average cycle time. Let us assume that this time can be reduced by half if the contractor exercises better material layout—piles the plywood closer. In effect, it is logically expected that the "carpenter walking to plywood" task can be reduced to 14.16 seconds per cycle. The recalculated select time is now as follows:

Select time = 156.67 second – 14.16 seconds = 142.51 seconds

Similarly the recalculated normal time is as follows:

Normal time = 142.51 seconds (1.1) = 156.76 seconds

The recalculated standard time then follows:

Standard time = 156.76 seconds + 156.76 seconds (.15)
= 156.76 seconds + 23.51 seconds
= 180.27 seconds (3 minutes plus)

Finally, this standard time can be converted to a productivity number standard for estimating by the following calculation:

$$\text{Standard productivity} = \frac{3600}{180.27} \cdot \frac{32}{4}$$

Standard productivity = 159.76 square feet/carpenter hour

This estimating of a productivity standard can be considered a "scientifically" calculated attainable work standard for the work process.

METHOD PRODUCTIVITY DELAY MODEL

In this section the techniques developed by the author will be used to model construction method productivity. Called the Method Productivity Delay Model (MPDM), it can be viewed as an alternative to the more traditional work sampling, time study, and motion analysis models. The objective of the MPDM model is to provide the construction firm with a practical way to measure, predict, and improve productivity, and thus develop productivity standards.

An overview of the MPDM is shown in Figure 2.12. The model has four elements. These consist of the collection of data, the processing of data, the structured model, and the implementation element. Each of these elements will now be explained separately.

Collection of Data: The purpose of the collection element is to collect method productivity data to be used as a basis for modeling the method

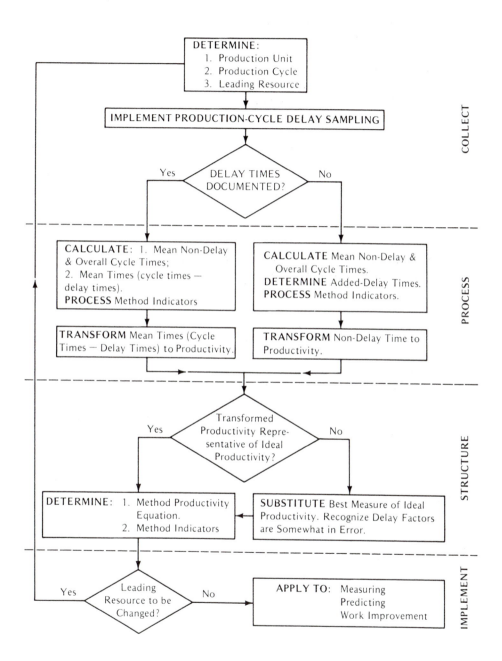

Figure 2.12 *Overview of the Method Productivity Delay Model*

productivity. (Note: the term "method" is used as a substitute for a work process.) This element identifies the determination and collection of the three following types of information:

1. The "production unit."
2. The "production cycle."
3. The time required for completion of production cycles and document any productivity delays.

The definition of a method's production unit serves as the basis for measuring, predicting, and improving method productivity. The "production unit" is an amount of work descriptive of the production which can easily be measured visually. Typical examples of production units:

a. Arrival of a scraper in a borrow-pit.
b. Release of concrete from a crane bucket.
c. Placement of one row of concrete blocks on a wall.
d. Placement of a structural member.

A "production cycle" is the time between consecutive occurrences of the production unit. The production unit cannot be defined independently of the other elements of the model. The definition of the production unit dictates the detail used to measure method productivity. If the production unit is defined in too broad a context the collected productivity information may be of little aid in productivity measurement, prediction, and improvement. For example, if the production unit is defined as the completion of the placing and finishing of concrete for a 200 square-foot wall section, the production cycle times and delay information thus collected may be too broad to focus on parameters that affect productivity. On the other hand, too detailed a definition of the production unit may curtail the model user's ability to identify either productivity delays or even completion of a production cycle. For example, if the placing of a single brick is identified as the production unit when four masons are simultaneously placing bricks, the model user will find it difficult to document the successive completion of production cycles, let alone cite productivity delays.

The data collected in documenting the cycle times and delays is the basis for the delay information essential to the productivity equation of the MPDM. The documented production cycle times will serve as a basis for the first part of the productivity equation.

The following data are documented:

1. Time required to complete production cycle.
2. Occurrence of productivity delay.

3. Total cycle delay allocated by approximate percentage or by documented times—if more than one productivity delay type is found in a given cycle.
4. Any unusual events that characterize a given production cycle.

Data Sampling Techniques: The procedure used by the author for collecting this data is called Production Cycle Delay Sampling (PCDS). A data collection form, along with example data is shown in Figure 2.13.

Documenting production cycle times is straightforward. The model user merely clocks the time between occurrences of the production unit. If the method is very complex or the production cycle times are very short, so that visual inspection is impossible or inaccurate, a filming procedure such as time lapse can be used to document the method productivity.

Citing and documenting production cycle delays, whether they are single or multiple type delays, requires a degree of skill on the part of the MPDM user. The model user's ability to single out productivity delay types normally increases with experience and decreases as the complexity of method increases. The types of delays identified may vary depending on the construction method. Construction method delays can normally be identified when they are observable and independent of one another. The following example of the MPDM will consider five delay types: environment, equipment, labor, material, and management. (A user may find it convenient to add or delete from these five types, the MPDM procedure will remain unchanged.) Examples typical of the five delay types follow:

1. **Environment**: Change in soil conditions, change in wall section, change in roadway alignment.
2. **Equipment**: Stationary production equipment in transit, equipment operating at less than capable production rate.
3. **Labor**: Workman waiting for another workman, workman loafing, worker fatigue, workman not productive because of lack of knowledge of his work.
4. **Material**: Material not available for equipment or labor demand, material defective.
5. **Management**: Poor planning of method resource combination and placement, secondary operation interfering with method productivity, poor method layout planning.

The processing element of the MPDM provides the connection between the model's collected data, the structured method productivity equation, and the accompanying measure of risk and variability. This processing consists of adding, subtracting, multiplying, and dividing. As a procedure, MPDM possesses the desired model attributes of ease and economy of implementation.

Productivity Records: Ways to Determine Standards for Construction Estimating

PRODUCTION CYCLE DELAY SAMPLING

Page 1 of 1   Unit: Second
Method: CRANE BUCKET CONCRETE POUR   Production Unit: Concrete Drop

Production Cycle	Production Cycle Time (sec.)	Environment Delay	Equipment Delay	Labor Delay	Material Delay	Management Delay	Notes	Minus Mean Non-delay Time
1	120			✓				27
2	126		✓					33
3	98						✓	5
4	112		✓					19
5	108						✓	15
6	1122					✓	crane move	1029
7	116		✓					23
8	214		✓					121
9	92						✓	1
10	88						✓	5
11	100						✓	7
12	312		✓					219
13	110			✓				17
14	666	90%	10%					573
15	146	25%	75%					53
16	120		✓					27
17	138		✓					45
18	144	20%	80%					51
19	598	✓					crane slip	505
20	118		✓					25
21	138		✓					45
22	108						✓	15
23	98						✓	5
24	120			✓				27
25	116			✓				23
26	368		✓					275
27	118		✓					25
28	140		✓					47
29	136			✓				43
30	138			✓				45

Figure 2.13 *Sampling Form Used for Data Collection for the MPDM*

PRODUCTION CYCLE DELAY SAMPLING

Unit: Second
Method: CRANE BUCKET CONCRETE POUR   Production Unit: Concrete Drop

Produc-tion Cycle	Produc-tion Cycle Time (sec.)	Envir-onment Delay	Equip-ment Delay	Labor Delay	Mater-ial De-lay	Manage-ment Delay	Notes	Minus Mean Non-delay Time
31	154			✓				61
32	396		✓					303
33	96						✓	3
34	286		✓					193
35	80						✓	13
36	82						✓	11
37	84						✓	9
38	212			✓				119
39	78						✓	15
40								

Figure 2.13 *(cont.)*

The data formulated from the collected data shown in Figure 2.13 are processed on a form such as the one shown in Figure 2.14. The form and entries are intended to be self-explanatory.

The first part of the structured MPDM as shown in Figure 2.15 is a **method productivity equation** that relates overall or actual method productivity to ideal method productivity as a function of the identified productivity delay types. This equation is used directly to aid in measuring, predicting and improving method productivity.

"Ideal productivity" is that productivity which occurs when productivity delays are absent. The non-delay production cycles documented in the collection of data (Figure 2.13) are, in fact, ideal productivity cycles if all delays have been cited. However, if conditions exist that indicate some delays are not being detected, the model user should select other techniques for determining ideal productivity. Even if the non-delay cycles appear adequate as a measure of ideal productivity, it may be beneficial to compare the calculated productivity with a historical record.

The non-delay production cycle time is transformed into a production per time period interval for which productivity is to be measured, predicted, and work improvement is to be performed. This time period interval will normally be in units per hour or day. The mean non-delay production cycle time for the example is 92.6 sec. The ideal productivity calculated on an hourly basis is

MPDM PROCESSING

Unit: second

Method: Crane Bucket Concrete Pour   Production Unit: Concrete Drop

Units	Total Production Time	Number of Cycles	Mean Cycle Time	$\Sigma[\mid (\text{Cycle time}) - (\text{Nondelay cycle time})\mid]/n$
A. Nondelayed production cycles	1112	12	92.6	8.7
B. Overall production cycles	7596	39	194.7	104.5

Delay Information

	Environment	Equipment	Labor	Material	Management
C. Occurrences	1	17	11	0	1
D. Total added time	505	1939	500	0	1029
E. Probability of occurrences[a]	.026	.436	.282	0	.026
F. Relative severity[b]	2.59	.58	.23	0	5.28
G. Expected % delay time per production cycle[c]	6.7	25.4	6.6	0	13.7

[a] Delay cycles/total number of cycles

[b] Mean added cycle time/mean overall cycle time

[c] Row E times row F times 100

Figure 2.14 *Example of Form Used to Process Data for the MPDM*

$$\frac{(60 \text{ min./hr.}) \times (60 \text{ sec./min.})}{92.6 \text{ sec./unit}} = 38.9 \text{ units/hr.}$$

The calculated 38.9 units/hr is in fact a scientific productivity standard. Given the quantity of work performed with each unit of work, the calculated 38.9 units/hr can be transformed to a measure of productivity common to the construction industry. For example, let us assume that each unit of work is equivalent to a cubic yard of concrete in place. In that case, the MPDM has indicated a possible productivity of 38.9 cubic yards of concrete per hour.

MPDM STRUCTURE

Crane Bucket Concrete Pour

Production Unit: Concrete Drop

I. Productivity Equation
    Overall method
    productivity   = (Ideal productivity) $(1 - Een - Eeq - Ela - Emt - Emn)$
    18.5 units/hr   = (38.9 units/hr) $(1 - .067 - .254 - .066 - 0.137)$

    where: Een = expected environmental delay time as a decimal fraction of total
                  production time

            Eeq = expected equipment delay time

            Ela = expected labor delay time

            Emt = expected material delay time

            Emn = expected management delay time

II. Method Indicators

    A. Variability of method productivity
        Ideal cycle variability = 8.7/92.6 = 0.09
        Overall cycle variability = 104.5/194.7 = 0.54

    B. Delay information

	Environ-ment	Equip-ment	Labor	Material	Manage-ment
Probability of occurrence	0.026	0.436	0.287	0	0.026
Relative severity	2.59	0.58	0.23	0	5.28
Expected % delay time per production cycle	6.7	25.4	6.6	0	13.7

Figure 2.15 *Example of Structured MPDM Data*

The 18.5 units per hour derived by dividing the productivity of 194.7 sec/cycle into 3600 sec/hour is in effect the overall or average productivity achieved by the contractor. In effect, this productivity is equivalent to the standard that would be obtained from the use of accounting data.

The difference between the calculated "scientific" standard of 38.9 and the overall or average productivity of 18.5 might be considered as the "potential for improvement." It is not realistic for the estimator to use the 38.9 unit/hr or cubic yards per hour in estimating a future project. However, knowing and being aware of this potential productivity, the contractor may seek means and methods of managing the construction to approach this "scientific" productivity standard. If this is done successfully, the estimator will be able to improve on the 18.5 accounting standard and in that way prepare a more competitive construction bid.

The factors in the right-hand side of the productivity equation shown in Figure 2.15 relate the method's ideal productivity to the method's overall productivity. These factors, Een = environmental Eeq = equipment Ela = labor Emt = material Emn = management, which can have values ranging from 0 to 1, are the decimal fraction of total production time caused by each delay type. The factor 1 minus the sum of Een, Eeq, Ela, Emt, and Emn relates the probability of productive work being performed. The left-hand side of the productivity equation in Figure 2.15 is the "overall method productivity".

Part 2 of the structured model shown in Figure 2.15 contains "method indicators." The information set out in this segment of the model will be used as indicators to predict and improve method productivity.

Four types of information are set out in the "method indicator" part of the structured MPDM. The first, "variability of method productivity", gives a measure of the variable nature of both the non-delay productivity cycles and the total overall productivity cycles. These are determined as follows:

$$\frac{\text{Ideal Cycle}}{\text{Variability}} = \frac{\text{Non-delay Cycle time} - \text{Mean non-delay cycle time}}{\text{Number of non-delay cycles}}$$

$$\frac{\text{Overall Cycle}}{\text{Variability}} = \frac{\text{Overall cycle time} - \text{Mean non-delay cycle time}}{\text{Number of total cycles}}$$

Non-delay cycle times are durations of cycles in which no delays are detected. Overall cycle times are the durations of all cycles without regard for detection of delays. For the data shown in Figure 2.13 and processed in Figure 2.14, the "variability of method productivity" is .09 for the ideal cycles and .54 for overall cycles.

The three other indicators in this part of the model relate to selected productivity delays. They are the "probability of occurrence," "relative severity," and the "expected percent of delay time per production cycle." These indicators are used in analysis of potential work improvement.

The MPDM calculated non-delay productivity rate and the overall productivity rate represent measured productivities. Measuring is one of the objectives of the model. The derived productivity equation provides the basis for method productivity prediction. The factors relating method ideal productivity to overall method productivity may be predicted or calculated from past data.

The constant in the method productivity equation, the "ideal productivity," should be fixed. For a given construction method some variation exists in the time of ideal production cycles. However, this variability is small compared to the variability in the cycle times for overall production cycles.

In effect the calculated overall productivity rate is equivalent to a determined standard based on accounting data. On the other hand, the calculated ideal productivity can be considered a scientific-based standard. Further, because the user of the MPDM documented correctable delays, it can be argued that the calculated ideal productivity is in fact an attainable standard.

The delay factors can all be considered variables. Although the delay factors or E's are given values as a result of collection of data from the previous performance(s) of the method in question, they may take on new values for a future occurrence of the construction method.

When a contractor is faced with a construction method for which he has collected MPDM data, he should focus attention on the calculated delay factors. He might determine that one or more of the factors will be lower or higher on the upcoming performance of the method. The previously determined delay factors should therefore be adjusted. This adjustment in delay factors will result in prediction of a new overall method productivity.

The MPDM user cannot assume that actual results will correspond to predicted values. If the method's leading resource is changed in quantity, a new method results and MPDM prediction becomes difficult. If only support resources are changed, prediction is still not precise. For example, three support laborers might be used on a construction method and the MPDM for the method may indicate a 30% labor delay factor. We could then ask: How much of a reduction in the delay factor will occur if an additional laborer is used for the method? The MPDM does not yield a single prediction, however, it does indicate that the method productivity should increase between 0% and 30%. Further consideration of the interdependencies between the method's resource parameters and the other delay factors should provide the MPDM user with a means of estimating the actual reduction in the delay factor and the corresponding increase in method productivity.

Once the construction productivity for the new method is documented, the prediction can be evaluated. This evaluation should serve as a means of better predicting future method productivity.

The contractor or estimator should also be concerned about the reliability of his prediction or the variability of productivity. As an aid in judging the variability of the productivity of a method, the MPDM user should focus on "variabilities of method productivity." In particular, the higher the overall cycle variability and the ideal cycle variability, the less dependable the productivity prediction. Ideally, these ratios should be small.

In addition to measuring productivity and thereby determining standards, the MPDM is useful in the work improvement task. Consider the structured MPDM in Figure 2.15 which was derived from the data collected in Figure 2.13 and processed in Figure 2.14.

Inspection of the MPDM structure for the method, indicates that the equipment delay and the management delay are critical as they result in a large percent of production delay time. Attention should probably be focused on these two types of delays when the contractor tries to improve productivity.

THE PRODUCTION FUNCTION

Simple types of models may be constructed to represent the many different possible relationships between the inputs to the construction activity and thereby determine a productivity standard. One of these models addresses the construction activity which has completely independent inputs. Others focus on the construction activity which has totally interdependent inputs. Of course, most actual construction activities contain some inputs which are totally independent in nature, and some inputs which have interdependent characteristics. Studying the simple independent and dependent input models shows that it is very probable that real construction activity consists of several components, these having either independent or interdependent characteristics.

Models with Independent Inputs

The construction activity with independent inputs may be modeled as shown in Figure 2.16. In the model, the X's represent the inputs to the activity, a and b are transformations which transform the inputs into the outputs, Y_1 and Y_2. The total output, Y, is merely the sum of the outputs, Y_1 and Y_2. The system may be modeled mathematically as follows:

$$Y = Y_1 + Y_2 = a(X_1) + b(X_2)$$

Note that the inputs and their resulting outputs are totally independent. If one of the inputs is zero—as for example, during a labor strike or equipment breakdown—output may still be obtained for the activity. In actuality, most

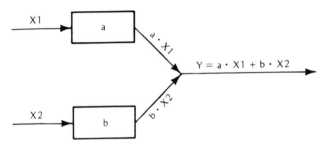

Figure 2.16 *Independent Resource Inputs*

construction activities may be visualized as containing more complex relations than those described in the independent model. Owing to their infre-
quency in the actual world, and because they may be easily analyzed, models
with independent inputs are not deserving of further study here.

Models with Interdependent Models

Most construction activities contain inputs which are interdependent.
Figure 2.17 shows a model containing interdependent production transformations. The model shows a single input being inputted to two production
processes, yielding a single output. In this system, the input to the second
process is the output of the first process. Mathematically, the process may be
modeled as follows:

$$Y = b1 \cdot b2 \cdot X$$

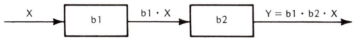

Figure 2.17   *Independent Resource Inputs*

X represents the input to the activity; the b's represent production processes.
It may be observed that if either of the production processes is inoperative—
for example, because of a machine breakdown—the production output is
zero. This model is characteristic of many types of industrial production,
such as assembly line production.

Another production system containing interdependent inputs is shown
in Figure 2.18. The construction activity can usually be represented by this
type of model. Paul Douglass, an economist, did much work on the formulation of this type of model. Working with a mathematician named Richard
Cobb, he derived a general type of mathematical model for this system. The
model takes the following form.

$$Y = dX_1^{\alpha_1} X_2^{\alpha_2}$$

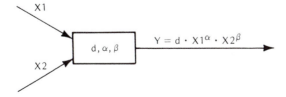

Figure 2.18   *Interdependent Resource Inputs*

The model is labeled the Cobb-Douglass production function. The d, α_1 and α_2 are constants for a particular production system. The α_1 and α_2 have significant economic importance. The sum of these constants indicates if the system has decreasing, constant, or increasing returns to scale. Decreasing returns indicate decreasing incremental output for increasing units of input; constant returns indicate constant incremental output for increasing units of input; and increasing returns indicate increasing incremental output for increasing units of input. Very few systems exhibit increasing, or even constant, returns to scale.

The Cobb-Douglass production function is often representative of construction activity production. Naturally, a construction activity may contain inputs other than interdependent inputs. A production function for such an activity may be expressed as follows:

$$Y = z + X_1^{\alpha_1} + C \cdot X_2^{\alpha_2} \cdot + d\, X_3 \cdot X_4$$

The model describes a very general type of production process. It contains a constant for output nonrelated to the model's inputs, independent input characteristics, and interdependent input characteristics.

The described production function's inputs and outputs may be either deterministic or stochastic in nature. The main difference between them is that in a stochastic system the same output may not result when identical inputs are put into the system, while in a deterministic system the resulting outputs would be the same. This section addresses itself to deterministic models. With a little effort, they may be extended to stochastic models.

The Cobb-Douglass production model may easily be extended to account for more than two system inputs. The more general form of the model follows:

$$Q = A I_1^{\alpha_1} \cdot I_2^{\alpha_2}\, I_3^{\alpha_3} \cdot I_4^{\alpha_4}$$

In the above model, Q is the output rate, I's are inputs, and A and α's are constants. If the I's are inputs for some given time interval, the model becomes a productivity model.

The production or productivity function (the terms will be used interchangeably in the remainder of this section) models the relationship between the quantities of various inputs used per period of time. Economists also define a total product of an input, an average product of an input, and a marginal product of an input. The total product of an input is the total output of the system obtained for varying amounts of the input, all other inputs being held constant. The average product of an input is defined as the total output of the system divided by the amount of input required to product this output, all other inputs being held constant. The marginal product of an input is the addition to total output, owing to the addition of the last unit of the input, all other inputs being held constant. The production function, in addition to these other concepts, will be demonstrated by an example.

Before looking at an example, mention should be made as to how measurements of production functions have been taken historically. Generally, three methods are identified when considering the measurement of production functions.

The first method is based on a time-series analysis of production in a particular industry, over many years. By studying the various inputs to the system and the resulting output, the production function is formulated. The difficulty of the method is that methods and inputs to a production system often change drastically over a large time period. These changes often make past data irrelevant to an existing production function.

A second method of measuring production functions is to gather production information from various firms manufacturing the same product, then formulate the information into a production function. This method also has problems, in that various firms may have differing capabilities: in particular, the varying sizes of their production systems may be an influencing factor.

A third method is to use technical information supplied by engineering studies. The difficulty with this approach is that historically most of these studies have addressed a specific part of the production system, while ignoring the total system.

For the purpose of presenting an example, let us assume that information necessary to produce a production function for a construction activity can be obtained. This data may be acquired in one or more of the described measurement methods. In particular, let us use a computer program which has been constructed to yield outputs (some numbers) for various combinations of inputs (numbers). The example will contain two possible inputs, X_1 and X_2, and the resulting output will be a single item, Y.

Several sets of combinations of inputs X_1 and X_2 were tried for the activity in question, and resulting outputs were obtained. It should again be pointed out that the computer was programmed to represent the activity production. The results from this production system are listed in Figure 2.19.

From the data, the system may be judged to be deterministic. This was tested by inputting identical sets of inputs to the system and observing identical outputs. Actually most activity production is not deterministic. However, it is not difficult to extend the production function to be derived, to account for the stochastic behavior of production.

The data indicate that the system has interdependent inputs, as opposed to independent inputs. This is determined by inputting zero for either X_1 or X_2, and observing the zero output. In other words, no output results if one of the inputs is absent, indicating an interdependency between the inputs.

Figures 2.20 and 2.21 show graphically the data tabulated in Figure 2.19. Note that in Figure 2.21, lines are drawn through outputs of equal production. These lines are often referred to as isoquants. An isoquant is a line or curve of points of equal output for various combinations of inputs. It should be observed that the figures actually represent the production function. A

Activity Production		
Input 1 $\longrightarrow$	Input 2 $\longrightarrow$	Output
0.0	0.0	0.0
0.0	0.0	0.0
1.0	1.0	51.12
5.0	0.0	0.0
5.0	10.0	240.96
5.0	15.0	271.07
5.0	20.0	287.13
5.0	25.0	300.23
6.0	10.0	283.99
1.0	20.0	93.07
2.0	20.0	151.19
3.0	20.0	200.81
4.0	20.0	245.61
6.0	20.0	326.21
10.0	0.0	0.00
10.0	10.0	406.06
10.0	15.0	440.36
10.0	20.0	466.44
10.0	25.0	487.73
11.0	10.0	434.08
11.0	20.0	498.62
15.0	0.0	0.0
15.0	10.0	539.33
15.0	15.0	584.89
15.0	20.0	619.53
15.0	25.0	647.80
16.0	10.0	564.25
16.0	20.0	648.16
20.0	10.0	659.65
20.0	15.0	715.37
20.0	20.0	757.74
20.0	25.0	792.32
21.0	10.0	682.56
21.0	20.0	784.06
25.0	10.0	771.17
25.0	15.0	836.31
25.0	20.0	885.84
26.0	10.0	792.63
26.0	20.0	910.50

Figure 2.19 *Computer Output for a Production System*

three-dimensional model could be constructed to show the various outputs obtained from different combinations of the two inputs.

Figures 2.22 and Figures 2.23 and 2.24 show calculations and graphs for the total product, average product, and marginal product of input X_1. In all these measures, input X_2 was held constant at 20 units per time interval. Observe that the marginal product of input X_1 decreases after an initial increasing stage. This indicates the production system, with respect to input

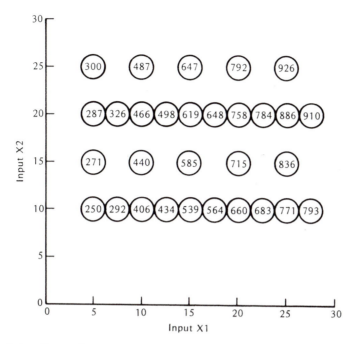

Figure 2.20 *Output from an Example Production System*

X_1, is a decreasing returns-to-scale system. Later, it will be shown that the entire production system studied is a decreasing returns-to-scale system.

Having observed that the production system is characterized solely by two interdependent inputs, the system may now be modeled mathematically, using the Cobb-Douglass production function. The general form of the production function for the activity is as follows:

$$Y = {}_dX_1{}^{\alpha_1}X_2{}^{\alpha_2}$$

To quantify the model, the constants d, α^1 and α^2 must be determined.

To determine the constant d in the model, inputs of X_1 = 1.0, and X_2 = 1.0 were input to the system. Because 1.0 raised to any exponent is 1.0, the model becomes Y = d. The resulting output obtained was Y = 51.12, and the coefficient d was thus determined as 51.12. The model, therefore, takes the following form:

$$Y = 51.12X_1{}^{\alpha_1}X_2{}^{\alpha_2}$$

To determine the values of the coefficients, α_1 and α_2, two sets of inputs and corresponding outputs were inputted so that two equations may be determined and solved. Following this method and checking various other

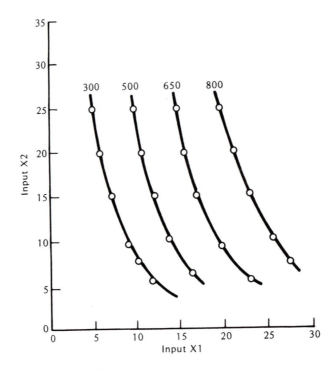

Figure 2.21 Output from a Production System

Input X_1	Output	Average Product	Marginal Product
0	0	0/0 = 0	
1	93	93/1 = 93	93/1 = 93
2	157	151/2 = 75	151 − 93 = 58
3	201	201/3 = 67	201 − 151 = 50
4	246	246/4 = 61	246 − 201 = 45
5	287	287/5 = 57	287 − 246 = 41
6	326	326/6 = 54	326 − 287 = 39
10	466	466/10 = 47	
11	498	498/11 = 45	498 − 466 = 32
15	619	619/15 = 41	
16	648	648/16 = 40	648 − 619 = 29
20	758	758/20 = 38	
21	784	784/21 = 37	784 − 758 = 26
25	886	886/25 = 35	
26	910	910/26 = 35	910 − 886 = 24

Figure 2.22 Output and Average and Marginal Products from a Production System (Inputs X_2 is Fixed at 20; Average Product is Rounded to the Nearest Integer)

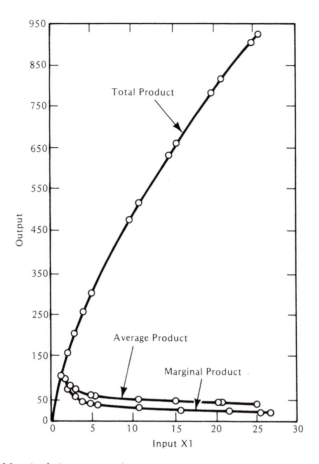

Figure 2.23 *Marginal, Average, and Total Products from a Production System*

points to insure correctness, α_1 is found to be equal to 0.70, and α_2 equal to 0.20. The production function may be completed and represented as follows:

$$Y = 51.12\,(X_1^{0.7})(X_2\alpha^{0.2})$$

The mathematical model may now be used to further study the construction activity it represents, and also to predict outputs for any combination of X_1 and X_2.

Earlier, it was stated that the sum of α_1 and α_2 indicates whether the production system has increasing, constant, or decreasing returns to scale. In particular, if the sum is less than one, as in this case (0.90), it indicates decreasing returns to scale.

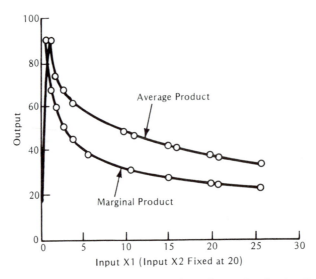

Figure 2.24 *Average and Marginal Products from a Production System*

Note, also, that the average product of an input and the marginal product of an input may be obtained directly from the derived mathematical model. The average product of an input may be found by merely dividing the production function by the input, the marginal product of an input may be found by taking the derivative of the production function with respect to the input in question.

To make further analysis of the construction activity meaningful, we must define our immediate objective for such a further analysis. Let us assume we are interested in determining the minimum cost ratio of the inputs to be used in obtaining various amounts of production. It may be shown that when the isoquants of production are parallel (as in this example), the minimum cost ratio of the inputs is constant for the various outputs produced. Minimum cost ratio of inputs refers to the ratio of input amounts used to obtain a desired production, given the unit costs of the various inputs. This, in effect, is the same as finding the ratio of inputs to be used to minimize the total costs of all of the necessary inputs.

The objective of minimizing cost is acutally a realistic objective for production systems. If the firm has the objective of maximizing profits, it is faced with a required production output, and the unit price of the output is constant—as is often the case—then the objective may only be accomplished by minimizing costs. Of course, if output is not fixed, or the unit price of output is a function of supply or demand, then the profit maximizer may not want to minimize costs.

The objective of minimizing costs may be a good starting point in dealing with construction activities. Later, in the development of a productivity model, it may be necessary to somewhat "relax" the objective, owing to target dates, penalty costs, and so on. Let us assume that the unit costs of each of the inputs is constant, regardless of quantity purchased or used. This is the case in a perfectly competitive market.

It may be easily shown that the firm will minimize its input costs if it distributes its expenditures among the various inputs, so that the marginal product of a dollar's worth of any one input is equal to the marginal product of a dollar's worth of any other input used. Thus, if a firm is using several inputs in a production system, it should employ them in a combination so that

$$MP_1/P_1 = MP_2/P_2 - MP_3/P_3 \ldots MP_n/P_n$$

where the MP's represent the marginal products of inputs 1, 2, 3...., n, and the P's represent unit prices in inputs 1,2,3,..., n. This is easily demonstrated to be the optimal policy. For example, assume one of the ratios is smaller than the rest. Since the marginal product per dollar spent on that input is less than on the other inputs, the firm should spend some of the money on the other inputs, thus obtaining a larger total product.

Knowing this optimal policy, how may it be applied to our previous example? The answer is simple. We merely have to know the prices of the inputs, their marginal products, and then equate the defined ratios. Let us define the price of a unit of input X_1 as PX_1, and the price of a unit of input X_2 as PX_2.

One of the easiest ways to solve for the optimal combination of resources to use is to use the Lagrangian function:

$$L = dX_1{}^{\alpha_1} X_2{}^{\alpha_2} - (X_1 PX_1 + X_2 PX_2 - C)$$

The first part of the function is merely the production function; the second part is the cost function for the system, multiplied by a constant. We now find the marginal product, with respect to cost, by taking the derivative of the function with respect to input X_1 and input X_2, and solving. Such information yields the following:

$$X_2 = (\alpha_2/\alpha_1)(PX_1/PX_2)(X_1)$$

We could have obtained the same relationship by merely determining the marginal product of each input (finding the derivative of production function) and dividing these by the respective prices of the inputs.

Having determined this ratio of inputs, we may now proceed to solve our example problem. Let us assume that the unit price of input X_1 is 1.0, and the unit price of input X_2 is 2.0. The optimal combination of inputs (or optimal method) to use in obtaining any desired output from our example production system may then be determined as follows:

$$X_2 = (0.2/0.7)(1/2)(X_1) = 0.14X_1$$

Therefore, 0.14 units of input X_2 should be used for every unit of input X_1. This optimal combination of inputs determines the optimal method for producing output. The ratio of optimal inputs is shown versus various amounts of production in Figure 2.25. Since the ratio is a constant (for parallel production isoquants), the optimal combination of ratios of inputs plots as a straight line. In economics, this line is often referred to as the expansion line of the firm. Given a required production or productivity, an optimal method of obtaining it may be derived.

Having determined the optimal combination of inputs (minimum total cost) to be used in obtaining various production amounts of the activity, economists extend the model to determine the optimal output, given an objective such as profit maximization. A production cost function may be determined by plotting the total cost of the inputs to the system against the corresponding output. Such a graph is shown in Figure 2.26. The optimal cost and output are then determined by equating marginal revenue to marginal cost. Marginal cost is the added cost associated with an additional unit of output. When the selling price is constant, marginal revenue is simply equal to price. In this case, the optimal policy is to equate the price to marginal cost.

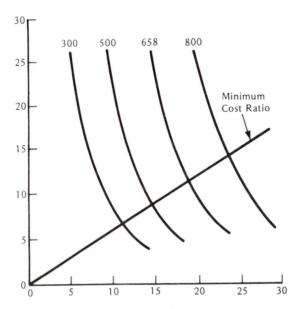

Figure 2.25 *Optimal Outputs from a Production System*

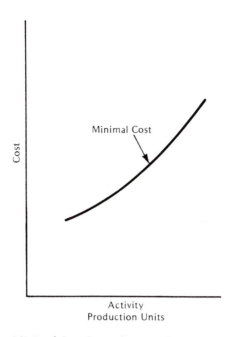

Figure 2.26   *Minimal Cost Curve for a Production System*

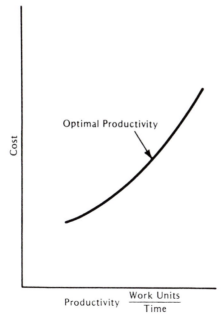

Figure 2.27   *Optimal Productivity Curve for a Production System*

When considering the production associated with a construction activity, the production cost graph shown in Figure 2.26 no longer has the same meaning. The total production (output) required for a construction activity is fixed by the project specifications. However, the amount of work per day, or hour (productivity), is not usually fixed, and must be determined by the contractor. If we change the ordinate in Figure 2.26 to productivity rather than output, as in Figure 2.27, we may visualize the graph as a productivity cost trade-off curve. On a given day, the firm may use one of the feasible productivities shown by the graph. To optimize such a selection with respect to the total project, the other activities within the project would have to be considered.

Undoubtedly, the benefits to be received from being able to map an activity into a production function are more or less unlimited. However, it should be observed that it may be difficult to gather the input-output information which is needed to generate an activity production function.

COMPARING ALTERNATIVE CONSTRUCTION METHODS

Construction estimating is in great part the determination of work to be performed and the determination of the cost or price of doing that work. Perhaps the most difficult and challenging of these two somewhat independent processes is the determination of cost. Skillful determination of the cost of doing work is not limited to the knowledge of costs of the labor, material, and other direct costs of doing the work. It is also dependent on the estimator's choice of alternative construction means and methods of doing the work.

For each type of work activity to be performed for a project, a constructor typically has several alternative means of performing the work. For example, for placing concrete in a suspended slab, the contractor might choose among different types and sizes of labor crews using either a concrete pump, or a crane and bucket, or perhaps even a conveyor system. Or a contractor may choose between forming a wall with a job-built system or may use one of several available steel-ply pre-built forming systems as an alternative. Each of these alternative construction means and methods may result in a different work duration and different unit and total cost.

The choice of a given construction method by a contractor is dependent to some degree on the available resources. For instance, a contractor who owns a crane is more likely to use that crane for placing concrete rather than a rented pump. But, regardless of whether available resources are owned or rented, the contractor should make an economic analysis of alternative means and methods of performing a specific type of work in order to estimate and do the work efficiently for the least cost.

The economic analysis of alternative means and methods of construction for a specific type and quantity of work can and should be considered part of the estimating function. If an estimator is to determine accurate and economical costs of doing work, that estimator must be in a position to determine the "best" method. What is best is in some measure dependent on the estimator and contractor's objectives. However, best usually means the lowest cost method.

The analysis of alternative means and methods for a specific type and quantity of work usually entails consideration of both direct labor cost and direct material cost. If equipment is needed to perform the work, that also has to be recognized in the analysis. Often when two different resources are part of the analysis—such as labor and material—one resource will cost more and one less for one method relative to the other. For example, a specific type of steel-ply pre-built form may itself cost more than job-built wood forms for a certain type of concrete wall forming that has to be done. However, the direct labor cost for placing the steel-ply pre-built forming system may be considerably lower than the direct labor cost for placing the job-built system. The estimator's recognition of the combined significance of the two costs is what is needed for determining the most economical method.

Figure 2.28 displays a good format for comparing a cost analysis of two alternative means of performing work. Additional means could be analyzed, using this same process. While the analysis shown in Figure 2.28 implies that the cost comparison is valid independent of the quantity of concrete forms to be placed, there might be instances where the relative costs of the two methods could change. For example, given the relatively high cost of a steel-ply pre-built forming system, its cost per work quantity may estimate higher than for a job-built system if the needed total square feet is relatively small.

METHOD ANALYSIS

Forming Method	Initial Quantity Estimate (sfca)	Revised Quantity Estimate (sfca)	Quantity Placed To Date (sfca)	Total Estimated Labor Hrs	Labor Hours To Date	Projected Labor Hrs Variance
System 1	2000	2500	1250	200	125	50 over
System 2	1000	1000	400	100	40	0

Figure 2.28   *An Example Job Cost System Report with Work Item Information*

This may be a result of too few or no multiple re-uses of the steel-ply forms; however, that same pre-built system may be cheaper than the job-built system when the square feet of contact area of forming increases. Then the contractor is able to take advantage of the multiple re-uses possible with the steel-ply system.

It is obvious that a competent estimator should know means and methods of construction in order to accurately determine the cost of construction work. To emphasize this point, each subsequent chapter on the preparation of a detailed contractor direct cost estimate includes an example of an economic analysis of at least two alternative construction means and methods for performing a specific type and quantity of work.

EXERCISE 2.1

A contractor has determined for an upcoming project that 2,500 cubic yards of concrete has to be placed, and is studying two alternative means. The contractor wants to perform the work at the least cost in order to maximize potential profits. The two possible construction methods are as follows:

Pumping Concrete with Rented Pump This way, the contractor rents a pump at a cost of $200 per day. The contractor estimates that he can place 100 cubic yards of concrete per day using this method. The following crew size will be needed to place and/or finish the concrete:

   6 laborers at $12 per hour each
   4 carpenters at $14 per hour each
   2 finishers at $15 per hour each
   1 operator at $16 per hour each

Placing Concrete with Own Crane Bucket This way the contractor uses its own crane with a bucket. The contractor estimates that he can place 50 cubic yards of concrete per day using this method. The contractor has also estimated the cost of owning and operating the crane at $12 per hour. (It is assumed that the crane can be used during the same time period on another project or be rented to another contractor for this hourly rate if it is not used on this project.) The following crew size will be needed to place the concrete:

   4 laborers at $12 per hour each
   2 carpenters at $14 per hour each
   2 finishers at $15 per hour each
   1 operator at $16 per hour each
   1 oiler at $14 per hour each

The contractor will only work 8-hour days and 5-day weeks. The contract documents for the project indicate that if the placing of concrete takes in excess of 40 days, the contractor will be penalized a cost of $50 per day for each day over the 40 days.

Given the contractor's above data regarding the productivity and cost of the resources for these two alternative construction methods, you are to assist him in selecting the better construction method to minimize the cost of the work. You should also calculate the work duration in hours for this better method, and the total cost of the work.

EXERCISE 2.2

Using time-study techniques, an observer studies the placement of 2' x 8' wall forms for a straight wall. Three carpenters are employed, and the observer breaks down the overall work process into the following operations:

Operation 1 Carpenters walking to get form.
Operation 2 Carpenters picking up form.
Operation 3 Carpenters taking form to placement location.
Operation 4 Carpenters positioning and putting form in place.

The following times in seconds are collected for each of 8 cycles:

	1	2	3	4	5	6	7	8
Operation 1	135	130	130	135	130	120	120	120
Operation 2	120	120	125	120	140	130	130	140
Operation 3	135	140	145	160	150	125	130	120
Operation 4	170	140	150	180	140	135	180	150
Total time	560	530	550	595	560	510	560	530

Assuming a rating factor of 1.1 and an allowance factor of .15 solve each of the following 3 problems:

a. Calculate a standard time in seconds per cycle.
b. Calculate the standard time productivity in terms of carpenter hours per square foot of contact area.
c. Assume that the estimator judges Operation 3 (taking form to placement location) can be reduced by one-half the time and recalculate the standard time for the process.

EXERCISE 2.3

The unit cost of any type of construction consists of several cost components, including labor craft costs, material costs, and equipment costs. Often the increase (or decrease) in the composite unit cost is distorted by any one single cost component. In other words, the mere fact that a unit cost increases 10% does not mean that each cost component increases by 10% only.

Let us assume the following data during a given year:

BEGINNING-OF-YEAR COST		PCT. INCREASE (DECREASE) AT YEAR END
Concrete Finisher	$14.00/hr.	15%
Laborer	$12.00/hr.	10%
Cubic Yard of Concrete	$50.00/yd.³	– 5%
Pumping Machine	$80.00/hr.	5%

Let us further assume that to place a cubic yard of concrete, 0.8 finisher hours, 0.4 laborer hours, and 0.4 pumping machine hours are required.

Calculate the unit cost at the beginning of the year for placing concrete, the unit cost at the end of the year, and the percentage increase (decrease) at year end.

3

Cost Accounting: Terminology Relevant to the Construction Estimate

Definition of a Cost: Its Importance in Estimating *
Different Types of Cost * *Job Order Costing* *
Standard Costs * *The Measure of Cost as a Function of*
Time * *Interest Formulas* * *Cost Indexes* * *Costs*
and Work Items

DEFINITION OF A COST: ITS IMPORTANCE IN ESTIMATING

As noted previously, the successful preparation of a construction estimate is essentially the combining of the determination of work to be performed, of the determination of the productivity of the resources used for the work, and of the determination of the cost of each of these resources. Therefore this chapter addresses resource costs just as Chapter 2 addressed the impact of productivity on estimating. (In part, Chapters 6 through 15 further address the first matter—the determination of work to be performed for a project.)

The importance of the estimator's ability to understand the concept of a cost and its components is highlighted by the fact that the cost estimate serves to initiate the project and to engage the contractor who will build that project. So the contractor must make a forecast of costs in order to bid a project. In effect, the cost estimator sets out to accurately judge the potential for profit on a project.

Determination of cost is the final step in the preparation of an estimate. Given a determination of work to be performed for a project, and given the resources to be utilized and their productivity, the determination of the cost of the work may seem somewhat easy. However, this is not necessarily the case. Even assuming a known amount of work and productivity, the estimate of the cost of doing the work is dependent on knowledge of different types of costs as well as the variability of costs as a function of time. We will consider both in detail in this chapter.

Cost accounting encompasses the study of costs. The very term "cost" is subject to many interpretations. One interpretation is that cost is anything one has to give up. This would include one's time and energy, in addition to dollars or non-monetary products exchanged. However, cost as it relates to accounting study is more limited. In an accounting sense cost means dollars exchanged for goods or services, and it is this meaning that will be used here. Dollars paid for material, for labor, or for a bond or insurance are examples of cost to the construction firm.

A properly designed cost accounting system assigns the various types of costs that accrue to certain defined cost objects. A cost object is an activity or part of an organization for which a separate determination of cost is needed, and they are defined in a manner consistent with the decision-making needs of management. For example, if the construction firm wants each project superintendent to be responsible for individual job overhead, then a job overhead cost object should be set up for each individual job. Similarly, segregating overtime hours from normal working hours may be the result of management's desire to analyze overtime policies. Thus management may choose to allocate such overtime labor costs to job overhead rather than to direct cost. Regardless of its decision or policy as to the given cost, the firm's very decision to segregate the cost results in the creation of a cost object. Various construction work items such as concrete slab work, footings, masonry walls, etc., are commonly segregated as to data collection. Each of the work items is thus identified as a cost object. Cost objects can be identified either as individuals, or work items, or even segments of one's business.

DIFFERENT TYPES OF COST

Costs in an accounting sense are often identified as to behavior—either as a function of time or as a function of relations to the makeup of the product produced or service performed. Variable or fixed, unit or total, and product or period are terms used for such purposes.

Costs that vary directly with changes in activity are called variable costs. Examples of variable costs are costs of material and labor used in the production process, and equipment hours expended in the manufacturing process. Fixed costs are necessary recurring costs. They ordinarily vary only indirectly with the passage of time despite changes in the level of activity. For example, fixed costs include a secretary's wages, and the salaries of estimators or a superintendent who are employed the full year regardless of whether or not company jobs are available.

It should be noted that a fixed cost may vary substantially if the level of activity changes substantially. For example, if a construction firm reorganizes so that its volume is reduced by 50%, it may be necessary to lay off an estimator or two. As such, the fixed cost is no longer "fixed." The end result

is that when speaking about fixed cost, one must recognize the existence of the relevant range of business activity. The relevant range is that level of activity for which the firm budgets and expects to operate. Thus, the definition of fixed cost is valid only for the firm's relevant range of activity. To illustrate the problem of fixed costs not being fixed over unexpected activity levels, the term semifixed is introduced. The concept of variable cost, fixed cost, and semifixed cost are shown graphically in Figure 3.1.

It should be noted that what is a variable cost to one firm, might in fact be a fixed cost to another firm. For example, one construction firm hires and fires job superintendents as a function of workload, whereas another firm may keep a given number of superintendents on its payroll irrespective of

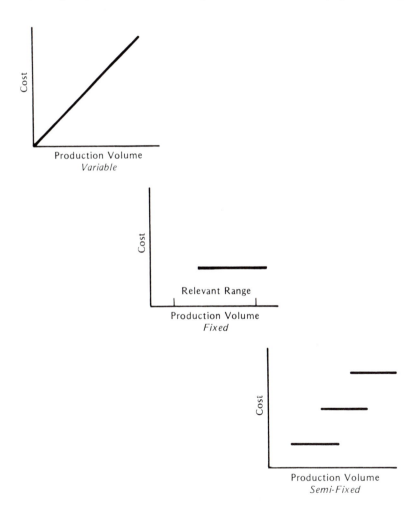

Figure 3.1 *Example of Schematic Showing Variable, Fixed, and Semifixed Costs*

workload level. The advantages and disadvantages of each of these policies is not at issue here. What is at issue is the fact that the costs of superintendents are variable to the first firm, whereas to the latter firm they are fixed.

Most firms operate by absorbing both variable and fixed costs. Many production costs—those directly related to making a product—tend to be variable, whereas selling and administrative costs tend to be fixed. The behavior of all a cost object's costs, including variable and fixed, taken together are expressed as cost functions. A simple cost function is illustrated in Figure 3.2. Needless to say, not all cost functions are as simple as that shown. The concept of cost functions, variable costs, and fixed costs are used in cost-volume-profit analysis and in variance analysis. These types of cost accounting analysis are studied in the following sections.

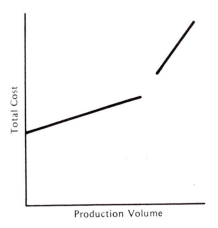

Figure 3.2   *An Example of a Simple Cost Function*

Costs associated with a product are classified as being either unit or total costs. Total costs merely represent the sum of all the costs associated with the type of production or cost object in question. For example, the sum of the costs for placing 25 cubic yards of slab concrete, 20 cubic yards of wall concrete, and 5 cubic yards of footing concrete might be $1,000. This $1,000 represents the total cost of placing the whole 50 cubic yards of concrete. Unit costs merely represent the total cost of the production in question divided by a defined unit of the total production, and can be thought of as average costs. For example, the common production unit in the example above can be identified as a cubic yard of concrete. As such, the unit cost of the concrete is calculated as $1,000 divided by 50 cubic yards, averaging $20 per cubic yard.

Note should be made of the importance of the identification of the base or common production unit in the unit cost calculation. Different bases can be defined for a given type of work and each different base will result in a different calculated unit cost. For example, three types of concrete placement were considered in the calculation of the unit cost of placing a cubic yard of concrete. If we were to limit the production unit to a given type of placement, such as the placing of slab concrete, it is likely that the unit cost for placing a cubic yard of concrete would be different. If $250 of the $1,000 total cost is associated with placing the 25 cubic yards of slab concrete, the unit cost per cubic yard of placing slab concrete is calculated as $250 divided by 25 cubic yards or $10 per cubic yard. The consideration of the base in the unit cost calculation is especially relevant when considering the several different possible bases that might be identified when singling out the various types of construction work performed. That is, an analyst might either single out total concrete or more detailed bases such as footing concrete, or 4-inch slab on grade concrete, or 6-inch suspended slab concrete, or wall concrete. In addition, the production unit might be identified as cubic yards of concrete, or square feet of concrete, or linear feet of concrete.

Another important way to identify costs in regard to a construction process is by classifying them as either product or period costs.

Product Costs in Construction

Costs incurred in transforming materials into a useful product are referred to as product costs. These types of costs can be thought of as "attaching" to the product. Concrete and steel costs, carpenter and ironworker wages, and even the job superintendent's wages—assuming he is allocated to supervision of the production process—are product costs. In general, production related costs are identified as product costs. Product costs are sometimes referred to as inventoriable costs.

Product costs are closely related to the concept of variable costs. While many product costs are in fact variable, not all product costs are variable costs. The wages of a superintendent may be a fixed cost to a construction firm. On the other hand, his time and wages associated with his supervision of a given project may be considered product costs for that project.

Period Costs in Construction

Period costs do not result directly from the manufacturing of the product. They are indirect costs related to the selling of the product and the overall general administrative costs of operating the firm. They do not "attach" to the product produced. Secretarial wages, estimator's wages, and office equipment are examples of construction firm period costs. (Note, however, that if the secretary or estimator is "assigned" to a given job, these labor costs can then be considered as product costs.)

It is also true that while most period costs are fixed in nature, there are exceptions. Personnel responsible for finding work may be paid a commission based upon the dollar volume of work contracted. This cost, while not attached to the work itself, is in fact a variable cost.

Tax Implications of Product and Period Cost Differences

The concepts of product cost and period cost have implications for both the financial statements and the tax liabilities of a construction firm. Whereas product costs are visualized as attached to the product, they are identified as being part of an inventory of products that the firm maintains or holds at the end of a reporting period. As such, these costs are not viewed as expenses to the firm until the inventory is sold. Instead, they are temporarily viewed as assets. On the other hand, period costs do not attach to the manufactured product and as such, are viewed as expenses in the period in which they occur. The difference between the two accounting procedures is very important in construction work because the rather long duration of projects is common to the construction industry. That is, the construction project, which can be viewed as inventory or work-in-progress, may have that status for two or three years. Thus rather significant differences in financial statements and tax implications result from the alternatives of viewing a cost as a product or period cost.

Types of Manufacturing Costs

Perhaps the most meaningful classification of cost for use in a cost accounting discussion is the classification of costs as elements of manufacturing costs. Three elements of the costs of manufacturing a product are identified as follows:

1. Direct material costs.
2. Direct labor costs.
3. Overhead costs.

Direct material costs are those that constitute an appreciable part of the finished product. Concrete, steel, and lumber costs are all direct costs of the construction firm. Certain minor materials such as nails or glue may sometimes be identified as indirect. That is, for convenience they may be called project overhead costs rather than direct costs, since it may be difficult to assign a given number of nails or glue to a given project or work item. However, difficulty by itself does not justify identifying a given cost as an indirect cost. If the amount of the item is significant, it is necessary, regardless of the difficulty, to identify it as direct and relate it to the work item in question.

Direct labor costs are those costs involved in the transformation of material into a product. Laborer, carpenter, ironworker, and foreman wages are all examples of direct labor costs. But overtime premiums are sometimes more appropriately assigned to overhead than to direct labor. This is the case when overtime results because of external conditions. For example, if laborers are pulled off Project A in order to finish Project B, the later occurrence of overtime on Project A (in order to finish it on schedule) does not justify penalizing Project A with the overtime premium. It is more appropriate either to charge this time directly to Project B, or to charge the premium costs to overhead.

Overhead includes all the costs necessary to the manufacturing operations of the firm that cannot be directly identified with a given product. Overhead is also referred to as factory overhead, manufacturing expense, or indirect manufacturing. The latter term is probably the most descriptive of the actual costs in question.

Overhead is further identified as either variable or fixed. Variable overhead includes such items as supplies and indirect labor. An estimator's time

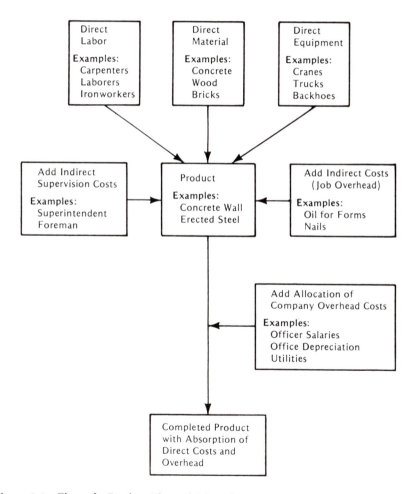

Figure 3.3   *Flow of a Product Through Manufacturing Process*

is most appropriately identified as indirect labor. Additional variable over-head would include a job superintendent's work since it is appropriately identified as indirect labor, rather than direct labor. Fixed overhead includes such construction costs as insurance for the firm's central office, deprecia-tion of its own building, and some office supervisory salaries.

Prime costs and conversion costs are two less familiar terms that are sometimes used in grouping direct material, direct labor, and overhead. Prime costs consist of the sum of direct material costs and direct labor costs. Conversion costs consist of the sum of direct labor costs and overhead costs.

An overview of the various manufacturing costs and other costs that have been discussed so far is brought together in Figure 3.3, which focuses on the flow of a firm's product through its production and selling cycles. The illustration shows a firm in the business of manufacturing a product. This is, of course, true of the construction firm. However, a retail firm, which does not manufacture a product but only buys and resells it, has no direct labor in-put to the manufacturing process and thus is not illustrated in Figure 3.3.

One other classification of costs should be mentioned before leaving our discussion of construction terminology. A cost is either a historic cost or a budgeted (predetermined) cost. A historic cost is one that is determined from the record of an actual cost. Most construction estimating systems are based on historic costs. Budgeted or predetermined costs are estimates and predic-tions of what costs should be, based on historic costs and on an analysis of the work performed.

JOB-ORDER COSTING

The construction industry differs from most other manufacturing industries in that the cost of its product usually has to be estimated before the product is even contracted for. Manufacturing industries such as the automobile indus-try measure their direct costs and distribute overhead to their finished auto-mobiles. This measured cost is then used as a basis of fixing the selling price. The unit direct costs and overhead costs are relatively constant because of the repetitive characteristics of their manufacturing process. Construction projects do not share this cost characteristic. Even if two projects are similar in their appearance, the different environments in which they are built usual-ly result in the builder having to predict future costs as a basis of his present contract price. This is especially the case in regard to a competitive bid lump sum contract.

Cost-plus fixed fee or cost-plus percentage contracts do provide the builder an opportunity to determine costs before billing the project owner for the work performed. With these, there is less financial risk for the builder since what he is paid is a function of his actual costs. He is reimbursed for his costs plus a negotiable profit.

In order to be able to determine product costs so as to properly determine the "cost" portion of a cost-plus contract, the construction firm needs a reliable means of formulating job costs. In other words, in the competitive lump sum bid contract process, the contractor needs to formulate job costs so that the firm can better estimate costs of a project.

The construction firm usually builds several projects simultaneously. During construction, each project is considered to be part of the firm's work-in-progress. The basic document that is used to accumulate costs for each project is part of the firm's work-in-progress and is called the job order or job-cost sheet. Such a document is shown in Figure 3.4. The file (referred to as subsidiary ledgers) of job-cost sheets for uncompleted projects makes up the firm's work-in-progress and is summed and controlled by a work-in-progress control ledger account.

If material is procured by means of field purchase requests, the purchase requests discussed previously are used to charge job costs sheets for direct material used. Work tickets or time cards prepared by a project timekeeper or foreman are used to charge jobs for direct labor used. This process of charging direct material and direct labor to jobs is shown in Figure 3.5.

Overhead costs are applied to a project's job-cost sheet as a function of a pre-determined base or bases. For example, job overhead (direct overhead) might be applied to a project on the basis of 10% of the direct labor cost expended. Similarly, general overhead (indirect overhead) might be applied at a rate of $1,000 per month of project duration. This allocation process is shown in Figure 3.5.

Actual job overhead costs are summarized in a job overhead control account. If the predetermined overhead base cost per unit, or percent of direct labor cost, is accurately determined and actual job overhead costs are controlled, the applied job overhead will equal the actual job overhead costs. Control of the job overhead costs during construction is facilitated by comparing the actual costs to those applied for a given time period. For example, assuming direct labor costs of $10,000 and a job overhead rate of 10% of project direct labor costs, $1,000 of job overhead would be applied to a job. However, if the actual job overhead cost is $2,000 at the time in question, management should probably investigate the reason for this cost overrun and take any required corrective action.

General overhead costs can be handled in a manner similar to job overhead costs. However, it is usually more difficult to allocate actual general overhead costs to a project. As such, the accuracy of the basis by which these costs are to be applied to the job cost sheets becomes very important. Because actual costs are difficult to compare to the applied costs, control of general costs becomes more difficult. If the allocation base unit is inaccurate, jobs may be underapplied or overapplied. For example, a firm's total annual general overhead cost is expected to be $1,000,000. Expecting 1,000 months duration of work throughout the year, the firm would apply general

Sample Construction Co., Inc.

Project 101

DATE	DES-CRIPTION	LABOR COST	MATERIAL COST	JOB OVERHEAD COST	APPLIED OVERHEAD COST	SUB-CONTRACTOR COST
6-24-X0		$5,000				
6-25-X0			$4,000			
6-25-X0	BONDS			$9,000		
6-30-X0					$4,500	

Figure 3.4 *Sample Entries on a Job Cost Ledger Sheet*

overhead on the basis of $1,000 for each month of project duration. However, if the firm underestimates its annual general overhead or overestimates its volume of work, too little general overhead will be applied to each and every project. The end result is that the firm will be making less than planned

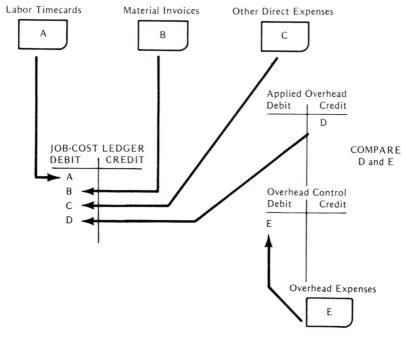

Figure 3.5 *Job Cost or Allocation by Project*

profits on their projects. In fact, if the general overhead is grossly under-stated, the firm may actually lose money on their projects.

Overhead is underapplied when the applied amount is less than the actual overhead costs. It is overapplied when the applied amounts exceed the actual overhead costs. Proper control should aid in keeping underapplied overhead to a minimum. However, even with adequate control, there is likely to be weekly or monthly differences in the amount of actual overhead costs and those applied. This is due to variations in the nature of such costs. For example, the actual purchase of small tools (a job overhead cost) may occur at the beginning of a project whereas the applied cost of this overhead is distributed more evenly over the duration of the project.

There are several accounting alternatives for treating differences between actual overhead costs and those applied. On the one hand, the difference can simply be ignored in recognition of the fact that it is likely to even out in a project's accounting period. On the other hand, if a project is complete, it is best to charge (or deduct) the difference in the actual overhead costs and the applied amount to the job-cost sheet.

Projects which are completed are transferred out of the work-in-progress accounts. This is done by eliminating them (crediting the account) from the work-in-progress account and placing them (debiting an account) in a completed contract account. This completed contract account is referred to as a finish goods account in many manufacturing industries. The entire job-costing procedure is summarized in Figure 3.5.

STANDARD COSTS

Work and cost standards are noticeably absent in the construction industry. This lack of standards was discussed in the previous chapter. Such standards are a necessary part of a budgeting and control system. Much of the construction industry's inability to increase its productivity substantially can be traced to its inability to establish work-and-cost standards methodically. There is a direct correlation between an industry's ability to increase its productivity and its ability to set work and cost standards. The better the industry's potential for setting such standards, the larger the potential for increased productivity.

Undoubtedly the uniqueness of each construction project and the variability of each project's environment combine to complicate the task of setting standards. However, the neglect of cost accounting concepts can also be cited as a significant reason for the failure to determine and implement meaningful standards in the construction industry.

The concept of standard work and cost is yet another way of measuring a cost. For instance, standards for a construction method outline how a given work package should be accomplished and how much it should cost. As work is performed, comparison of actual costs and production with standard production costs reveals variances that may be economically dangerous.

Standard costs can be considered as target costs. They are carefully predetermined costs and should be attainable. They should not be set as ideal standards that require "perfection" on the part of management or the individual worker. However, standards should be set high enough for their achievement to signify a satisfying accomplishment.

For direct costing, two different types of standards should be set up. Each construction method should have a standard for direct material and a standard for direct labor. Let us first consider the determination of standards and the analysis of variances for direct material.

A standard amount of material for a given construction method and a standard unit price for the material are determined first. The amount of material required is probably best determined by the foremen or similar personnel. In a large, more sophisticated firm, a method's department would analyze method material requirements and set the standards. Cost-oriented personnel are the likely persons to set unit price standards for material. Typically, accounting or estimating personnel are best qualified to make such determinations.

The amount of material actually used and the cost of the material used are compared to the predetermined standards. For example, let us assume we are considering pouring concrete for a 5-3/4 inch on-grade slab. The slab is 100 feet square. Assume that the standard amount of concrete for this slab is 185 cubic yards. This amount is based on the actual volume and a small wastage percentage based on it. The estimating department sets $20 as the standard cost per cubic yard of concrete delivered to the site. Comparison and analysis of any variances between actual material and costs and the standards is then made by analyzing data in a columnar format, as in Figure 3.6. The figures entered in Figure 3.6 indicate that 192 cubic yards of actual concrete was used for the slab and that the unit price of the concrete was $19.50 per cubic yard.

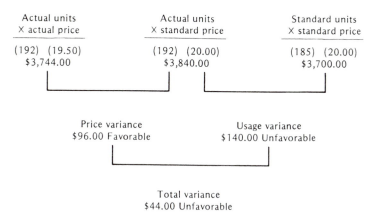

Figure 3.6 *Analysis of Material and Cost Variances*

The analysis is carried out by focusing on two variances. One is the purchase-price variance, the other is the usage variance. The purchase-price variance is the difference between the product of the actual amount of material used and the unit purchase price and the product of the actual amount of material used and the standard unit price. This difference is $96 in the example shown. If the amount of material used times the unit purchase price cost more than the amount of material used times the standard unit price, the variance determined would be considered unfavorable. That is, the variance would indicate that the material was purchased at an unfavorable price. Responsibility for such a price variance should probably focus on the purchasing or procurement department. This same department should be favorably recognized when the variance is favorable, as occurs when the purchase price is less than the predetermined standard unit price. In the Figure 3.6 example, the $96 difference is a favorable purchase-price variance.

The second variance to be determined in a material-cost analysis is the usage variance. It is calculated by taking the difference between the product of the actual amount of material used and the standard unit price, and the product of the predetermined standard amount of material for the work and the standard unit price. In the example in Figure 3.6 this difference is $140. If the amount of material used times the standard price is more than the standard amount of material times the standard unit price, the usage variance is unfavorable. Such an unfavorable variance can usually be traced to excessive material wastage and to poor handling and placing procedures. However, a favorable usage variance will exist when the amount of material used times the standard unit price is less than the standard amount of material times the standard unit price. An unfavorable usage variance of $140 is shown in the Figure 3.6 example. Responsibility for favorable or unfavorable usage variances should be credited to the project foreman or job superintendent.

The total material variance can be determined by summing the purchase-price variance and the usage variance. In the example considered there is a $96 favorable purchase price variance and a $140 unfavorable usage variance resulting in a $44 differential or an unfavorable total variance. It should be noted that merely focusing on the difference between total actual cost and a total standard cost (that is, the total variance) can be misleading in regard to method efficiency. In this example, the favorable efforts of the purchasing personnel are negated by inefficient material usage at the job site. Similarly, much of the inefficient material usage is hidden because the favorable purchase price of the material offsets some of the failure to maintain predetermined standards. Adequate control can only be determined by focusing on the purchase-price and usage variances.

The determination and analysis of direct labor standards is performed in a manner similar to that for direct material standards. A labor rate variance and a labor efficiency variance are determined by methods similar to those used for material purchase-price and usage variances.

Construction labor rate standards are usually set by labor union agreements. As such, they are somewhat outside of the control of management. However, while the rate standard might be set, actual wages in dollars spent may exceed the standard. Overtime pay would ordinarily be the primary reason for such a difference. In addition, an unplanned rate change during a project will cause a difference in the predetermined standard and the amount actually paid to workers. This can happen if a union agreement expires during a project and a higher wage is subsequently negotiated.

Labor efficiency standards are controllable. However, they are difficult to set. Disputes over labor efficiency standards are much more likely to arise than are disputes over material usage standards. Determining appropriate fatigue limits and rest time for workers has to be considered. Union work rules have to be considered also. Time and motion studies are the most widely used method of setting labor efficiency standards. These are usually made by a staff engineer or someone at that level.

The construction industry has not always been receptive to the use of time and motion studies. Management has sometimes failed to apply such techniques for fear labor unions would not accept them. In addition, construction managers have argued that it is difficult to set labor efficiency standards for workers who are performing varying types of work in varying types of environment. There is no question that difficulties arise. However, this does not mean that adequately trained personnel cannot determine such standards. Moreover, without such standards, the industry has little means of budgeting and controlling total direct labor costs. Current standard practice in the more informed construction firms indicates that it is possible to set a meaningful labor efficiency standard. Once such a labor efficiency standard is determined, project foremen should be held responsible for any efficiency related variance. Time cards and time card summaries discussed previously provide the means of comparing actual labor rates and efficiency with the predetermined standards.

The analysis of direct labor variances is done in a columnar format similar to the one used for analysis of direct material variances. The total direct labor variance is separated into two variances: the rate variance and the efficiency variance.

The rate variance for a given work package such as a construction method is the difference between the product of the actual labor hours used for the work and the actual price paid for each hour (that is, the actual rate), and the product of the actual labor hours for the work and the predetermined standard rate. If the actual hours times the actual rate exceeds the actual hours times the standard rate, the variance is unfavorable. Similarly, if the actual hours times the actual rate is less than the actual hours times the standard rate, the variance is favorable.

Let us assume that 70 actual carpenter hours are used to form 3,000 square feet of vertical walls. In addition, the predetermined labor efficiency standard is set as 2.5 carpenter hours per 100 square feet. At the beginning of the project a rate standard of $8 per hour was used in setting the budget. However, because the union work agreement terminated during the work and a higher rate was negotiated, the average actual rate used for paying labor for the method in question is $9.50.

Given the above time/rate information on forming vertical walls, the rate variance is calculated as shown in Figure 3.7. There is an unfavorable rate variance of $105.

The labor efficiency variance is calculated as the difference between the product of the actual labor used to perform the work in question and the standard predetermined wage rate, and the product of the standard hours for the work in question and the standard predetermined wage rate. A higher actual hour times standard rate than standard hours times standard rate

Cost Accounting: Terminology Relevant to the Construction Estimate

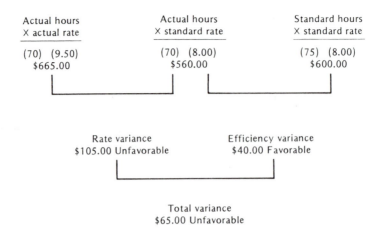

Actual hours X actual rate	Actual hours X standard rate	Standard hours X standard rate
(70) (9.50) $665.00	(70) (8.00) $560.00	(75) (8.00) $600.00

Rate variance $105.00 Unfavorable   Efficiency variance $40.00 Favorable

Total variance $65.00 Unfavorable

Figure 3.7 *Analysis of Labor Cost Variances*

indicates an unfavorable variance. A higher standard hours times standard rate than actual hours times standard rate indicates a favorable variance. As shown in Figure 3.7, there is a $40 favorable labor efficiency variance for the form work method previously described. This is based on the fact that for 3,000 square feet of formwork the standard number of hours is calculated as 2.5 times 30 or 75 hours.

The total direct labor variance is the sum of the rate and labor efficiency variances. For the example considered, there is a $105 unfavorable rate variance and a favorable $40 labor efficiency variance. Thus a $65 unfavorable variance is evident. As is true when analyzing the direct material variance, it is necessary to focus on each of the two ingredients of the total variance in order to assure adequate control.

The determination of standards and the analysis of variances for overhead recognizes the different nature of job overhead and general overhead. For the most part, job overhead is variable in that its amount is dependent on the volume of work the construction firm performs. But general overhead tends to be fixed in that costs such as secretarial salaries and central office maintenance are more a function of time than of volume, so far as construction projects are concerned.

In the previous section on job order accounting we discussed use of an overhead control account in order to compare overhead costs to the amount of overhead applied. When making use of standard costs, this control account should be replaced with two new accounts, one for variable overhead (job overhead) and one for fixed overhead (general overhead).

The method for determination and analysis of variances for variable overhead is similar to the method we recommend for calculation of direct material and direct labor. As we saw earlier in the chapter, construction job overhead is applied to a project as a function of a predetermined base such as total direct cost, direct labor hours, or direct material cost. Let us assume that a construction firm applies its job overhead to projects as a function of the total direct labor hours estimated for a project. In particular, let us assume that it applies job overhead at a rate of $4 per direct labor hour. The $4 rate is, in effect, a standard variable overhead rate. Determination of this rate is the responsibility of cost accounting and engineering personnel.

In addition to setting a standard variable overhead rate, a standard number of direct labor hours must be determined for performing the method in question. Because of the definition of the allocation basis, this determination is the same as the determination of the labor efficiency standard previously discussed.

Unlike the direct material and labor variances, job overhead variances relate more to the sum of several ongoing activities. For example, actual monthly job overhead costs might be compared to a predetermined, standard monthly overhead cost. If three methods are being performed during the month, the total of the direct labor hours incurred on each of the projects is summed and used as a basis for comparing the actual job overhead costs to the standard.

The total variable overhead variance consists of a spending variance and an efficiency variance. The spending variance is merely the difference in the actual variable overhead (job overhead) incurred during the period in question and the product of the actual direct units of the basis—such as the actual direct labor hours—and the standard variable overhead rate. Let us assume that our example firm (that is, the firm that applied job overhead at a rate of $4 per direct labor hour) incurs $8,200 job overhead during a month in which 2,000 direct labor hours of work are performed. As shown in Figure 3.8, the spending variance is calculated at $200. This variance is unfavorable in that

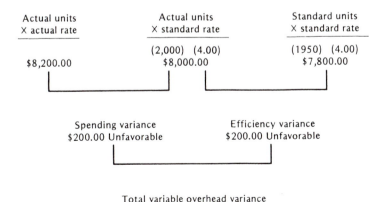

Figure 3.8   *Analysis of Variable Overhead Variances*

the actual job overhead incurred is greater than the actual direct labor hours incurred times the standard application rate.

Variable overhead efficiency variance is calculated as the difference between the product of the actual direct units of the basis—the actual direct labor hours—and the standard variable overhead rate, and the product of the standard or budgeted units of the basis and the standard variable overhead rate. For the previously discussed example firm, let us next assume that the firm set a standard of 1,950 direct labor hours. We find that an efficiency variance of $200 exists (also shown in Figure 3.8). This efficiency variance is unfavorable in that the actual direct hours incurred times the standard rate exceeds the standard direct hours times the standard rate.

The total variable overhead variance is the sum of the spending and efficiency variances. For the example firm considered, both the spending and efficiency variances were unfavorable. As such, a $400 unfavorable total variable overhead variance was evident. This variance is partly due to the fact that more direct hours were used for the work in question than were budgeted. Called the efficiency variance, it resulted from a lack of labor production control. The other half of the variance can be traced to higher than expected job overhead costs per direct labor hour incurred. This is called the spending variance; job overhead control should focus on this part of the total variance.

Fixed overhead costs by definition are not variable. As such, no variation occurs in them as a function of time. Since actual and budgeted or standard units of the basis (for example, direct labor hours incurred or total direct costs incurred) may differ during the time period, the effective unit of cost applied to the basis unit or production may vary.

The total fixed overhead variance consists of a budget variance and a volume variance. The budget variance is the difference between the actual fixed overhead costs incurred in a given period of time (such as a month) and the budgeted amount of overhead for the time period. For example, let us assume the example firm in Figure 3.9 expects to incur an annual general overhead cost of $600,000. Accordingly $50,000 of fixed overhead is budgeted for each month. However, if $62,000 of fixed overhead is incurred during a given month, there will be an unfavorable $12,000 budget variance.

As discussed earlier, construction fixed overhead is appropriately allocated on a duration basis. However, within a given month several projects may be ongoing. The monthly overhead budget ($50,000 in the example considered) may be distributed to individual ongoing projects as a function of their total duration or as a function of the amount of work performed on each project during the month. For example, a rate of $5 per direct labor hour might be used. Assume the firm is expecting 120,000 total labor hours during the year, which averages 10,000 per month.

A volume variance is calculated as the difference in the monthly budgeted overhead and the product of an actual unit of the basis—such as 12,000 estimated direct labor hours for the example month—and the standard rate.

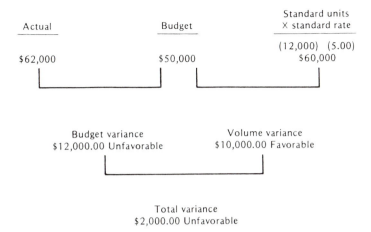

Figure 3.9 *Fixed Overhead Variances*

As shown in Figure 3.9, the volume variance is calculated as the difference between $50,000 and 12,000 hours times $5 per hour. Since the budget amount is less than the budgeted hours times the standard rate (that is, the amount applied to jobs), a favorable volume variance occurs. It amounts to $10,000 in this case.

Summing the $12,000 unfavorable budget variance and the $10,000 favorable volume variance, a total unfavorable fixed variance of $2,000 is calculated. However, unlike the total variable overhead variance, not all of the fixed overhead variance is controllable—for instance the volume variance. This variance occurs because of seasonal variations in production. The variance is only relevant to product costing. In particular, a volume variance indicates that fixed overhead costs are not properly matched to units of production.

Regardless of whether it is a material, labor, job overhead or general overhead standard, the determination of the standard and the analysis of any variations provide the firm with both a means of budgeting (that is, project costing) and controlling the budgeted costs. Actual costs are compared to standards. The investigation of the differences provides the indication for corrective management procedures, change of plans or objectives, or even a change in the standard.

THE MEASURE OF COST AS A FUNCTION OF TIME

As we explained earlier in this chapter, a cost is measured as a money value. In the United States that money is in turn measured in dollars. The value of money, and therefore the measure of a cost, is not static. Instead the value is dependent upon when the money or cost is incurred or expended as a function of time.

Associated with borrowed money is an interest rate. Interest is a rent charged for using money for a period of time. A borrower pays a lender interest for using the lender's money. The interest rate charged by the lender is the ratio of rental charges to the amount of money lent or borrowed for a given time period. When determining the interest rate to be charged, the lender considers the risk of loss of his money, his opportunity costs for the money, and such things as administrative expenses of drawing up and closing the loan. When determining what interest rate he is willing to pay for using money, the borrower compares his expected rate of return from the proposed investment of the money to the interest rate associated with borrowing the money.

The interest rate as it pertains to the availability of money from banks, or savings and loan companies, is related to the state of the national economy. In theory the federal government can dictate interest rates to some degree, through monetary policies aimed at controlling the amount of money in circulation and the issuance of debt. In practice, the modeling of interest rates as a function of distinct factors is indeed complex.

Whenever an economic analysis is made of several possible alternatives involving revenues and costs, a borrower not only must consider the interest rate he will pay for borrowing money, but also the interest rate he will earn by investing its income. Ordinarily, the interest rate charged the borrower is higher than the interest rate paid the investor.

The interest rate not only applies to the incomes and costs associated with the proposal the entity is considering, but also to the opportunity incomes or costs as they pertain to the proposal in question. For example, even though a project owner may have enough money available to finance a project without borrowing, the fact that he could safely invest his own money for 10% a year instead (and not use it) should be charged against the cost of doing the project.

Similarly, when a contractor is choosing between such alternatives as either buying a piece of equipment or renting it, the opportunity interest income that could be obtained if equipment were rented, should be recognized in the analysis. Failure to recognize such opportunity income, can result in a less than optimal decision.

INTEREST FORMULAS

Owing to the existence of an interest rate associated with investing or borrowing money, the value of money to an individual is dependent upon time. That is, the value of a proposed future sum of money must be discounted to determine its present value. When purchasing equipment with borrowed money, both the interest rate paid for the money and the time associated with paying it back must be considered to determine the equipment's "true" cost.

In this section, we will explain several formulas for determining the value of money or a cost at different points in time, given an existing interest rate. In presenting these interest formulas the following symbols will be used:

i   interest rate per defined time period
n   number of time periods
P   present sum of money
S   future sum of money
R   one of a uniform series of end of period payments

When the interest paid by the borrower to a lender is proportional to the length of time the money is borrowed, it is called simple interest. The interest paid per period is equal to the amount of money borrowed, multiplied by the interest rate per time period. The total amount of interest paid by the borrower to the lender is equal to the amount paid per period, multiplied by the number of time periods for which the money is borrowed. Having borrowed an amount of money (P) from a lender, the borrower pays back the lender a total sum on money (S) as follows.

$$S = P + Pin = P(1 + in)$$

For example, if a person borrows $1,000 from a lender for a 3-year period, at an interest rate of 6%, he pays $60 interest at the end of each of the 3 years, in addition to the $1,000 at the end of the last year. These payments are shown on a time scale in Figure 3.10. In this illustration, a negative payment refers to the money that the borrower pays the lender.

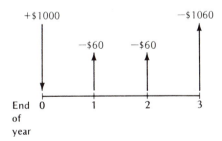

Figure 3.10   *Simple Interest for Loaned Money*

Very few loans between two parties are drawn so that simple interest applies to the loaned money. The more common type of loan uses compound interest. Suppose that you lend the bank $1,000. In this case, the bank is the borrower and you are the lender, and the bank agrees to pay 6% yearly compounded interest for the use of your money. At the end of the first year the bank credits $60.00 (6% of $1,000.00) to your account. Assuming you withdraw no money from this account, the bank credits your account $63.60 (6% of $1,060.00) at the end of the second year. And, at the end of the third year the bank credits your account $67.42 (6% of $1,123.60). That is, at the end of each time period the bank pays you interest on the actual new (increased) balance in your account. If after n pay periods, you decide to make a withdrawal of your initial loan and all your accumulated interest, you would receive a sum of money compounded by the following formula:

$$S = P(1 + i)^n$$

For example, if you withdrew your accumulated sum of money after a 5-year period, you would receive $1,338.00 ($1,000 $(1 + 0.06)^5$. The payments are shown in Figure 3.11: the positive payment represents the money you receive, the negative payment the money you loaned the bank.

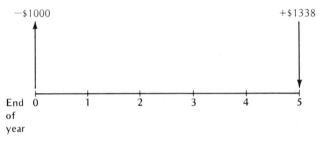

Figure 3.11   *Single Payment Compound Amount for Loaned Money*

It should be obvious that there is a great difference between simple and compound interest. For this reason, it is very important for the borrower and the lender to fully understand the type of interest, the relevant time periods, and the interest rate per time period written into a particular loan contract. Many people, including contractors, have taken out a loan expecting to make certain interest payments, only to discover later they have committed themselves to pay a larger sum of money. Historically, lenders have been misleading in stating the "real" interest rate associated with a loan to a borrower. For example, a lender may state an interest rate of 1% monthly, leading some individuals to believe that they will only have to pay an interest charge of $10 a year on a $1,000 loan. Of course, this is not correct. A 1% monthly interest rate is not equal to 1% annual interest rate. The government has recently taken steps to alleviate this "information" problem by requiring a lender to state the "true" annual interest rate charged on a loan, rather than the weekly, monthly, or quarterly interest rate. The lender is also required to state whether the interest charged is to be simple or compound. Even with these preventive actions by the government, it cannot be overemphasized that the borrower should be aware of all liabilities associated with borrowing money.

Regarding the time value of money, an individual may be interested in knowing the worth of some future sum of money to him at the present time. For example, let us imagine an individual who is promised $1,000 five years from the present. The individual realizes that if he presently has the $1,000, he could invest it at the present interest rate and accumulate a sum greater than $1,000 by the end of n years (at which time he would receive the promised $1,000). As a result of this opportunity to presently invest the money at some positive interest rate, the future sum of $1,000 is worth less in 5 years than $1,000 is to him at present.

The present value of a future monetary sum is often referred to as the present worth of the sum of money. To determine a future sum's present worth, we need to know how far into the future the sum is to be paid or received, and also the interest rate at which the money could be invested, or in the case of borrowing money, the interest rate associated with borrowing the money. The present worth (P) of a future sum of money (S) may be found by using the following formula:

$$P = S (1 + i)^{-n}$$

The term $(1 + i)^{-n}$ is known as the single-payment present worth formula or factor. It assumes a constant i over the relevant time period n. It may be observed that the single-payment present worth factor is merely the inverse of the single-payment compound amount factor. That is to say, finding the future worth of a present sum of money is the reverse of finding the present value of a future sum.

Returning to the problem of finding the present worth of $1,000 to be received by an individual 5 years from the present, let us assume the individual could invest the money at an interest rate of 6%. The present worth of the money is equal to $747.30 ($1,000 $(1 + 0.06)^{-5}$). Figure 3.12 shows the finding of the present worth on a 5-year time scale.

By using the single-payment compound amount factor and the single-payment present worth factor, the value of a sum of money may be found at any desired point in time. However, to facilitate the actual calculation

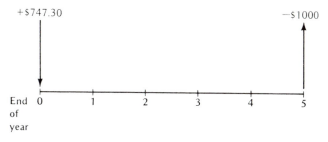

Figure 3.12 *Single Payment Present Worth for Loaned Money*

procedure, which may involve shifting money on a time scale when there are several payments, several other interest formulas or factors have been derived. One of these factors is the capital recovery interest factor. This is useful in determining the equal payments required in a loan in which the borrower is to only pay interest on the unpaid balance. The equal payments he is to make (R) may be obtained from the following formula:

$$R = P[i(1 + i)^n/((1 + i)^n-1)]$$

The term in brackets is known as the capital recovery factor. Note that the factor is merely the result of summing several single payment compound amount factors. In the above example, the interest factor is found to be 0.2374. The equal payments to be made (R) are $1,000 x (0.2374) or $237.40. In Figure 3.13 the payments involved are shown on a 5-year time scale.

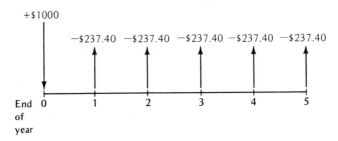

Figure 3.13   *Equal Payment Calculations with Capital Recovery Factor*

Observe that in using the capital recovery factor, we are actually finding a uniform series of payments which are equivalent to a present monetary sum. We may well be interested in reversing the problem, that is, we may want instead to know the present value of a uniform series of payments. This may be found by using the following formula:

$$P = R [((1 + i)^n - 1)/(i(1 + i)^n)]$$

Note that the term in the brackets, referred to as the uniform series present worth factor, is merely the inverse of the capital recovery factor. The reason should be obvious.

The uniform-series present worth factor is useful in determining the present worth of a uniform series of payments. For example, let us suppose an individual is promised $1,000 at the end of every year for 5 years. Since he knows he can invest available money at a 6% interest rate, he knows this promised sum of $5,000 (5 x $1,000) is, in fact, worth less than $5,000 at the present time. The uniform series present worth factor for i = 0.06, n = 5, is found to be equal to 4.2124. Thus, the present worth of the payments is equal

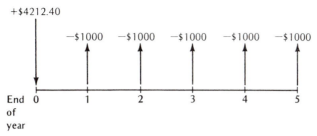

Figure 3.14 *Payments Calculated with Uniform Series Present Worth Factor*

to $4,212.40 ($1,000 × 4.2.24). These payments are shown in Figure 3.14 on a 5-year time scale.

Two other interest formulas or factors are of immediate interest here. These interest factors define the relationships which exist between a future sum of money (S), and a set of uniform series of payments (R). These relationships may be derived from the single-payment compound amount interest factors. One of these interest factors, the uniform-series compound amount factor, determines what equal value end-of-period payments, invested at an interest rate i and earning interest immediately upon investment, will sum to after n payment periods. The amount accumulated, which consists of the sum of the uniform payments and the earned interest, may be determined from the following formula:

$$S = R[(((1 + i)^n - 1)/i)]$$

The term in the brackets is referred to as the uniform-series compound amount factor. Suppose an individual invests $1,000 at the end of every year into a savings account, which pays 6% compounded interest. The money starts drawing interest as soon as it is invested. To determine the amount of money accumulated after 5 years, we calculate the uniform-series compound amount factor for i = 0.06, n = 5, and obtain a value of 4.637. Thus, the sum of money accumulated is found to be the product of $1,000 times 5.637, or $5,637.00. This is shown on the 5-year time scale in Figure 3.15.

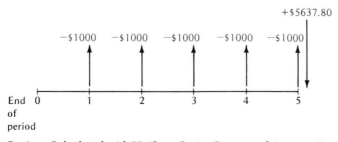

Figure 3.15 *Savings Calculated with Uniform Series Compound Amount Factor*

The inverse of the uniform series compound amount factor is the sinking fund factor. The sinking fund factor determines which uniform series of payments must be made at a given interest rate, to accumulate a stipulated future sum (S) after a given number of payment periods. For example, assume an individual desires to invest a uniform series of payments (each to be made at the end of the year at an interest rate of 6%), so that he has $1,000 in savings at the end of 5 years. This uniform series of payments may be derived from the following formula:

$$R = S[(i/((1 + i)^n - 1))]$$

The term in the brackets is referred to as the sinking fund factor. For the above example, the sinking fund factor equals 0.1774. Thus, the uniform end-of-the-year payments which the individual must invest in order to accumulate $1,000 after a 5-year period is equal to $1,000 multiplied by 0.1774 ($177.40). In Figure 3.16 these payments are shown on a 5-year time scale.

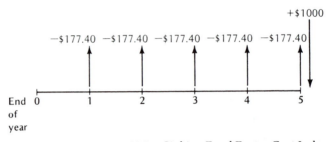

Figure 3.16 *End-of-Year Payments Using Sinking Fund Factor Cost Index*

It should be observed that the value of any interest factor described, whether single payment compound amount, capital recovery, sinking fund, or the like, depends only on the relevant interest rate, i, and on the number of payments, n. The computed values of the various interest factors as a function of the interest rate, i, and the number of payment periods, n, are arranged in standard interest tables. Their use eliminates the need to mathematically solve the interest formula. A complete set of interest tables is presented in Appendix A.

COST INDEXES

The time value of money interest formulas discussed in the prior section calculate the varying value of money or a cost that arises because of the existance of the impact of interest. The formulas do not take into account that inflation itself often results in an escalating cost as a function of time. For example over a relatively short period of time a common laborer's wage rate may be increased from $8.00 per hour to in excess of $12.00.

Even if the estimator has historical data regarding cost or productivity, the fact that a cost changes because of inflation results in the estimator's need to adjust historical cost or productivity data for inflation. One means for such an adjustment is through the use of indexes.

A cost index provides a comparison of cost or price changes from year to year for a fixed quantity of work or services. It enables the estimator to forecast the cost of a similar type of work from the past to the present or future period without going through detailed costing. Provided the estimator uses discretion in choosing the proper index, a reasonable approximation of cost should result. Extrapolation through time-series analysis of cost indexes is possible for future periods. Index numbers have been used for these purposes for a long time.

Index numbers are useful in other respects. With time and money for estimating usually scarce, the cost estimator is forced to make immediate use of previous costs which are based on outdated conditions. Because costs vary with time due to changes in demand, in economic conditions, and in prices, indexes convert costs applicable at a past date to equivalent costs now or in the future. A cost index is merely a dimensionless number for a given year showing the cost at that time relative to a certain base year. If a construction cost at a previous period is known, present cost can be determined by multiplying the original cost by the ratio of the present index value to the index value applicable when the original cost was obtained. This may be stated formally as

$$C_c = C_r \frac{(I_c)}{(I_r)}$$

Where C_c = present cost in dollars
C_r = original reference cost in dollars
I_c = index value at present time
I_r = index value at time reference cost was obtained

In selecting an index to upgrade an estimate, the estimator should consider the construction region, elements of the index, and the individual elements indicated in the original estimate. If major items are ignored in the index as compared to the estimate, adjustments in the composition of the index are necessary. An index seldom considers all factors such as technological progress or local and special conditions.

Consider the following example. Construction of a 10,000-square-foot warehouse is planned for a future period. Several years ago a similar warehouse was constructed for a unit estimate of $32.50 when the index was 118.

The index for the construction period is forecast as 143, and construction costs per square foot will be

$$C_c = \$32.50 \, \frac{(143)}{(118)} = \$39.39$$

While many indexes are published and are widely used, the estimator needs to recognize that it may be better for his firm to develop its own index. Therefore, he needs to understand the basics of constructing, altering, and applying an index. Arithmetic development of indexes are of six types: (1) adding costs together and dividing by their number, (2) adding the cost reciprocals together and dividing into their number, (3) multiplying the costs together and extracting the root indicated by their number, (4) ranking the costs and selecting the median value, (5) selecting the mean cost, and (6) adding together actual costs of each year and taking the ratio of these sums. The weighted arithmetic method (1) is the most popular. Other versions incorporating refinements are also popular.

A cost relative to n items is

$$I_c = \frac{(C_{11}/C_{01}) + (C_{12}/C_{02}) + \ldots + (C_{1n}/C_{0n})}{n}$$

Where C_{1n} = cost of nth item for the first year. Year zero would be the base year. If three prices—say for concrete, steel, and lumber—rise 4.4%, 3.6%, and 10% respectively, their average rise is 6%, and the index number is 106 compared with the original price level of 100 taken as a base of comparison.

The equation above treats all items equally, and consequently the index is called simple. One item may be considered more important—as though it were two or three items—thus giving it two or three times as much weight as the other item. A weighted approach is

$$I_c = \frac{W_1(C_{11}/C_{01}) + W_2(C_{12}/C_{02}) + \ldots + W_n(C_{1n}/C_{0n})}{W_1 + W_2 + \ldots + W_n}$$

Where $W_{1,2,\ldots,n}$ = weight 1, 2, and n of items 1, 2, and n. Successive years would substitute values for C_{m1}, C_{m2},..., C_{mn} in the numerator, and calculation would then be relative to the base year, 0 in our case. While the formulas are simple enough, determination of cost indexes bears little resemblance because of the variety, number, and complications involved.

Several characteristics distinguish indexes. In the construction of an index, there is a choice in the selection of its parameters. Typical alternate choices are wholesale prices or retail prices, wages or volume of production, proportion of labor to materials, and the number of separate statistics used. Indexes apply to a place and time; that is, to a period covered or a region considered, to a base year, and to the yearly or monthly interval between successive indexes. Additionally, indexes are varied as to the data compiler and to the data sources used. A variety of objectives create diversity for the many indexes.

While the formulas suggest a straightforward procedure, the finding of weights and their calculation is not so simple. Consider the following data:

ITEM	QUANTITY UNIT	UNIT COST	TOTAL QUANTITY	TOTAL DOLLARS	PCT. OF TOTAL
Steel	lb.	$0.04	2500 lb	$100.00	12.94%
Cement	bbl.	4.00	6 bbl	24.00	3.10
Southern pine	fbm	0.04	10,000 fbm	400.00	51.66
Labor/hour	hr	2.50	100	250.00	32.30
				$774.00	100.00%

Successive years would substitute different unit costs. For example, let us assume a following year exhibited the following data:

ITEM	QUANTITY UNIT	UNIT COST	TOTAL QUANTITY	TOTAL DOLLARS	PCT. OF TOTAL
Steel	lb.	$ 0.16	2500 lb	$ 400.00	8.94%
Cement	bbl	12.00	6 bbl	72.00	1.62
Southern pine	fbm	0.25	10,000 fbm	2,500.00	55.90
Labor/hours	hr	15.00	100	1,500.00	33.54
				$4,472.00	100.00%

During this period the steel component increased by 4.00 times, while hourly wages for common labor rose 6.0 times. While 100 manhours of labor might have been required in the earlier year to place the quantities of lumber, cement, and steel, a lesser number of labor hours may be required in the subsequent years in view of technological progress. In effect it might be argued that the 33.54% labor component and 55.90% lumber component are not representative of the work process in the subsequent years. This, of course, is the difficulty of the component index. It does not reflect the change in the component mix of the work process.

A variety of cost indexes are available to the estimator. Perhaps the most well known index is the *Wholesale Price Index* published by the Bureau of Labor Statistics.

In addition to these national indexes there are several indexes that specifically address building construction, including the Boeckh and Marshal and Steven Indexes. Perhaps the most widely known construction indexes are those published by *Engineering News Record. ENR* maintains both a building cost index and a construction cost index. The building index charts costs for building type projects, whereas the construction index charts the

Building Cost Index

1913 = 100

1961 — 568	1966 — 650	1971 — 948	1976 — 1425
1962 — 580	1967 — 672	1972 — 1048	1977 — 1545
1963 — 594	1968 — 721	1973 — 1138	1978 — 1674
1964 — 612	1969 — 790	1974 — 1204	1979 — 1826
1965 — 627	1970 — 836	1975 — 1306	1980 — 1940
			1981 — 2130

Construction Cost Index

1913 = 100

1961 — 847	1966 — 1019	1971 — 1581	1976 — 2413
1962 — 872	1967 — 1070	1972 — 1753	1977 — 2583
1963 — 901	1968 — 1155	1973 — 1895	1978 — 2822
1964 — 936	1969 — 1269	1974 — 2062	1979 — 3050
1965 — 971	1970 — 1385	1975 — 2246	1980 — 3350
			1981 — 3580

Engineering News Record Cost Index

Figure 3.17   *Example Cost Index*

costs of engineering and heavy and highway type of projects. Figure 3.17 illustrates the significant escalation of costs contained in both of these indexes.

COSTS AND WORK ITEMS

The concept of a work item is similar to the concept of a cost object discussed earlier in this chapter. Both concepts are basic to an estimating system. An example construction firm performs many types of work and projects. In particular, the problem often arises as to what is and what is not a work item. That is, whereas one firm might choose to define all work related to concrete including forming and stripping as a work item, another firm may choose to identify only placing concrete for all types of concrete work as a work item, and yet another firm may identify placing concrete in a wall as a work item. Clearly the problem of identifying specific work items is one of establishing the degree or magnitude of the work involved.

The definition of the work items dictates the manner in which the estimator takes off quantities and field data are collected. For a given set of drawings, the number of work items the firm establishes as part of the take-off procedures dictates a specific groupings of the items to be identified from the drawings. The broader the scale of a work item, the fewer work items the estimator will single out when taking off quantities for a project. On the other hand, the broader the scale or scope of work items, the less accurate the take-off will be that determines the cost of performing the project's work. That is, if the estimator takes off all the concrete placement as one work item, the fact that it costs the construction firm more to place concrete in an elevated wall than placing a similar amount in a slab on grade will not be reflected in the estimate. This difference in costs is one major reason why we would want to delineate separate work items for the two different work processes. No single rule or principle can be used to determine what is and what is not a work item. However, if we keep in mind why we are listing separate work items, this will aid us in determining them.

There is a relationship between our costing or pricing of work and determining work packages. If two different types of work require different resources to perform the work, they probably will have different costs for a given unit of work performed on each of the two types. Thus, an overall guideline is that whenever one or more different resources are required to perform work, we probably are justified in identifying each as a separate work item so as to reflect the different resources required. However, even this general guideline will not be practical for solving work-item identification in all cases.

In establishing a set of work items that dictate the quantity take-off, one should remember that the work items can prove to be a means of integrating several construction-management functions. If work items are properly defined, they can serve as a means of interrelating project estimating, project planning, project control, and project accounting and payroll functions. (Review our treatment of these topics in Chapter 1.)

The importance of the relationship between project estimating and the definition of work items is shown in Figure 3.18. In this figure there is evidence that project estimating depends on the compatibility of the work item definitions and the cost data file that a firm has gathered from its past projects. To illustrate what can happen if the systems are not compatible, let us assume that a construction firm has defined two separate work items: (1) placing concrete footings, and (2) placing concrete walls. The estimator then proceeds to take-off the cubic yards of concrete for these two different types of concrete work. This is done as recognition that each type of work may require different costs in regard to pricing a unit of work. Let us further assume that the firm has gathered and structured cost data on placing concrete, but with no differentiation as to type of concrete placement. However, when the necessary quantities for placing the concrete footings and for placing the concrete walls are determined, the firm will be forced to make calculations with a less than accurate price for the specific work items. The result will probably be an inaccurate estimate.

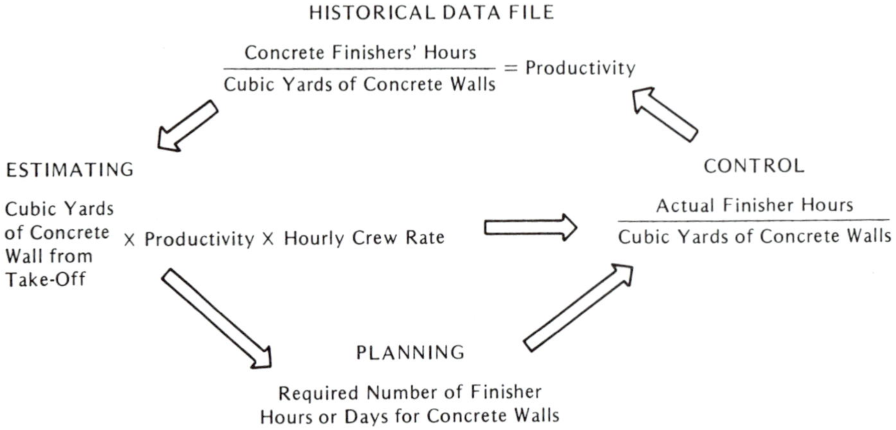

Cost Accounting: Terminology Relevant to the Construction Estimate

HISTORICAL DATA FILE

$$\frac{\text{Concrete Finishers' Hours}}{\text{Cubic Yards of Concrete Walls}} = \text{Productivity}$$

ESTIMATING

Cubic Yards
of Concrete
Wall from X Productivity X Hourly Crew Rate
Take-Off

CONTROL

$$\frac{\text{Actual Finisher Hours}}{\text{Cubic Yards of Concrete Walls}}$$

PLANNING

Required Number of Finisher
Hours or Days for Concrete Walls

Figure 3.18 *Relationship of Work Item History and Estimating*

The construction industry is burdened by fragmentation. Work package standards are absent. Moreover, units of measure differ for the designer, builder, and material supplier. Use of the uniform index as defined by the Construction Specifications Institute (CSI) is a move toward providing a common communication base for all the parties in building construction. This Uniform Construction Index is reproduced in Appendix B. Definition of work items for an estimating and control system can be compatible with the CSI terms codified in the Index. By using each division as a broad classification, specific work items can be identified for each. For example, Division 3 is identified as "Concrete."

Defining work items in terms consistent with this uniform code will benefit all firms since they will probably be better able to relate their data to published cost data averages and systems used by designers. Already in fact, building construction specifications are increasingly prepared so as to be compatible with the uniform code.

The CSI uniform code is designed primarily for building construction. Heavy and highway contractors should define work items with a data base common to heavy and highway construction. Once again the benefits are better communication.

As an alternative to using work items as a means of project control, some construction firms trace costs to a specific room or wing of a building. While such a process may aid in controlling the project in question, it offers little help in estimating future projects because usually the design of the building components varies significantly from one building to the next.

As with tracing costs to rooms, tracing and summing labor costs by crafts can be used to help control the project in question. However, as is also true of tracing costs to rooms, summed labor craft costs have little value for future estimating efforts.

In conclusion, defining work items as the cost data base or object offers the most potential for both project estimating and control. The concept of utilizing a common set of work items for estimating along with other project management functions is emphasized in the following chapters, which address the preparation of a detailed contractor estimate.

EXERCISE 3.1

A given formwork system results in a productivity of 3 manhours per 100 square feet of contact area. The productivity for placing the concrete in the wall is 2 manhours per cubic yard of concrete placed. There are 2.5 cubic yards of concrete in 100 square feet of wall. The cost for a manhour is $10. The cost of a cubic yard of concrete is $30. The formwork system is made up of panels each one foot in width by eight feet high. Each panel costs $40 and has an expected life of 30 re-uses. Calculate in an orderly procedure, the unit cost of 1 *square foot of finished wall.*

EXERCISE 3.2

Vary Construction Company is attempting to implement the use of past project data and variances to improve the accuracy of its estimating system. The firm typically analyzes monthly data to aid it in the estimating of projects to be bid in the following month.

It currently is the end of April and Vary Company has asked you to aid it in the determination of various costs to be applied to projects to be bid and performed in the month of May.

On January 1, the beginning of its fiscal year, Vary Company estimated its total annual company overhead (which is considered fixed) to be $240,000. While budgeting it equally over each of the twelve months, Vary Company applies it to specific jobs on a basis of $.60 per direct labor hour. This application rate has been determined from an estimate of total direct labor hours for the upcoming year. While variations in direct labor hours have occurred in the months of January, February, March, and April, the firm believes its estimate for the year to be accurate. The actual fixed overhead expenditures by month have been as follows:

MONTH	OVERHEAD
January	$25,000
February	$25,000
March	$25,000
April	$25,000

On the basis of the first four month costs, it is expected that the monthly fixed overhead cost will remain at $25,000. The estimate of direct labor hours for May is 50,000 hours.

The firm has applied job overhead costs on the basis of $.40 per estimated direct labor hour. April has been the first month that Vary Company has gathered comparative data as to estimated direct labor hours and actual direct labor hours. For the month of April, 40,000 hours were estimated, however 45,000 labor hours were actually used. The firm actually spent $18,400 on job overhead costs in the month of April.

The 40,000 estimated direct labor hours for April consisted of 30,000 laborer hours and 10,000 carpenter hours. The estimated wage rate for April for the two crafts were $8 and $10, respectively. Payroll records indicate that a total of $300,000 was paid to laborers and $98,000 to carpenters during the month of April. This was on the basis of 33,100 actual laborer hours and 11,900 carpenter hours.

The two main materials that the firm uses are lumber and concrete. During April, project drawings specified lumber use of 8,000 board feet of lumber and 1,450 cubic yards of concrete. Actual estimated prices are $2.50 per board foot of lumber, and $34 per cubic yard of concrete. Actual price paid averaged $2.25 and $35.25, respectively. Total material costs in April for lumber and concrete were $21,000 and $69,000, respectively.

The firm expects to bid and perform a single project during the month of May. On the basis of calculations made before consideration of April's data, the estimated carpenter hours and labor hours are 20,000 and 30,000 respectively. The project drawings specify 7,000 board feet of lumber and 1,200 cubic yards of concrete. The project will start the first day of the month and is estimated for completion on the last day of the month.

Using the variance analysis method you are to aid Vary Company in determining its bid price for the May project. The firm desires to include a 4% profit margin on total direct cost.

EXERCISE 3.3

One way of defining inflation is to measure the difference between the increase in the cost of a worker and the lesser increase in the productivity of that worker. (Obviously, if a worker is able to increase his productivity at the same rate as his increase in labor wage there is no inflation.) To illustrate the impact of low productivity on the cost of construction—which must be recognized in an estimate—consider this hypothetical data next.

Let us assume that in 19**XO** a mason's labor rate was $8.00 per hour and received a 12% annual increase each of the following 5 years. Thus, his wages in years 19**X1** through 19**X5**, increased at a rate of 12% a year. Let us further assume that a mason in year 19**XO** placed 120 blocks per day or 15 blocks per hour and that his annual increase in productivity for the following 5 years was 1% a year. This productivity increase may have occurred because of either better technology or better management, or because the mason worked harder.

    a. Based on the above information, calculate the labor cost per block placed for each of the 6 years.

    b. Calculate the inflation rate for each of the last 5 years; that is, the percentage increase in the labor cost per block for each year from the previous year.

4

Project Owner's Estimate

*Importance of the Feasibility Estimate   *   Measuring*
*Project Benefits   *   Owner's Estimated Project Costs   ***
*Comparative Methods for Reckoning Cost   *   Income Tax*
*Related Factors   *   Time Value of Money and the*
*Feasibility Estimate   *   Office Building Example*

IMPORTANCE OF THE FEASIBILITY ESTIMATE

In Chapter 1 we saw that several kinds of cost estimates must be made for a construction project—from its conception as a project through its completion. Often attention focuses on the contractor's detailed estimate. However, several pre-construction estimates also play a significant role in the construction process. One of these is the project owner's feasibility estimate.

Owners initiate construction projects for various reasons. Public construction is built to satisfy the social service needs of the public at large. Private construction, such as a commercial building or an apartment complex, is often undertaken as an economic investment to yield the owner future monetary rewards. Yet other projects, such as schools, are undertaken to fulfill both social and economic needs.

Regardless of the underlying reason for the initiation of a project, each and every project has measurable benefits and costs. The ability to measure these benefits and costs at an early stage of a project can decrease the potential risk and cost associated with it. The rapidly escalating costs of construction have drawn attention to the need for a comparison of the anticipated benefits and costs of a project at the time of its conception. This comparison is commonly referred to as the **feasibility estimate**. A feasibility estimate might be considered a project owner's estimate.

This chapter presents the mechanics of the preparation of feasibility estimate. Tax depreciation, investment tax credit, debt service, and capital gains are also discussed. It is intended that this knowledge, along with construction and project design knowledge, will enable an individual to prepare an accurate, complete, and reliable feasibility estimate.

The project drawings and even preliminary plans are usually non-existent at the time the feasibility estimate is made. As such, the extent of detail and means used for a feasibility estimate differ from those for a contractor's detailed estimate. On the one hand, the owner's feasibility estimate has to place a strong emphasis on the accurate prediction of project benefits in that these benefits, which have great weight in deciding overall feasibility, are in fact more difficult to forecast than project costs. Tax considerations are also important in a feasibility estimate. On the other hand, the contractor's detailed estimate will commonly address itself only to costs.

The project owner's time of involvement with a project commonly far exceeds that of the construction contractor. While an owner may sell his ownership in its newly constructed project, this transfer of ownership may not occur for several years after construction, if in fact it does occur. Depreciation, debt service, and life-cycle costs are three concepts that are major considerations in any feasibility estimate.

The accuracy, degree of completeness, and reliability of a feasibility estimate or study for a proposed project are dependent on the use of several skills. These skills include careful pricing, knowledge of the correct use of the expected life of construction materials, training in accounting and related tax laws, and a good grasp of project design.

Absent one of the identified skills, the feasibility estimator's work will likely be less than optimal in regard to accuracy or completeness. For example, an accountant, lacking construction knowledge, may fail to utilize correct initial project costs and annual maintenance and repair costs in his feasibility estimate. The result will be a correct process but invalid results. Similarly the accountant's lack of knowledge regarding the taxable lives of various building components may result in his failing to recognize optimal tax benefits in his analysis.

The constructor without either accounting or tax knowledge will not be in a position to do a complete or accurate feasibility estimate. The same is true of an individual whose skills are limited only to construction or design knowledge.

MEASURING PROJECT BENEFITS

Objective estimating of a project's predictable benefits is fundamental to the timely evaluation of the feasibility of a construction project. Nevertheless, this prediction is often somewhat subjective since it is based on factors that are sensitive to future events.

Future housing demand, industrial production levels, cost of capital, lifestyles, and shifting population trends are all factors that affect the measurement of a project's future benefits.

The types of benefits to be predicted depend upon the source and type of construction project in question. Figure 4.1 is an abbreviated list of types of

Highway or Roadway	*Prime Benefits* Motor vehicle usage	*How Measured* Vehicles/hour
	Decreased safety risk	Decreased accident cost
	Secondary Added stimulus to a depressed labor market	Jobs created
Office Building & Apartment Complexes	*Prime* Lease receipts	Occupancy rates of newly competing units
	Tax advantages	Using depreciation methods, investment tax credit, capital gains, and so on
	Secondary Increase in market value	Historical trends, capital gains treatment
Manufacturing Plant	*Prime* Increased productive capacity	Output per square foot
	Lower unit cost of production	Estimated cost per unit
	Secondary Decreased rental costs	Purchase versus rental analysis
Residential	*Prime* Elimination of rental costs	Purchase versus rental analysis
	Increase in market value	Historical trends, market analysis
	Secondary Homeowner prestige	Status in community, neighborhood, and so on

Figure 4.1 *Primary and Secondary Benefits from Typical Construction Projects*

construction projects and the primary and secondary benefits relevant to the measurement of future use or ownership.

A forecast is needed to establish the amount of benefits to be obtained for the types of projects listed in Figure 4.1. In order to make a reliable forecast, it may be necessary for the project owner to secure a consultant who has expertise in making such forecasts. Real estate appraisers, marketing analysts, management service departments of certified public accounting firms, management accountants, and real estate developers offer such services.

The basic approach or analysis used by the project owner or the consultant is referred to as **market research**. Market research includes the study of population trends, age and size of family households, industrial development, production demand, real estate occupancy rates, personal income, and the like. Marketing research firms typically make forecasts based on correlation between the various parameters whose values determine project characteristics. For example, the market research used by a potential housing developer may relate a forecast of near-term employment opportunities in a given location to the need for work-force housing. Such a hypothetical analysis is illustrated in Figure 4.2.

Subject	This year	Next year
Employment	24,000	25,200
Employment participation rate	35.3%	35.8%
Population	68,000	70,382
Household population ratio	97.1%	97.1%
Household population	66,000	38,350
Average household size	3.30	3.29
Households	20,000	20,800
Overall occupancy rate	95.2%	95.4%
Housing units	21,000	21,800

Figure 4.2 *Correlation Analysis for a Proposed Housing Project*

In order to be able to evaluate both a project's benefits and costs on the same basis, the benefits of a project are normally converted to a monetary value per year. This is done by transforming increased production rates or anticipated rental or leasing receipts into an annual cash inflow.

OWNER'S ESTIMATED PROJECT COSTS

Perhaps easier than estimating the project benefits but no less significant to a project's feasibility is the reckoning of project costs. These include initial construction costs, design costs, finance costs, and maintenance and repair costs. While construction and design costs are one-time initial costs, others such as finance, maintenance, and repair costs are annual costs.

Pros and Cons of Using Cost Books

The feasibility estimate usually includes an approximate estimate for initial construction costs. In order to make this approximate estimate reasonably close, the estimator normally turns to the use of cost books or his own historical cost data. These cost books or periodic publications give the costs of various types of buildings per square foot of building, or per volume of building (typically measured in cubic feet). For example, such a book may indicate a cost of $45.50 per square foot for building an apartment complex. Some of these estimating books also provide data for various quality building types. For example, they may indicate a cost per square foot for cheap construction, average construction, or high quality construction. It is wise for the user of these cost books to determine where a given type of project fits in terms of these defined quality categories. Needless to say, the estimator has to be careful using such cost books. In particular, most costs per square foot tabulated in them are indeed approximate—for several different reasons. One difficulty is that the costs are usually reported for a rather large geographic area. That is, average costs are reported for numerous buildings of the same type built in many different locations. Thus the cost of construction (labor and material prices) for an actual building in a specific location may differ substantially from the reported average cost. The second difficulty with using the average cost reported in these cost books is that few actual buildings tend to be exactly like an averaged ideal building. So, even when using cost books that give different qualities of construction, it is unlikely that the proposed building will in fact have identical design, material, and other price parameters, as those factors were considered for determination of the numbers published in these cost books. Nevertheless, the fact that there are difficulties with using standard cost books to reckon future square foot or cubic foot costs of various buildings does not take away from the value of these cost books for approximate estimates. In particular, the cautious use of these books provides the estimator a quick way of determining an approximate cost for building the project in question. Such approximate costs are usually accurate enough to be able to determine the feasibility of a project. This fact—taken along with the usual unavailability of complete sets of drawings and specifications when a feasibility estimate is prepared—can result in valuable service from the approximate cost numbers published in cost books.

COMPARATIVE METHODS FOR RECKONING COSTS

One solution to the difficulty of taking the cost book too literally is for the estimator to compare the cost per square foot or cubic foot for the type of building considered with those in two or three other similar cost books. By comparing various standard costs the estimator can better assure himself that his estimate is in fact close to reality. This is especially true if the cost per square foot or cubic foot varies little, as published in two or three reliable cost books. Among the best in a long list of cost books and periodicals designed for this sort of approximate estimating are *Richardson's Service*,[1] *The Dodge Systems Book*,[2] and *Mean's Building Systems Cost Guide*.[3] Real

1. Richardson Engineering Services, Inc., Solona Beach, California, (published bi-annually)
2. Dodge Systems Book, McGraw Hill, New York, N.Y. (published annually)
3. Means Building Systems Cost Guide, Robert Snow Means Company, Inc., Kingston, MA (published annually)

estate appraising books also commonly have such cost data. In addition, as part of business promotion in some geographic areas, various large cities or groups of cities publish public cost data for buildings on a square-foot basis or a cubic-foot basis for construction in those geographic areas.

Caution should also characterize a feasibility estimator's determination of estimate of construction costs so that he recognizes the time differences between the estimating and the actual constructing. Too many owners and estimators have been surprised by the fact that the bid for a particular project has come in as much as 50% or more higher than the initial feasibility estimate by the time of actual construction. This may be due to incorrect estimates. However it is likely also to result from a too-long time lag between the estimating and the letting of the project.

Phases and Timing of Projects

The construction process can be broken into three logical phases. That is, ordinarily each project goes through the *feasibility and design* phase, the *bidding and letting* phase, and the *construction* phase. On some projects, especially public construction projects, it is not uncommon for each of these phases to be of approximately the same duration. For example, a project that finally takes 6 months of constructing time may first take 6 months of feasibility and design time, and 6 months of bidding and letting time. Thus, during the 6 months between the initial feasibility phase and the letting phase, material and labor costs may have increased as much as 5 or 10%. The feasibility estimate therefore must take such cost escalation into account initially.

The estimator must also consider design costs in the feasibility estimate. Almost every construction project requires considerable input from either an architect or an engineer for design of the project. Ordinarily building construction is much more dependent on the services of an architect, than are public works projects. Those—dams, highways, and the like—require the services of an engineer. Of course, some projects require both the services of an architect and engineer. For example, the overall design of a large office building, including the selection of exterior materials, interior finishes, and overall interior layout may be the responsibility of the architect. But the structural frame for the same building, be it concrete or steel, may be the responsibility of an engineer. Most states require that both architects and engineers be registered with the state, to certify both their professional qualifications and knowledge of local building and environmental codes.

The fees charged by the architect/engineer depend on several variables, including the degree of services provided. That is, an architect may only provide design services such as preparation of the contract documents that furnish drawings and specifications, or the architect or engineer may also provide construction supervision and perhaps letting services.

The professional fees charged by architects/engineers also depend on the type and the dollar size of the project. Unique one-time projects, such as an elaborate church or high-rise office building, may warrant a fee that is relatively higher than for a simpler project with a similar dollar value. In addition, it is common that an architect's or engineer's fee decreases as the cost of the project increases. That is, while the fee for a $1,000,000 project may be 5% of the cost of the project, the fee for a $5,000,000 project may be only 3%.

Like construction costs and design costs, the potential owner's land cost is a one-time initial cost. It is this type of cost that places a heavy burden on the calculator of the initial feasibility estimate. High initial land costs are often prohibitive to the feasibility of a project. Land costs in the recent years have typically outpaced the increased cost of labor and materials. The cost of land relative to the construction cost of a given project depends on several factors. Included in these are the following.

1. The scarcity of land in the geographic area of a project.
2. Total acreage of the land purchased.
3. The relative demand for the land purchased as a function of time.
4. The negotiating powers of the seller and the buyer.

Other unique factors for a given land acquisition may also affect the relative cost of the land as compared to the construction cost of the project in question. In general, land cost for any construction project can be estimated to vary anywhere from as little as 10% to as much as 30% of the construction cost for the project. High land costs are especially prohibitive when one considers the fact that the land is not a depreciable asset and therefore will not reward the project owner with a tax savings through depreciation that the construction improvement costs will return to him. (Depreciation benefits are discussed separately later in this chapter.)

Land costs are usually more determinative of feasibility than construction costs. That is to say, land is an observable commodity at the exact time that the estimator is making his feasibility estimate. Therefore, it is possible to get an appraised value assigned to the land when it is being considered for purchase. On the other hand future construction does not exist at the point in time that the feasibility estimate is being made. As such these costs are much less determinable and therefore much more difficult for the estimator to trace for the feasibility estimate.

Finance Costs

Finance costs may run as little as 5% of the construction costs but more often can run as high as 10 to 15% of these costs over the life of the project. That is, each and every year that it is under construction a project may have finance costs equal to 10 to 15% of the initial construction cost. It is not unusual for the total finance cost of a project over the life of the project to equal or exceed the initial construction cost. Finance costs are annual costs, and this stream of costs must be recognized appropriately in the feasibility estimate.

The capital that an owner uses to fund a particular construction project comes from one of two broad classifications of funds. First, the owner may use either his own capital or the capital that he raises from additional stockholders. This is called **equity capital**. When using equity capital the project owner does not pay out a finance cost to an external party. Therefore, it is not uncommon for an estimator or owner to neglect to figure a finance cost in his feasibility estimate when he uses equity capital for a project. However, it should be noted that had the project owner not in fact built the project with his own funds, the funds could have been invested at some favorable rate of return. This potential rate of return should be charged against the new project in the feasibility estimate. Specifically this loss of rate of return is referred to as an **opportunity cost**. While this opportunity cost will normally be less than a finance charge it can affect the feasibility estimate.

More often than not an owner is dependent on funds other than his own for funding a construction project. In this case the funds are called **debt capital**. Debt capital is referred to as a loan. When a project owner builds a project he may need both construction and mortgage loans. The first type, called a **construction loan**, is needed to fund the project during construction. Sources of construction loans include banks, savings and loans institutions, and increasingly a Real Estate Investment Trust. REIT is merely a pool of funds raised by a group of investors to fund a building construction project. The group of investors give the potential project owners this money in return for an interest charge. As is true of most construction loans, Real Estate Investment Trust loans typically have a high interest rate associated with it. Part of the reason for this high interest rate charged to the project owner for the construction loan is the fact that should the project owner go bankrupt during the construction, the project in its uncompleted stage could not return a great deal of monetary value to the lender.

In addition to the construction loan, the project owner later commonly needs a second type called a **mortgage loan**. The mortgage loan is used by the project owner to pay off the construction loan. The source of mortgage loans include saving and loan institutions, insurance companies, bond issues, and to a lesser degree commercial banks. To obtain a mortgage loan the project owner puts up the completed project as security. Because the completed project has an identifiable market value if the owner defaults on the loan, the lender of the mortgage loan is usually willing to provide such a loan at a somewhat lower interest charge than the construction loan. In other words, this lower interest rate reflects a lower degree of risk.

Mortgage loans are usually amortized. This merely means that the owner pays back the principal by means of payments which extinguish (or deaden) part of the capital sum of the loan and part of the interest or service charge for it. Initially the owner's payment includes a large amount of interest and a small amount of principal as pay-back. Over the years of the loan the principal pay-back increases systematically while the interest charge decreases. While construction loans are usually written for a period to equal actual construction, mortgage loans typically have a duration from 20 to as long as 40 years.

Other ways of classifying the loans that the project owner takes on when building a project are conventional and non-conventional. Sometimes referred to as a subsidized loan, a **non-conventional loan** typically is tied to a government-related program. For example, FHA and VA mortgages are non-conventional or subsidized loans. Their availability is dependent upon the project owner's ability to meet certain qualifications. For example, the borrower's income normally can not be greater than some pre-determined maximum, and he may have to show the unavailability of conventional loans. The **conventional loan** is merely a loan that is not subsidized. More often than not the savings and loan institution is the primary source of such conventional loans. The cost of a loan is dependent on the following characteristics:

1. The availability of money in the money market.
2. The monetary policy of the government (including the Federal Reserve System's charge to banks for capital).
3. The financial soundness of the borrower—the potential project owner.
4. The negotiating ability of the project owner.
5. The degree of risk the money lender associates with the project.
6. The lender's forecast of the future availability of money and the cost of such money.

Other factors such as the ability of the owner or estimator to market the project to lenders also bears on the cost of financing. The point to be made is that the finance cost associated with equity capital or debt capital can be considerable and due to the fact that it occurs annually it can weigh heavily on the total feasibility of the project. One point satisfying to the potential project owner is that interest or finance cost is a taxable expense and as such yields certain benefits to the project owner. These benefits will be discussed in a following section.

Operating Costs

Of all the costs relevant to the feasibility estimate, perhaps the most difficult to estimate is the annual operating expenses of the project. Sometimes referred to as maintenance and repair costs, these operating expenses are annual costs. The difficulty associated with their estimation is the fact that, while their various components are not only hard to identify, operating expenses can and do increase at varying rates over the life of the project.

Unlike the finance cost that is usually determined for the entire project when the project is initiated, the operating costs can fluctuate dramatically from one year to the next. Components of total operating expenses include everyday maintenance and repair costs such as repairs on heating and ventilating components, repairs on exterior and interior finishes, monthly utility bills, custodian fees, and yearly landscaping costs. Numerous other maintenance and repair costs can also be traced to a particular project. Thus their correct identification can be instrumental in determining the annual operating cost for a project.

The difficulty associated with being able to reckon each of the operating costs as a function of time to the life of the project can be resolved in a couple of different ways. For one, based on past history an attempt can be made to project these costs over the life of the project. This procedure assumes that future cost patterns will follow past cost patterns. Recent years have shown this assumption to be less than totally accurate. A second and more acceptable procedure for recognizing variable operating costs as a function of time is to merely estimate the first-year operating cost and assume that benefits (returns) will be increased to cover increases in operating costs. This is the assumption that investors often make in feasibility estimates for apartment complexes and office complexes. That is, if the operating cost of an apartment complex increases 10% in the second year of its life, rental rates will be increased in an amount to cover that increased operating amount. This makes the estimation of operating costs more determinative. However the assumption that the owner will be able to raise the benefits to the level of increase in operating expenses is thus implicit. Competitive factors may prohibit the owner from increasing benefits somewhat at an equivalent rate. Nonetheless, the potential project owner does have to recognize the variability of the operating expenses as a function of time. Failure to do this would likely result in an underestimation of operating costs and therefore a less than conservative feasibility estimate.

Operating costs for various types of projects are less documented than construction costs. Most cost books on the market do not attempt to identify operating cost for a particular type of project. Undoubtedly part of the reason for this failure to document operating costs is due to uniqueness of the cost for a particular project. In addition, the energy costs that affect utility costs for a project vary substantially from one part of the country to another. Finally, operating costs are closely related to construction costs. A building made with inferior construction materials will likely have higher operating costs than one made with superior materials. These higher operating costs would ordinarily be reflected in increased maintenance and repair cost and in higher utility costs.

Perhaps the best source of cost data in regard to operating costs is the information available from the owners of similar types of projects. Additionally, based on his own experience, the estimator should be able to give the owner some assistance in estimating such future operating costs. While they are difficult to estimate, the significance of operating costs to the feasibility estimate cannot be overemphasized. Annual operating costs can exceed annual finance costs, and the total operating cost for a project over its life can far exceed the initial construction cost.

The significance of operating costs has been highlighted in recent years by those significant increases in energy costs which are reflected in significantly higher operating costs. There is evidence that operating costs are now often underestimated. While some of this underestimation is undoubtedly due to the difficulty of predicting recent escalation of costs for labor, materials, and energy, part of the fault also lies in overlooking several of the factors that make up the operating cost. One related to the operating cost of a project is that sooner or later a project will approach its total estimated life.

If a proposed project clearly has a finite life, then another cost referred to as a **replacement cost** should be recognized in the feasibility estimate. This is the cost of replacing the project after it has become non-functional. This replacement cost will likely exceed the initial construction cost owing to increasing costs of construction labor and materials. More often than not an estimator fails to recognize this replacement cost in the feasibility estimate. This failure can only be justified if in fact the owner or estimator is looking at a project from the point of view of the returns for the project for the finite life of the original design. Depreciation accounting is an attempt at setting a reserve for the replacement of a project. However, depreciation only recognizes the construction cost as originally determined for the project. But in fact, depreciation does not provide a fund or reserve great enough to replace the project in times of inflation. More will be said about depreciation in following sections.

INCOME TAX RELATED FACTORS

The factors relevant to the feasibility estimate are not limited to the cash benefits and cash costs discussed previously. In particular, depreciation effects, capital gains considerations, and deductions for operating interest expenses can all dramatically affect the feasibility estimate. In fact some construction projects are built with the stimulus deriving from the positive tax effects that include depreciation benefits.

Depreciation

The Internal Revenue System recognizes that buildings as capital assets yield returns to their owner over long periods of time. Since most buildings are built to include the business function of producing profit, the I.R.S. allows recognition of a depreciation expense to negate the income-producing revenue related to the particular project. However, rather than let the project owner expense the entire project when built, the I.R.S. requires him to distribute the expense of building a project over a defined period. The process of doing this can be referred to as **depreciation**.

Depreciation provides the potential project owner a value-lowering method for reducing the cost of building the project. In effect, depreciation is a benefit that can be used to reduce taxable income received from the project or from any other source for the project owner. This is especially beneficial if the project owner is in a high income tax bracket. For example, should the project owner be in a 40% income tax bracket, each and every dollar of depreciation he can claim for his project can reduce his tax liability by $.40.

Government guidelines indicate the life over which a particular building can be depreciated and the method of depreciation that the project owner can use to depreciate the building. Naturally an owner would like to depreciate a project as quickly as possible. This fast depreciation provides the benefits of the time value of money. In particular, the sooner benefits are received, the higher the monetary value associated with them.

Tables provided in various I.R.S. tax regulations specify the minimum number of years in which a building can be depreciated. Such a table is shown in Figure 4.3. However, the I.R.S. will usually allow the project owner to depreciate a project in a number of years greater than the minimum number if the project owner sees fit. Here again it is unlikely that the project owner would want to depreciate for any number of years greater than the minimum.

	Asset Depreciation Period (years)
Factories	45
Garages	45
Apartments	40
Dwellings	45
Office Buildings	45
Warehouses	60
Banks	50
Hotels	40
Office Furniture	10
Computers	6

Figure 4.3 *I.R.S. Guidelines for Minimum Depreciation Periods*

I.R.S. regulations provide the project owner much flexibility in choosing a depreciation method. Methods such as straight-line depreciation, declining-balance, double-declining-balance, and others are available to the taxpayer who is depreciating capital assets.

The **straight-line method** is perhaps the easiest to utilize. The taxpayer merely takes the total cost of his assets minus any salvage value and divides that by the number of periods or years for the life of the asset. The result is the annual depreciation charge that is recognized in calculating income tax liability. While the easiest to apply, the straight-line method is perhaps the least advantageous in terms of time-value money concepts. Fast write-off depreciation methods are preferred. However, the project owner is somewhat limited in his choice of depreciation methods. In particular, the I.R.S. limits the number that can be used for depreciating buildings.

The effects of recognizing the depreciation expense as a benefit are tabulated in Figure 4.4. On the one hand, depreciation itself is a non-cash expense to the project owner. As such there is no cash outflow for this expense. On the other hand, as previously indicated, for each dollar recognized as a depreciation expense, the project owner receives a benefit through a reduced tax liability. This reduced tax liability increases as the taxpayer's income tax bracket increases.

Year	Initial Cost	Depreciation	Cash Inflow Depreciation[a]
0	$100,000		
1		$30,000	$12,000
2		21,000	8,400
3		16,333[b]	6,533
4		16,333	6,533
5		16,333	6,533

[a] Assumes 40% income tax bracket

[b] Depreciation method changed to straight line

Figure 4.4 *Depreciation Benefits for a $100,000 Initial Cost*

Investment Tax Credit

The tax advantages of a new building to an owner are not limited to depreciation. Elements of an owner's building may be eligible for investment tax credit benefits.

From time to time government has attempted to stimulate economic growth by granting an incentive to firms that construct or purchase capital assets. The existence of this incentive and the degree of the incentive have varied over time. It is referred to as an investment tax credit. When such a credit exists or is expected to, it should be recognized in the capital budgeting analysis as well as in the project design.

In its present form, the investment tax credit is defined as taken only on "qualified" investments. Basically, qualified investment means that the asset is to be used in a production process.

The investment tax credit differs from a depreciation expense in that the credit is subtracted from the tax entity's tax liability rather than used to determine the tax liability. Whereas a dollar depreciation may result in something less than a 50% lessening of the tax liability, a dollar investment tax credit results in a lessening of a dollar of tax liability. Additionally, the use of the credit has no subsequent effect on the depreciation deduction.

Normally a building in itself is not eligible for investment tax credit benefits because it is not directly part of the production process. However, analysis of the elements or components of a building may result in the identification of several "production-type" elements or components.

As an example of a possible investment tax credit element, consider a manufacturing plant with several pieces of production equipment on the first floor. The slab that holds the equipment might be considered part of the building and therefore not eligible for investment tax credit. However, it may be that the slab would only have to be 4 inches thick if it did not need to hold the production equipment. However, because the heavy equipment does exist, a 10-inch slab of concrete is required to maintain the structural integrity of the building. In effect, the majority of the component slab is there because of the production equipment. It can therefore be argued that part, if not all, of the concrete slab is eligible for investment tax credit benefits.

Other examples of building elements or components that may be eligible for investment tax credit benefits include plumbing and mechanical equipment that supports work in a butcher shop within a grocery store–shopping center, elevators in a building used to transport production material, and even certain sidewalks and driveways utilized in transporting production material to and from the manufacturing process.

If a building element can be traced to a production process it may be eligible for a government subsidy in the form of an investment tax credit. The ability to identify these elements is dependent on an individual's creativity, analytical skills, and construction and production knowledge. Given such attributes the estimator can assist a potential project owner to optimize his investment tax credit benefits and thus lower the cost of the proposed project.

Investment tax credit considerations should also serve as input to design decisions. For example, the fact that moveable partition walls are eligible for investment tax credit benefits whereas fixed walls are likely to be ineligible, can affect the architect/engineer's design. This is not to say that the designer should design all moveable partition walls simply because they may receive better tax treatment. However, the knowledge that they may should be part of the architect's design considerations. It is obvious then that the knowledge of tax implications should be part of the design process.

Capital Gain Treatment

Capital gain treatment identifies the reduced tax rate that applies to the gain received by the project owner on selling a capital asset. In particular, if he as a taxpayer sells a building used for his business, that building may qualify for favorable capital gain treatment. If a gain is realized when he sells a qualified building, the taxpayer/project owner may be able to recognize much of the gain as a capital gain. This type of gain is taxed at a reduced rate.

The gain a project owner would have from the sale of a building would be the difference between the sales price and the building's tax base. Essentially the tax base is the initial construction cost or purchase cost minus the depreciation taken on the building. Because of the relatively high resale value of buildings, the taxable gain to be reported may have a significant impact on the economic analysis or feasibility estimate of a project.

Tax Impact of Operating Expenses

Operating expenses and finance related interest costs are both expenses that can be used to lessen the project owner's income. Finance expenses can be deducted from taxable income in the year that the expenses are incurred. Therefore, while the interest charge for a construction project may be $1,000 a year, in effect the actual interest charge is less when one recognizes the positive effects of being able to identify the expense in calculating taxable income. For the potential project owner in the 40% tax bracket incurrence of $1,000 in interest for the year will result in a $400 cash inflow. The result would be an interest charge less than the $1,000 for the year. In particular, the effective interest charge for the year would be $600.00.

IMPORTANCE OF TIME-VALUE-OF-MONEY CONCEPTS TO THE FEASIBILITY ESTIMATE

Time-value-of-money concepts along with the interest factors that can be used to determine the economic worth of money as a function of time were examined in Chapter 3. Interest tables for these formulas are presented in Appendix A.

Time-value-of-money concepts and formulas are utilized in an economic analysis referred to as the **discounted-cash-flow method**. This is the most general technique available to an estimator and project owner for evaluating project feasibility.

Time-value-of-money concepts provide the basis for the discounted-cash-flow method. Using this method, a capital expenditure is viewed as the acquisition of a series of future net cash flows consisting of two elements: the return of the original outlay, and the net income or cost savings from the capital expenditure. If the present value of the net cash inflows as discounted at a desired rate of return is in excess of the initial capital expenditure required, the investment may then be warranted. Alternately, two proposals may be compared on the basis of their respective differences in discounted net cash inflows versus required initial investment.

Let us assume a potential project owner is considering a project with an initial cost of $60,000. The estimated annual cash flow from rent is $9,000, and the increased cash flow from the depreciation expense is $5,000. Thus the total annual cash inflow is $14,000. In order to determine the feasibility of the capital expenditure using the discounted-cash-flow method, the future incomes or cash flows are discounted back to the present time and compared to the initial capital expenditure of $60,000. The future cash flows are discounted by means of calculating their present worths at a specified interest rate. The interest rate to be used in the calculation is the one at which the firm could have invested the initial capital expenditure had it not built the project. Alternatively, it is the interest rate at which the firm borrows the money to build the building. Let us assume that the latter is true in the example, and the effective interest rate for borrowing the money is 15%, and that includes closing costs and so forth. The net cash inflow discounted at 15% is calculated in Figure 4.5.

Year	Net Cash Flow	Present Value of $1 at 15%	Present Value of Net Cash Flow
1	$14,000	0.869	$12,166
2	$14,000	0.756	$10,584
3	$14,000	0.657	$ 9,198
4	$14,000	0.572	$ 8,008
5	$14,000	0.497	$ 6,958
6	$14,000	0.432	$ 6,048
7	$14,000	0.376	$ 5,264
8	$14,000	0.327	$ 4,578
9	$14,000	0.284	$ 3,976
10	$14,000	0.247	$ 3,458
11	$14,000	0.215	$ 3,010
12	$14,000	0.187	$ 2,618
			$75,866

Figure 4.5 *Discounted Cash Flows at a 15% Interest Rate*

The discounted sum of the inflows exceeds the initial cost of $60,000. This in effect means that there is a positive economic return associated with the building.

The use of the discounted-cash-flow or rate-of-return method is not limited to determining whether the discounted net cash inflows exceed the initial expenditure for an investment alternative. This method can be used to determine the actual rate of return associated with an investment such as the project in question. For example, let us assume that the owner contemplating the $60,000 project would like to know the true rate of return associated with the $60,000 expenditure.

Because the total net cash inflows sum to $168,000 whereas the initial expenditure is only $60,000, it follows that the rate of return is greater than zero. It can also be concluded from the calculations in Figure 4.5 that the rate of return on the initial expenditure is in excess of 15%. This conclusion follows from the fact that the sum of the net cash inflows, discounted at an assumed interest rate of 15%, exceeded the initial expenditure of $60,000. In effect, this means that had the owner invested the $60,000 at a compounded 15% annual interest rate, it would have received less cash inflows than those that it received from investment. Thus the return associated with the $60,000 project must be greater than 15%.

Having decided that the actual rate of return is in excess of 15%, the next step in determining the actual rate of return centers around the use of trial and error. The complex mathematical nature of interest formulas (i.e., they are not linear) results in the necessity for use of the trial-and-error method. Using such a method, a rate of return is selected on the basis of eliminating unfeasible values (i.e., in this case all values less than or equal to 15% have been determined unfeasible), and analyzing previously calculated discounted net cash inflows versus initial expenditures. For example, from Figure 4.5 it was determined that the sum of the discounted net cash inflows equals $75,866. This sum, discounted at 15%, is quite a bit in excess of the $60,000 investment. On this basis it follows that the next trial-and-error iteration should be much in excess of 15%. Perhaps 25% would be a good estimate. On this basis, the sum of the discounted net cash inflows is calculated in Figure 4.6.

Year	Net Cash Flow	Present Value of $1 at 25%	Present Value of Net Cash Flow
1	$14,000	0.800	$11,200
2	$14,000	0.640	$ 8,960
3	$14,000	0.510	$ 7,168
4	$14,000	0.410	$ 5,734
5	$14,000	0.328	$ 4,588
6	$14,000	0.262	$ 3,669
7	$14,000	0.210	$ 2,936
8	$14,000	0.168	$ 2,349
9	$14,000	0.134	$ 1,879
10	$14,000	0.107	$ 1,504
11	$14,000	0.086	$ 1,203
12	$14,000	0.069	$ 962
			$52,152

Figure 4.6 *Discounted Cash Flows—25%*

From the calculation in Figure 4.6 it is determined that the sum of the net cash inflows discounted at 25% is less than $60,000. This in effect means that had the firm invested $52,152 in the project and received the annual cash inflows of $14,000 for 12 years, the effective rate of return on the investment would be 25%. However, because the firm had to invest more than $52,152 (i.e., the project required an expenditure of $60,000), its actual rate of return is less than the 25%.

We have now bounded the solution. The actual rate of return is between 15 and 25%. If it is desirable to further focus on the actual rate of return, the next iteration would be carried out at a rate somewhere between 15 and 25%. The next trial might be made at 20%. This is done in Figure 4.7.

Year	Net Cash Flow	Present Value of $1 at 20%	Present Value of Net Cash Flow
1	$14,000	0.833	$11,662
2	$14,000	0.694	$ 9,716
3	$14,000	0.579	$ 8,106
4	$14,000	0.482	$ 6,748
5	$14,000	0.402	$ 5,628
6	$14,000	0.335	$ 4,690
7	$14,000	0.279	$ 3,906
8	$14,000	0.233	$ 3,262
9	$14,000	0.194	$ 2,716
10	$14,000	0.162	$ 1,890
11	$14,000	0.135	$ 2,268
12	$14,000	0.112	$ 1,568
			$62,160

Figure 4.7   *Discounted Cash Flows—20%*

While the sum of the net cash inflows discounted at 20% does not exactly equal $60,000, it is relatively close. Any further centering in on the actual rate of return may not be beneficial in that there is a degree of uncertainty associated with the expected annual net cash inflows. More likely than not, the rate is close to 21%.

The trial-and-error method has resulted in the determination of approximately a 21% rate of return for the project. This rate of return can be compared to the expected rate of return for alternative investments to determine whether or not the project is warranted.

It should be noted that while the actual rate of return was determined by means of a trial-and-error method, it could have been determined in a more straightforward procedure. In particular, because the expected annual cash inflows were equal in amount, the rate of return could have been solved from the following equation:

$$\$60,000 = (\$14,000) \times (\text{uniform series present worth factor})(\text{for } n = 12, i = ?)$$

Using interest formula tables, the value of i (i.e., the interest rate) could be found that results in the equation being valid. This interest rate would in effect be the rate of return.

This direct application of finding the rate of return can only be used if the expected inflows are equal in amount. More often than not they will vary (that is, both expected revenues and expenses may vary from year to year); when this is the case, the trial-and-error method must be used. Because of this likelihood, the calculations in Figures 4.5 to 4.7 are carried out in a manner that is general (i.e., the inflows may be equal or unequal).

FEASIBILITY ESTIMATE FOR AN OFFICE BUILDING

The purpose of this section is to illustrate a feasibility estimate. The relevant factors and terminology that are part of a feasibility estimate are somewhat a function of the type of project being considered; thus it is impossible to present a single example that encompasses all factors and terminology. However, the procedure and objectives of the feasibility study are essentially the same regardless of the project's type or size, so the techniques illustrated here should prove applicable to numerous types of projects.

Let us assume that an individual or firm is considering constructing an office building with the intention of leasing office space. Based on his available budget, the lot size, and his familiarity with similar projects, the potential investor has decided on a three-story building with approximately 12,500 square feet. At the time of the feasibility estimate, little has been determined about building quality other than the fact that a brick building is preferred. The project is to be conventionally financed and is to be built in a metropolitan area.

The initial cost of the project can be approximated by using office-building data published in cost books. On the basis of the square footage desired, at $80 per square foot, initial cost for the *improvements* on the land would be approximately $1,000,000. To this must be added the land cost. This cost is usually easier to calculate than the improvement costs because it can be determined by means of an **option contract** with the landowner. Such a contract gives the potential investor a right to purchase the land in question for a specific amount of money within a given time period—say 2 months. For this right the investor gives the landowner an amount of money that will be forfeited if the investor fails to exercise his purchase right. If he purchases the land, the option money usually is credited to his purchase price. For purposes of our example, let us assume that the investor has a right to purchase the required land for $100,000. Thus the initial office building costs are approximated as follows:

Land	$ 100,000
Improvements	$1,000,000
Initial Costs	$1,100,000

Let us assume that the cost of money is such that it will be possible for the investor to secure an 80% conventional financing loan at 15% interest per annum payable over 20 years. The other 20 percent of the initial costs are to be financed by means of equity capital of the investor. Thus the investor would obtain a loan for 80% of the $1,100,000, or $880,000.

By using interest tables such as the ones illustrated in Appendix A (the tables can be developed using the time-value-of-money formulas discussed in the previous section), it is possible to calculate the equal monthly or yearly payment that retires the loan and reflects the interest charged by the lender. In a feasibility estimate this monthly or yearly payment is commonly referred to as the **debt service**. Using the tables in Appendix A, an annual debt service of $140,588.80 can be determined as the annual payment required by the company to repay the $880,000 and the 15% interest.

The operating-expense component of the project is commonly expressed as a percentage of the construction costs of erecting the building. Operating expenses normally range from 6 to 30% of construction cost, depending on the type of building being erected.

Let us assume that the initial operating expense (maintenance and repair) for the office building is expected to be 10% of the construction costs. Based on anticipated construction costs of $1,000,000, the initial operating expenses are estimated to be $100,000.

A **market analysis** must be made of the immediate project vicinity in order to establish an estimate of the rental area. A primary part of the market analysis consists of a comparison of rental or leasing rates of comparable types of buildings in the surrounding geographic area. Let us assume that the market analysis for our example yields an anticipated leasing rate of $2.50 per square foot per month.

Although the gross square footage of the proposed office building is 12,500, interior walls, doorways, columns, and so forth will reduce the leasable space. The rule of thumb for office buildings is that a building is efficiently designed if the leasable area approaches 80% of the gross area.

Assuming 80% leasable space, a 7% vacancy allowance, and an expected leasing rate of $2.50, the gross possible income is calculated as follows:

Gross area = 12,500 square feet × .80 leasable
= 10,000 square feet − .12 vacancy allowance
= 9,300 square feet × $2.50 per square foot
= $23,250 per month × 12 months
= $279,000

As noted earlier, operating expenses will be assumed as $100,000 for the first year. Naturally the operating expenses will increase from one year to the next. However, they will be compensated by equal relative increases in leasing rates.

The most accurate means of recognizing all relevant factors concerning the financial rate of return of a project is to apply the **discounted-rate-of-return method**. The fundamentals of this method were discussed earlier.

For an *accurate* rate-of-return calculation, one must have sufficient information. For example, for our office-building project, analysis must include such factors as the owner's decision to sell the project after a number of years, the investor's tax rate, and the particular depreciation method to be used.

Let us assume that the office-building investor is in the 40% tax bracket; that he is planning to sell the project after 7 years; and that he would like to utilize the straight line depreciation method. The fact that the owner plans to sell the project before it has no value means that the selling price must be estimated. The estimated future value of a building is often determined by assuming a constant appreciation (or depreciation) rate for the building. This rate is probably unrelated to the depreciation rate used to reduce the value of the building for tax purposes. For our example, we will assume an annual appreciation rate of 1%.

We will now proceed to determine the annual cash inflow or outflow for each of the years of the analysis. As indicated earlier, the gross possible income (i.e., the rental income) and the operating expense will likely vary from year to year. However, the investor will probably increase rental rates to reflect increased operating expenses. Thus, it is the net of the rental incomes and operating expenses that is of concern to us here; this amount can be assumed constant throughout the relevant time period.

Initially (i.e., at time period 0), the investor in the office building has an outflow of $220,000. This represents the invested equity in the project. The annual cash inflows from owning and operating the office building for the first year is illustrated in Figure 4.8. The cash flow calculation is initiated by first listing the gross possible income (i.e., the lease income). The $279,000 is shown at the net of the vacancy allowance.

The operating expenses shown include annual utilities, maintenance and repairs, insurance, real estate taxes, advertising, legal and accounting costs, and caretaker/manager fees. On the basis of the previous assumed operating expense amount of 10% times initial construction costs, $100,000 is indicated in Figure 4.8 as the first-year operating expense. The $100,000 difference between the gross possible income and the operating expense is shown as the **net operating cash income**. It is the cash inflow to the investor before consideration of the debt service, depreciation, and tax effects.

The debt service amount of $140,588.80 is subtracted from the net operating cash income. The constant of .15976 used to determine the debt service is a factor which equates to a 15% interest rate and a pay-back period of 20 years. Subtracting the $140,588.80 from the $179,000.00 leaves a net cash flow of $38,411.20. This amount does not, however, reflect depreciation, interest, or tax effects.

Pro Forma
Annual Cash Results for Office Building

A. Gross Possible Income (including vacancy allowance)		$279,000.00
B. Operating Expenses		−100,000.00
C. Net Operating Cash Income		179,000.00
D. Less Mortgage of $880,000 times Debt Service Constant of .15976		−140,588.80
E. Net Cash Flow		$ 38,411.20
F. Depreciation, Interest, and Tax Effects		
1. Gross Possible Income	$279,000.00	
2. Operating Expenses	−100,000.00	
3. Depreciation-Construction costs of $1 million at 150% straight-line depreciation with life of 15 years	− 66,666.67	
4. Interest Cost (i.e., part of the debt service)	−132,000.00	
Net Operating Gain	$−19,666.67	
Tax Liability (40% bracket) ($19,667.67)		+ 7,866.07
G. Total Cash Inflow After Taxes		$ 46,277.27

Figure 4.8 *First Year Cash Inflow*

Part F of Figure 4.8 reflects the effect of depreciation, interest, and taxes. The calculation starts with a restatement of the gross possible income: that is, the $279,000 is the previously determined inflow from rental receipts. In order to calculate the actual operating gain or loss for purposes of taxes, the operating expenses, depreciation, and interest effects must be subtracted from the gross possible income. (The operating expense, previously calculated as $100,000, is shown in line 2.)

Depreciation is an expense that is allowable for purposes of taxes; however, it is not a cash expense in regard to the firm. Depreciation provides the potential investor a benefit in that while not expending cash, the depreciation expense has the effect of reducing tax liability. Earlier it was stated that the office-building investor wanted to use the straight line depreciation method. Current tax laws also dictate the minimum number of years over which an asset can be depreciated. Current tax laws permit many buildings to be depreciated in 15 years. A 15 year depreciation period is assumed in the example. Thus, the calculation of depreciation for the office building illustrated in Figure 4.8 uses straight-line depreciation over a 15-year life. Given these assumptions, the calculated depreciation for the first year of the office building is $66,666.67.

We previously calculated an annual debt service of $140,588.80 to pay back the mortgage that was taken on the office building. This debt service was made up of a pay-back of principal and interest. Only the interest cost or expense is a deductible expense for purposes of calculating taxable income. Therefore, the interest cost must be broken out of the debt service so we can calculate the effect of interest on cash flow. In early years the interest expense is large, while the pay-back on the principal is small. In particular, the interest cost for the first year is 15% times the mortgage. This yields an interest expense of $132,000 for the first year.

When we subtract the operating expense, depreciation, and interest expense from the gross possible income, we achieve a calculated net operating tax loss of $19,667.67. The effect of the depreciation and interest expense result in a net operating tax *loss*. This is in fact desirable, because it results in a reduction of taxes to the point at which the tax effect actually adds to the net cash flow for the investor. In our example the effect of depreciation and interest does reduce the net operating income to a negative number. The depreciation and interest expenses have the effect of reducing the tax liability.

Earlier it was indicated that the investor was in the 40% income tax bracket. His tax liability (gain) is calculated by multiplying 40% times the net operating loss of $19,666.67; this results in a tax gain of $7,866.07. This tax gain is added to the net cash flow in order to calculate the total cash inflow after taxes. As is shown in Figure 4.8, this results in a total cash inflow after taxes of $46,277.27. This is the total cash inflow at the end of the first year for the office building. This cash inflow will change in following years because the depreciation and interest expense for following years will also change. Thus the tax liability will change, also, so that total cash inflow decreases over the years.

In order to calculate the total cash inflow for years 2 through 7 we must be able to determine the actual amount of depreciation and the actual interest expense for each of these years. These calculations are made in Figures 4.9 and 4.10, respectively. In Figure 4.9, the depreciation charge for any one year is determined as $66,666.67 because the straight line depreciation method is used.

The calculation of the interest payment is shown in Figure 4.10. The debt service charge for each year is constant. This payment is $140,588.80, as shown in the figure. Note that this debt service includes an interest payment and a principal payment. The interest payment is calculated by multiplying the interest rate of 15% times the mortgage at the end of the previous year. The remainder after subtracting the interest payment from the debt service is the principal payment. As can be seen in Figure 4.10, the interest payment decreases each year, while the principal payment increases yearly.

End of Year	Depreciation Charge	Book Value
0	$66,666.67	$1,000,000.00
1	66,666.67	933,333.33
2	66,666.67	866,666.66
3	66,666.67	799,999.99
4	66,666.67	733,333.32
5	66,666.67	666,666.65
6	66,666.67	599,999.98
7	66,666.67	533,333.31

Figure 4.9 *Depreciation Charge and Book Value for Improvements*

End of Year	Debt Service	Interest Payment	Principal Payment	Mortgage
0				$880,000.00
1	$140,588.80	$132,000.00	$ 8,588.80	871,411.20
2	140,588.80	130,711.68	9,877.12	861,534.08
3	140,588.80	129,230.11	11,358.69	850,175.39
4	140,588.80	127,526.30	13,062.50	837,112.89
5	140,588.80	125,566.93	15,021.87	822,091.02
6	140,588.80	123,313.65	17,275.15	804,815.87
7	140,588.80	120,722.33	19,866.47	784,949.40

Figure 4.10 *Debt Service Breakdown*

Having calculated the depreciation expense and interest expense for each of the years 1 through 7, we can now calculate the total cash inflow for each of these years. Previously we determined the total cash inflow at the end of year 1. Note that the depreciation expense and the interest expense have the effect of changing the tax liabilty for each of the years. The operating expense and the gross possible income or rental income will be assumed constant because the net of these two amounts will remain unchanged throughout the year. The net operating loss for each of the years shown in Figure 4.11 has been calculated by subtracting the operating expense and depreciation expense from the gross possible income for each of these years. The tax cash gain shown in Figure 4.11 was determined by multiplying 40% by the indicated net operating loss for that year. Finally, the total cash inflow is calculated by adding the tax cash gain shown to the net cash flow previously calculated as $38,411.20.

End of Year	Operating Expense	Depreciation	Interest	Net Operating Gain	Tax	Total Cash Inflow
1	$100,000.00	$66,666.67	$132,000.00	$19,666.67	$7,866.67	$46,277.27
2	100,000.00	66,666.67	130,711.68	18,378.35	7,351.34	45,762.54
3	100,000.00	66,666.67	129,230.11	16,896.76	6,758.70	45,169.90
4	100,000.00	66,666.67	127,526.30	15,192.97	6,077.19	44,488.39
5	100,000.00	66,666.67	125,566.93	13,233.60	5,293.44	43,704.33
6	100,000.00	66,666.67	123,313.65	10,980.32	4,392.13	42,803.33
7	100,000.00	66,666.67	120,722.38	8,389.05	3,355.62	41,766.82

Figure 4.11 *Cash Inflows for Years 1-7*

An additional net cash inflow occurs at the end of the 7th year owing to the planned sale of the office building at that point in time. This calculation is shown in Figure 4.12.

A. Cash for Sale	
1. Sales price (1% appreciation)	$1,179,348.80
2. Less 5% commission	58,967.40
3. Gross cash	$1,120,381.40
4. Less mortgage balance	784,949.40
5. Adjusted cash income	$ 335,432.00
B. Taxes to be Paid	
1. Gross cash	$1,120,381.40
2. Less book value	533,333.31
3. Taxable gain	$ 587,048.09
4. Total tax due	$ 117,409.61
C. Net Cash from Sale	
1. Adjusted cash income	$ 335,432.00
2. Less taxes due	117,409.61
3. Net cash	$ 218,022.39

Figure 4.12 *Cash Flow from Sale*

The calculation starts with the listing of the sale price. The anticipated sale price is determined from our previous assumption of a 1% appreciation rate per year for the land and the improvements; it results in an anticipated sales price of $1,179,348.80. It is anticipated that the owner of the office-building complex will incur a 5% commission cost in order to sell the building. When this figure is subtracted from the anticipated sales price, we arrive at the gross amount of $1,120,381.40. The mortgage balance at the end of the 7th year for the loan is calculated in Figure 4.10 as $784,949.40. When the office building is sold, this mortgage balance will have to be paid to the lending institution. After paying this mortgage balance, the adjusted cash flow to the investor is $335,432, as shown in Figure 4.12.

Because taxes must be paid on the gain from the sale of the office building, the adjusted cash income will be reduced. The calculation of the tax liability is shown in Figure 4.12. The calculation is made by subtracting the book value at the end of the 7th year from the gross cash received from the sale of the office building at that time. The book value at the end of the 7th year was calculated in Figure 4.9. The result is a taxable gain of $587,048.09. This tax gain is taxed as a capital gain. A capital gain has the advantage of being taxed at a reduced rate relative to an ordinary gain.

Let us assume a capital gain rate of 20%. The tax liability is determined by multiplying the taxable gain times the capital gain rate. Finally, the net cash flow to the investor from the sale of the office building at the end of the 7th year is determined by subtracting the tax liability from the adjusted cash income previously calculated. This calculation is shown in Figure 4.12.

We have now calculated the cash outflows and inflows for the period in which the investor is affected by the project. These cash outflows and inflows are best illustrated by means of arrows on a time scale, as shown in Figure 4.13. The arrow pointing upward indicates a cash outflow. For the example in question, there is only one net cash outflow—resulting from the initial cash investment in the project. The arrows pointing downward represent net cash inflow to the office-building investor. Note should be made of the fact that after the 7th year two arrows are shown, indicating two net cash inflows. They represent the cash inflow from operating the office complex in the 7th year and the cash inflow from the sale of the office building after the 7th year.

At this point in time, the analysis has yielded a net cash inflow or outflow for each year. This in itself provides the potential investor with a tool for evaluating the feasibility of the project. The rate of return for any one

46,277.87	45,762.54	45,169.90	44,488.39	43,704.33	42,803.33	41,766.82	218,022.39
1	2	3	4	5	6	7	

220,000

Figure 4.13 *Time Scale of Cash Inflows and Outflows*

year can be determined by dividing the cash inflow for that year by the equity investment. For example, for the third year a net cash inflow was determined as $45,169.90. Thus the rate of return for this third year is $45,169.90 divided by $220,000.00. This calculation yields a rate of return for the third year of approximately 20.53%. However, this is not the *true* rate of return.

The error in calculating the rate of return in this manner for any given year is that the time value of money is not taken into account. In particular, future net inflows of money are not worth as much as if the cash inflows were received at the present time. A true rate-of-return calculation should *discount future net cash inflows* to reflect the period of time at which they are received.

In order to discount future cash inflows or outflows in a rate of return calculation, the analysis must include the interest formulas discussed in a previous section. Essentially, the analysis consists of determining the interest rate at which future sums of money are discounted back to the present time to equal the initial investment. Figure 4.13 shows that there are cash inflows at the end of each of the years 1 through 7. If we can determine the interest rate, i, at which the discounted value of their sum equates to the initial investment of $220,000, then we will have found the true rate of return for this investment.

As indicated in an earlier section, the analysis that reflects discounted cash flows is essentially a trial-and-error analysis. Initially, it consists of making a guess at the rate of return. The initial guess can be guided if one looks at the magnitude of the future cash inflows versus the initial investment. If one were to sum all of the future cash inflows, they would equal $527,995.57. This is considerably larger than the initial investment of $220,000. It follows that because the cash inflows are considerably in excess of the initial investment, a fairly large rate of return seems evident.

Let us start the analysis by assuming a rate of return as 20%. The cash inflows illustrated in Figure 4.14 are multiplied by interest factors that reflect the year in which the cash inflow is received. The discounted cash inflows from Figure 4.14 yield a sum of $222,345.54—this is considered the present worth of the future cash inflows. This value of $222,345.54 is in excess of the $220,000 initial investment. A comparison of these two numbers leads us to conclude that the true rate of return is in excess of 20% for the office building. This conclusion is determined by reasoning that if the investor in the office building contributed $222,345.54 in equity for the office building, he would in fact have received the cash inflow illustrated in Figure 4.14. However, he did not have to invest $222,345.54 in order to receive the cash inflow; he only had to invest $220,000. So it follows that the rate of return is in excess of 20%.

We have now determined that the interest rate or rate of return is greater than 20% for the office building. However, the question remains, how much

Year	Cash Inflows	SPPW i — 20% n = x	Discounted Value
1	$ 46,277.87	0.8333	$ 38,563.35
2	45,762.54	0.6944	31,777.51
3	45,169.90	0.5787	26,139.82
4	44,488.39	0.4823	21,456.75
5	43,704.33	0.4019	17,564.77
6	42,807.33	0.3349	14,336.17
7	259,789.21	0.2791	72,507.17
			$222,345.54

Figure 4.14 *Discounted Cash Inflows at i = 20%*

greater than 20% is the true rate of return? Again we require another guess. Let us assume an interest rate of 25%. By so doing, it is likely that we will bound a solution. That is, if we find that an interest rate of 25% is too great, at least we will have established that the rate of return will be between 20% and 25%.

The discounted cash inflows at 25% are shown in Figure 4.15. The mathematics is the same as shown in Figure 4.14; the only difference is that the interest factors used now reflect an interest rate of 25%. As is shown in Figure 4.15, the sum of the present worths of the discounted cash inflows is equal to $187,679.27. This is less than the initial investment of $220,000. Had the owner invested in equity $187,679.27, he would have obtained the cash inflow shown in Figure 4.15. However, our office-building owner had to invest more than this amount—in fact, he had to invest $220,000. This means that the true rate of return is *less than* 25%.

Year	Cash Inflows	SPPW i — 20% n = x	Discounted Value
1	$ 46,277.87	0.8000	$ 37,022.30
2	45,762.54	0.6400	29,288.03
3	45,169.90	0.5120	23,126.99
4	44,488.39	0.4096	18,222.44
5	43,704.33	0.3277	14,321.91
6	42,807.33	0.2621	11,219.80
7	259,789.21	0.2097	54,477.80
			$187,679.27

Figure 4.15 *Discounted Cash Inflows at i = 25%*

We have now bounded a solution. Naturally the next guess at a true rate of return should be between 20% and 25%. On the basis of the calculations shown in Figure 4.14 and Figure 4.15, we will estimate an interest rate of 21%. Figure 4.16 shows an analysis for an interest rate of 21%; the sum or present worth of the discounted cash inflows discounted at 21% results in a sum of $214,641.01. This is close to the initial investment of $220,000. Thus the true rate of return is approximately 20 to 21%. However, it is not necessary to carry the solution any closer at this point. Recognizing all the uncertainties of the various cash inflows, further analysis is not justified.

Year	Cash Inflows	SPPW i − 20% n = x	Discounted Value
1	$ 46,277.87	0.8264	$ 38,244.03
2	45,762.54	0.6830	31,255.81
3	45,169.90	0.5645	25,498.41
4	44,488.39	0.4665	20,753.83
5	43,704.33	0.3855	16,848.02
6	42,807.33	0.3186	13,638.41
7	259,789.21	0.2633	68,402.50
			$214,641.01

Figure 4.16   *Discounted Cash Inflows at i = 21%*

The discounted-cash-flow analysis of the office-building project is now complete. The result of the analysis has been the determination of the rate of return the investor will receive from his equity investment in the project. This figure can be compared to alternative investments that may be available to him, including such things as interest or rate of return received from investment in stocks, treasury bills, bonds, and simple savings and loan passbook savings. Other investments might include business ventures such as investments in new products, new lines of business, or perhaps other types of construction projects. The analysis can also be used as a basis for evaluating design modifications. The point to be made is that the rate of return that has been calculated reflects all relevant financial considerations so that the calculated rate of return is in fact the *true* rate of return to the investor.

EXERCISE 4.1

A company builds a new plant for $1,500,000 (land and building). The building itself is valued at $1,200,000. The company invests $300,000 cash and obtains a mortgage for $1,200,000 at 10% payable in 20 end-of-year payments. The firm expects revenue during the first year (assume it is at the end of the year) of $250,000 and expects maintenance and repair independent of depreciation of $100,000 (assume it is at the end of the year). The building is to be depreciated over 15 years. Using a 1.75 times straight line depreciation method for the building in question, determine the after-tax cash flow for the firm for the end of year one. Assume the company is in a 40% tax bracket. The capital recovery factor for an interest rate of 10% and 20 periods is 0.11746. Ignore investment tax credit.

EXERCISE 4.2

Assume that a project owner is evaluating the alternatives of depreciating a new building over either a life of 15 years or 30 years. The straight depreciation method is to be used.

The project owner is to build a $1,000,000 project and invest equity of $200,000. The owner is in the 40 percent income tax bracket.

In order to determine the impact of a 15 year depreciation period versus a 30 year depreciation period, determine the rate of return of the depreciation cash flows relative to the equity investment for the two alternatives.

EXERCISE 4.3

Assume that based on an analysis of cash revenues and expenses, an estimator has concluded that a proposed project yields the following cash flow:

YEAR OF YIELD	CASH INFLOW (CASH OUTFLOW)
0	$(80,000)
1	18,000
2	19,000
3	22,000
4	17,000
5	12,000
6	28,000

Assume that each amount is received at the end of the year indicated. For example, the $80,000 cash outflow occurs at the end of year 0 (which is equal to the beginning of year 1), etc. Determine the rate of return for the proposed project.

5

Pre-Construction
Design Estimates

The Importance of Pre-construction (Preliminary)
*Estimates * Pre-construction Estimates: When and Why*
*Needed * Unit-Cost Estimates * Parameter*
*Estimating * Factor Estimating * Range Estimating*

THE IMPORTANCE OF PRE-CONSTRUCTION
(PRELIMINARY) ESTIMATES

As part of its services, the project designer (sometimes referred to as the A/E
in reference to the fact that an architect or engineer prepares the estimate) is
often called upon to give the owner a preliminary cost estimate of the proposed project. This estimate may be revised throughout the design phase as
materials and aesthetics are determined. These preliminary cost estimates
are referred to as pre-construction estimates in this chapter.

The purpose of the pre-construction estimate can vary depending on the
owner's demands and the type and size of the project. Frequently the estimate is used to support the feasibility estimate (discussed in the previous
chapter), to evaluate possible design modifications to keep the project within
the owner's budget, to evaluate contractor bids, and as an aid in budgeting
cash-flow needs throughout the project.

There is benefit to be gained from an accurate pre-construction estimate,
but factors such as the lack of detailed drawings, lack of thorough knowledge
of construction methods, and limitations on the fees restrict the attainable
accuracy. Although it is not expected that the pre-construction estimate be
precise, it should be reasonably accurate.

The fact that the pre-construction estimate is necessarily done before
project drawings and contract drawings are complete means that these estimates are less accurate than the subsequent cost estimates prepared by contractors. It also follows that the procedures and source documents used by
the designer in preparing pre-construction estimates differ from those used

by the construction contractor. Often the use of a given set of procedures for estimating the cost of a project has resulted in the procedures and the related source documents being identified as a specific estimating technique or approach. For example, parameter estimating, factor estimating, and range estimating are each characterized by a specific process and procedure for preparing a pre-construction cost estimate. These and other pre-construction estimating techniques are discussed in this chapter.

PRE-CONSTRUCTION ESTIMATES: WHEN AND WHY NEEDED

The pre-construction preliminary cost estimate is made without working drawings or detailed specifications. In fact, the estimator may have to make such an estimate from rough design sketches, without dimensions or details, and from an outline specification and a schedule of the owner's space requirements. Recent developments in the construction process have focused attention on the need for accurate pre-construction estimates. The increasing complexity of construction projects, their rapidly increasing costs, and the owner's difficulty with the high cost of obtaining funds all draw attention to the importance of the pre-construction or design estimate. The use of fast-track or phased construction also increases this need. If the estimate fails to provide a reliable pre-construction estimate, an owner may start a project only to find that there is not capital to complete it.

The types of skills needed by the estimator for reliable preliminary estimates include the following:

1. Knowledge of construction materials.
2. Understanding of building design.
3. Ability to conceive design details.
4. Knowledge of construction trades.
5. Acquaintances with construction labor productivity.

Although this list is not all-inclusive, it is fundamental to the determination of an accurate estimate.

The achievement of an accurate preliminary estimate is tempered by the unavailability of sufficient design detail and the time and fee restrictions on the estimator. A pre-construction estimate that is within 5% of the actual construction cost should be looked upon with favor. Ignoring the fact that there are insufficient source documents available to the estimator at the time of the preliminary estimate, the more time the estimator allocates to this estimate, the more accurate the estimate will tend to be. However, the estimator must operate within budget—time is an overhead cost to the owner.

The marginal increase in the accuracy of the preliminary cost estimate as a function of time spent determining the estimate is difficult to determine.

This difficulty does not remove the relevance of the accuracy versus time trade-off.

It is difficult to set a number of hours for the determination of the pre-construction estimate. Instead, the estimator has to keep in mind the objectives of the estimate, the accuracy and completeness of drawings, and the relative marginal accuracy and benefits obtained for additional estimating time. These factors, along with the owner's willingness to pay for estimating time, should dictate the time expended for the preliminary cost estimates.

The pre-construction cost estimate can serve several purposes, including the following:

1. It supplements or serves as the owner's feasibility estimate.
2. It aids the A/E in designing to a specific budget.
3. It assists in the establishment of the owner's funding.
4. It serves as a means of evaluating contractor bids.
5. It provides the basis for determining progress payments to be made to contractors.

Normally a single pre-construction cost estimate serves several of these purposes. However, if the estimate is prepared to serve a single purpose, the amount of detail and the time at which the estimate is prepared can vary. For example, if an estimate is to be made only to monitor progress payments, the estimate may be made as late as the start of construction. But, because actual payments are keyed to the estimate, this pre-construction estimate will likely have to be more accurate than if it had been prepared to serve as a feasibility estimate.

The pre-construction cost estimate often varies in form and scope depending on whether the project is a building or "heavy and highway" construction project. Building construction consists of projects such as office buildings, industrial plants, commercial buildings, and manufacturing plants. Heavy and highway construction projects include roadways, sewage treatment plants, and bridges.

The project drawings and specifications prepared by the A/E for building construction projects typically leave out the quantities of work the contractor is to perform, nor do they address specific work items or quantities. In contrast, heavy and highway pre-construction estimates like the drawings for heavy and highway projects typically have the quantities of the various project materials set out. This is especially true in regard to roadway or bridge construction. In fact, such quantities often appear as part of the drawings.

The reasons for setting out the quantities for heavy and highway projects are several. One reason is the repetitive nature of work components in such projects. This means fewer work items and materials relative to building construction, and facilitates the A/E's setting out of the quantities. In addition, there is less variability in the way the contractor can break out work packages and in the technology available to perform heavy and highway work. This in itself makes the quantities more useful to the contractor in estimating.

In addition to making pre-construction cost calculations, an estimator may also be called upon to perform a **progress estimate** or a **final estimate**. Involvement in such procedures is dependent upon the owner's needs and the type of contract used in the agreement with the contracting firm.

The progress estimate involves inspection and determination of work completed at the job site. It is required periodically so that progress payments to the contractor can be established. It is common for progress estimates to be made monthly.

The final estimate is merely the last progress estimate. It serves the purpose of verifying that all work set out in the contract documents is in fact completed. If this is the case, then and only then is final payment to the contractor authorized. The final estimate can also serve the purpose of determining final quantities of work performed should the final quantities not be set out in the contract documents. For example, the final estimate may be necessary in order to determine the exact length of piling driven by the contractor for which the contractor is to be paid.

UNIT-COST ESTIMATES

Unit-cost estimates are the quickest and easiest pre-construction estimates to prepare. If based on reliable data, and if used properly, such estimates provide the potential for accuracy within an acceptable time allowance. Unit-cost estimates vary as a function of the base element considered. Perhaps the broadest base unit-cost estimate is the **function estimate**. The function estimate measures the cost of a building relative to its use or function. The following are examples of function estimates:

TYPE OF PROJECT	FUNCTION ESTIMATE
School building	Cost per student
Hospital	Cost per bed
Theater	Cost per seat
Parking deck	Cost per parking space

In establishing a preliminary cost estimate by means of a function estimate, the estimator is expected to know or approximate the cost per functional element. Multiplying this cost times the quantity of the functional element yields the total cost estimate.

Perhaps the most commonly used unit-cost estimate is the **cost-per-square-foot-estimate**. This is based on historical cost-per-square-foot data. By multiplying the historical square-foot cost times the calculated square feet of floor area for the proposed building, a pre-construction preliminary cost estimate for the building can be determined.

The reason for the acceptance of the unit cost-per-square-foot estimate is that many building costs are very closely related to the square feet of floor area. As indicated in the previous chapter, this type of estimate is also widely used in the preparation of feasibility estimates.

Unit cost-per-square-foot data are readily available in industry cost books (see Figure 5.1). It is common to list the cost per square foot for building type and different qualities of construction. That is, the cost per square foot for a hospital may be shown assuming cheap construction, average construction, or high-quality construction. The purpose of listing several quality grades of construction is to enable the estimator to more accurately estimate the cost of the project in question. Obviously, in order to use such data the estimator must first determine what quality of construction the project is to have.

Type of Building	Cost per Square Foot
Apartments	$34.50
Banks	62.40
Churches	38.40
Department stores	29.80
Dormitories	45.60
Factories	28.40
Hospitals	68.30
Office buildings	44.50
Schools	39.60
Shopping centers	31.40
Warehouses	25.20

Figure 5.1 *Sample Data for Unit Cost per Square Foot*

Similar to the cost-per-square-foot estimate is the **cost-per-cubic-foot estimate**. This type of estimate relates the cost of a building to its volume. Historical data are collected regarding the cost as a function of the enclosed volume of the building. For example, if the volume of the proposed building is 50,000 cubic feet, this number would be multiplied by a cost per cubic foot as determined from previous similar projects. Cost-per-cubic-foot estimates are usually rather unreliable unless virtually identical buildings are compared. There is not much relationship between the volume of the building and its cost.

Cost-per-cubic-foot estimates are primarily used for structures such as warehouses. Because such structures have widely varying floor heights, cost-per-*square*-foot estimates can become unreliable—thus cost-per-cubic-foot estimates are more accurate and used more often.

Yet another type of unit-cost estimate is the ***cost-per-enclosure-area estimate***. This is based on the area of all the horizontal and vertical planes of the building. The interior area of the floors or decks is added to the interior areas of the walls. Using this as a base, this area is divided into the total cost of the building to develop a cost per enclosure area. For similar projects, these data can be used to determine the preliminary cost estimate. The cost-per-enclosure-area estimate is usually reliable because the costs of the building are related to the interior horizontal and vertical planes.

Going one step further in detail is the ***cost-per-element or trade-unit estimate***. Such an estimate breaks down the total building into basic parts or trades. Past unit-cost data of these parts are structured to be used for future preliminary cost estimates. An example of such historical cost data for this type of estimate is shown in Figure 5.2. Such an estimate has a potential for accuracy not obtainable from the functional or cost-per-square-foot estimates.

The estimator may also determine a preliminary cost estimate by relating the quantities of work to each of the 16 uniform code divisions of work. Such an estimate (shown in Figure 5.3) is often used by the A/E because it enables the A/E firm to use the divisions of work for writing the specifications. In addition, this division of work category is compatible with the knowledge of other architects and engineers.

If used properly, the unit-cost estimate described can satisfy the purpose of preliminary cost estimate. However, it remains an *approximate* estimate. The estimator has to recognize that numerous factors, such as the following, can change the cost of a proposed building:

1. Changes in quality of construction.
2. Changes in aesthetics or design.
3. Site conditions of the project.
4. Weather conditions affecting the project.
5. The skills of the labor force.
6. The morale of the labor force.
7. The material prices and wage rates.
8. The experience and skill of the project management team.

Other factors also can change the cost of the proposed project. However, given the time constraint usually placed on the preliminary estimate, the unit-cost estimate serves well as a first cost estimate. It should be remembered that it is difficult to obtain total accuracy if all input data are not known. Even so, the unit-cost estimate may in fact be the most accurate, given the time at which it is made.

Excavation		$ 34,500
Formed concrete		70,000
Exterior masonry		170,500
Interior masonry		64,700
Structural steel		152,300
Ornamental metal		13,600
Carpentry		10,500
Waterproofing		7,400
Roofing		21,400
Doors		16,200
Windows		10,900
Curtain walls		24,800
Drywall		42,300
Plumbing		44,200
Sprinklers		44,000
HAVC		58,200
Electrical		39,800
		$825,300

Figure 5.2   *Trade Estimate for a Typical Building*

Division	Work	Estimate
1	General requirements	$ 40,000
2	Site work	30,000
3	Concrete	145,000
4	Masonry	85,500
5	Metals	31,200
6	Wood & plastics	52,400
7	Moisture protection	11,400
8	Doors, windows, & glass	8,600
9	Finishes	32,400
10	Specialties	18,300
11	Equipment	38,500
12	Furnishings	9,200
13	Special construction	19,800
14	Conveying systems	14,800
15	Mechanical	63,200
16	Electrical	21,200
		$621,500

Figure 5.3   *Sample 16-Division Estimate for a Building*

PARAMETER ESTIMATING

A recently evolved pre-construction cost estimate method is the **parameter estimate**. This type can be used to give the owner an approximate cost of a project and also to enable the estimator to evaluate contractor bids.

Engineering *News Record* first published parameter cost estimates in 1966. From that time, the periodical has expanded the concept of parameter estimating. Whereas at first only the gross floor area was used as a parameter, today there are several parameter measures. Contractors, owners, designers, or construction managers using *Engineering News Record's* parameter cost breakouts allocate lump-sum costs to trades or component systems of a building's construction. Parameter measures are then chosen to divide into the trade cost, one parameter to a trade. The choice relates the parameter measure to the function of the trade. For example, structural steel cost may be related to the gross area supported, and dry-wall cost to interior area. The parameter costs can yield guidelines to estimate the cost of a similar project or to judge the validity of contractors' quotes.

Type of building	Office
Location	Troy, Mich.
Construction start/complete	Aug./Oct.
Type of owner	Private
Frame	Structural steel
Exterior walls	Metal curtain wall
Special site work	None
Fire rating	2-A
PARAMETER MEASURES	
1. Gross enclosed floor area	650,000 sf
2. Gross area supported (excluding slab on grade)	608,400 sf
3. Total basement floor area	41,600 sf
4. Roof area	70,000 sf
5. Net finished area	500,000 sf
6. Number of floors including basements	26
7. Number of floors excluding basements	25
8. Area of face brick	0
9. Area of other exterior wall	150,000 sf
10. Area of curtain wall including blass	221,000 sf
11. Store-front perimeter	300 sf
12. Interior partitions	30,000 sf
13. HVAC	1,700 tons
14. Parking area	890,000 sf
OTHER MEASURES	
Area of typical floor	19,400 sf
Story height, typical floor	152 in
Lobby area	2,000 sf
Number of plumbing fixtures	460
Number of elevators	12
DESIGN RATIOS:	
A/C ton per building sq. ft.	0.0026
Parking square feet per building	1.3692

Figure 5.4 *History of Typical Parameter Data*

Let us look at an example of collected parameter cost data. These data could easily be collected by the estimator or could be taken from published cost data such as those in *Engineering News Record*. Parameter cost data for an office building (shown in Figure 5.4) includes such information as the type of project, parameter measures, and trade data. The descriptive information regarding the type of project is used merely to classify similar types of projects. That is, should the estimator attempt to estimate an office building, he would attempt to match the future project with parameter cost data for a similar type of project. Information such as the type of frame, the exterior walls, and the design ratios would aid the estimator in selecting a similar type of project from its parameter cost library.

Area or Trade	Parameter Cost			Total Cost	
	Code	Unit	Cost	$	%
General conditions and fee	1	sf	2.19	1,425,000	9.07
Sitework (clearing and grubbing)	1	sf	0.35	226,000	1.44
Excavation	3	sf	4.57	190,000	1.21
Foundation	2	sf	0.67	410,000	2.61
Formed concrete	2	sf	2.71	1,650,000	10.50
Interior masonry	12	lf	6.20	186,000	1.18
Structural steel	2	sf	1.23	750,000	4.77
Miscellaneous metal, including stairs	2	sf	0.38	230,000	1.46
Carpentry	5	sf	0.18	90,000	0.57
Waterproofing and dampproofing	10	sf	0.10	22,000	0.14
Roofing and flashing	4	sf	1.64	115,000	0.73
Metal doors, frames, windows	5	sf	0.11	55,000	0.35
Hardware	5	sf	0.12	60,000	0.38
Glazing·	11	lf	333.33	100,000	0.64
Curtain wall	10	sf	3.96	875,000	5.57
Lath and plaster	5	sf	0.08	40,000	0.25
Dry wall	12	lf	28.90	867,000	5.52
Tile work	5	sf	0.10	49,000	0.31
Accoustical ceiling	5	sf	0.18	90,000	0.57
Resilient flooring	5	sf	0.04	18,000	0.11
Carpet	5	sf	0.03	17,000	0.11
Painting	5	sf	0.17	87,000	0.55
Toilet partitions	5	sf	0.03	13,000	0.08
Elevators	om	ea	70,000	70,000	0.44
Plumbing	1	sf	5.15	3,350,000	21.35
HVAC	1	sf	5.15	3,350,000	21.35
Electrical	1	sf	1.77	1,150,000	7.32
Parking, paved	14	sf	0.25	223,000	1.42
TOTAL	1	sf	24.17	15,708,000	100.00

Figure 5.4 *Continued*

The parameter measures shown in Figure 5.4 indicate the actual measures for a previously constructed office building. That project has a gross enclosed floor area of 650,000 square feet and the frame is structural steel. Other parameter measures shown in the figure may be used to relate the actual cost of the project by segments or trades.

Let us look at the data regarding the trades. In the first column a code is indicated which relates the types of work in question to one of the parameters. For example, the general conditions and fee for the office building are indicated to be a measure or function of the parameter of gross enclosed floor area. The interior masonry shown in the trade has a code of 12, indicating that the interior masonry shown is most a function of the linear feet of partition wall. The measured linear feet of partition wall was 30,000. Dividing this quantity into the total cost amount develops a parameter cost of $6.20.

Similar parameter unit costs are developed and shown in Figure 5.4. Another column relates the percentage of trade cost amount to the total cost of the project: for example, the interior masonry is shown as 1.18 percent. This is developed by dividing the total costs for the interior masonry— $186,000—by the total cost of the project—$15,708,000.

If the value of the parameter cost data shown in Figure 5.4 is to be used in determining a measure of preliminary cost, the estimator merely has to take off the parameter-measure quantities to obtain a historical library of parameter costs by trade. For the office building this would mean that the estimator would only have to take off 14 measures of project quantities. This compares to several hundred quantities that are characteristic of a contractor's detailed estimate. The unit-cost estimate is in reality a one-parameter estimate. Typically, the parameter estimate used by an estimator would be restricted to 8 to 15 parameters.

Let us now look at how a parameter estimate can be made from previous parameter cost data. Assume we are faced with determining a preliminary cost estimate for another office project. Further assume that the office building in question is similar in quality and construction to the office building for which we have collected data in Figure 5.4. The *type of project* is described in Figure 5.5. The *parameter measures* are shown; for example, the office building we wish to estimate has gross enclosed floor area of 395,000 square feet. Other parameter measures are shown in the figure. The parameter cost data for the trades as determined from the previous office-building project are also shown. In reality, the parameter cost data we would use for estimating would likely be an accumulation of several past projects. However, for purposes of demonstration, let us assume that these historical accumulated cost data are only the ones shown in the previous example.

Now let us consider the determination of the masonry cost. The unit parameter is identified as a function of linear feet of partition walls. We would multiply the 35,500 by $6.20 to determine the total estimated masonry cost of $220,100. We determine the column of percentages by simply taking

Trade	Parameter Cost			Total Cost	
	Code	Unit	Cost	Amount	%
General conditions and fee	1	sf	2.19	$ 865,050	8.62
Sitework (clearing and grubbing	1	sf	0.35	138,250	1.38
Excavation	3	sf	4.57	205,650	2.05
Foundation	2	sf	0.67	234,500	2.34
Formed concrete	2	sf	2.71	948,500	9.45
Interior masonry	12	lf	6.20	220,100	2.20
Structural steel	2	sf	1.23	430,500	4.29
Miscellaneous metal, including stairs	2	sf	0.38	133,000	1.33
Carpentry	5	sf	0.18	57,600	0.57
Waterproofing and dampproofing	10	sf	0.10	11,000	0.11
Roofing and flashing	4	sf	1.64	65,600	0.65
Metal doors, frames, windows	5	sf	0.11	35,200	0.35
Hardware	5	sf	0.12	38,400	0.38
Glazing	11	lf	333.33	100,000	1.00
Curtain wall	10	sf	3.96	435,600	4.34
Lath and plaster	5	sf	0.08	25,600	0.25
Drywall	12	lf	28.90	1,025,950	10.22
Tile work	5	sf	0.10	32,000	0.32
Accoustical ceiling	5	sf	0.18	57,600	0.57
Resilient flooring	5	sf	0.04	12,800	0.13
Carpet	5	sf	0.03	9,600	0.10
Painting	5	sf	0.17	54,400	0.54
Toilet partitions	5	sf	0.03	9,600	0.10
Elevators	om	ea	70,000	70,000	0.70
Plumbing	1	sf	5.15	2,034,250	20.27
HVAC	1	sf	5.15	2,034,250	20.27
Electrical	1	sf	1.77	699,150	6.97
Parking, paved	14	sf	0.25	50,000	0.50
TOTAL	1	sf	25.40	10,034,150	100.00

Parameter Measures

1.	Gross enclosed floor area	395,000 sf
2.	Gross area supported (excluding slab on grade)	350,000 sf
3.	Total basement floor area	45,000 sf
4.	Roof area	40,000 sf
5.	Net finished area	320,000 sf
6.	Number of floors including basements	12
7.	Number of floors excluding basements	11
8.	Area of face brick	0
9.	Area of other exterior wall	80,000 sf
10.	Area of curtain wall including glass	110,000 sf
11.	Store-front perimeter	300 sf
12.	Interior partitions	35,500 lf
13.	HVAC	1,200 tons
14.	Parking area	200,000 sf

Figure 5.5 *Use of Parameter Data for Estimating*

the estimated cost of a given trade and dividing it by the calculated sum of the total estimated cost. In this manner we derive the total estimated cost of the proposed project and the components of the cost identified by trade. Thus it is possible for the estimator to determine an approximate cost of the project and also develop component trade costs that can be used to evaluate contractor bids.

FACTOR ESTIMATING

Factor estimating can be used by an estimator to make a pre-construction cost estimate for various types of projects. However, the factor estimate is best used for projects with a predominant cost component. Often this component is the purchased equipment for the building. Examples of such projects are oil refineries and foundries. These and other types of buildings are referred to as *process plants*.

The factor estimate is also based on historical data. However, unlike other estimating models that gather and structure past cost data, the factor estimate develops factors for each component as a function of a predominant cost.

Figure 5.6 shows collected factor cost data. It can be observed that the total project is broken up into a series of components—for example, equipment costs, equipment installation cost, process piping, instrumentation cost, and so on. After listing the items or components of work, a factor for each is listed. In this factor estimate the purchased equipment cost is given a factor of 1.00. The factors for such items as the equipment installation are developed by observing the actual cost as determined from previous projects or the actual construction costs. For example, the process piping cost for the project in question is $700,000. The purchased equipment cost was $1 million. Thus the process piping cost was 70% of the purchased equipment cost, or 0.70. The other factors shown are calculated in the same manner. That is, the actual cost is divided by the purchased equipment cost to determine the factor for that particular component.

By collecting data for numerous similar projects, the reliability of the factors becomes greater. That is, if the data represent a rather wide assortment of different project conditions, then we can be fairly assured that the process piping cost will indeed be 70% of the actual purchased equipment cost. The data that are kept for future estimating are the factors and not the actual costs from past projects. It is these factors that are used to aid the estimator in *future* estimating projects.

The theory behind factor estimating is that components of a given type of project will have the same relative cost as a function of a key or predominant cost for each and every project; for example, for a steel mill, the processing equipment often dictates the cost of the building components. It should be noted also that component factors can be greater than one. This is because the purchased equipment cost, assuming it is the key cost, may not be in excess of every other component cost for that type of project.

The collected factor cost data can be used to determine a preliminary cost estimate for a project. Let us assume that we are building a project similar to the one in Figure 5.6. However, this project will not necessarily be of the same magnitude. Let us assume that we estimate from vendor bids that the purchased equipment cost for the proposed project is $600,000. We can now use the factors we developed in the previous example to estimate the component cost of the project we are proposing. This is done in Figure 5.7.

Type of Work	Factor	Projected Cost
General conditions	0.09	$ 54,000
Excavation	0.07	42,000
Framing system	0.22	132,000
Equipment	1.00	600,000
Equipment installation	0.18	108,000
Process piping	0.70	420,000
Instrumentation costs	0.20	120,000
Finish material	0.15	90,000
Electrical	0.10	60,000
Plumbing	0.18	108,000
Mechanical	0.44	264,000
Total Project Cost		$1,998,000

Figure 5.6   *Factor Estimate Data from an Actual Project*

Type of Work	Cost	Factors
General Conditions	$ 90,000	0.09
Excavation	70,000	0.07
Framing System	220,000	0.22
Equipment	1,000,000	1.00
Equipment Installation	180,000	0.18
Process Piping	700,000	0.70
Instrumentation Costs	200,000	0.20
Finish Material	150,000	0.15
Electrical	100,000	0.10
Plumbing	180,000	0.18
Mechanical	440,000	0.44
Total Project Cost	$3,330,000	

Figure 5.7   *Projected Building Cost Using Factor Estimating*

Any one component cost of the proposed project is determined by multiplying the historical factor for that component by the estimated purchased equipment cost for the proposed project. For example, the electrical factor from the previous project was 10% or 0.10. So the estimated cost of electrical work for the proposed project is 10% times the purchased equipment cost of $600,000, or $60,000. The other cost components of the project are determined in a similar manner. Summing all of the estimated costs yields an estimated project cost. Assuming that the collected factors are reliable and that the components have the same cost relationship to the predominant cost as in previous projects, then the estimate is fairly reliable.

RANGE ESTIMATING

An estimate by definition is uncertain. The amount of uncertainty of a cost estimate of a building is dependent on the skill and knowledge of the preparer as well as the characteristics of the building being proposed and the timing of the estimate. One means of recognizing and evaluating the uncertainty of an estimate is through the use of **range estimating**. This has proved an effective means of determining accurate pre-construction estimates.

Range estimating consists, in part, of a computer software program. Therefore, the process cannot be totally illustrated without the presentation of the probability theory that is part of the program. However, because of the growing popularity of range estimating and its effective use by construction estimators, an overview of the process is given in this section.

Range estimating has the objective of setting out a range of possible project costs or probabilities of various project costs within this range. In conventional estimating, the estimator ultimately represents the anticipated cost of the project by a single number. This single dollar number is set out even though the estimator knows that hundreds, if not thousands, of actual project costs are possible. The range estimating process calculates an expected range of total project or building costs as a function of the range of expected costs of the project's individual work phases or packages. The result is that the user of the range estimating process can equate risks with various possible project budgets. For example, a range estimate may indicate that there is an 80% probability that a project will cost less than $1 million, an 85% probability that it will cost less than $1.1 million, a 90% probability that it will cost less than $1.2 million, and so on. Range estimating also sets out critical phases or packages that can greatly affect the project either adversely or favorably.

Let us illustrate the range estimating process by means of an example. Figure 5.8 shows an estimate of a building's cost as might be prepared by an architect/engineer. The estimated cost is $2 million. This amount might be determined with a parameter estimate, a factor estimate, or another estimat-

Work Package	Cost
General conditions and fee	$ 165,400
Sitework	26,200
Excavation	22,000
Foundation	47,600
Formed concrete	75,400
Interior masonry	21,600
Structural steel	319,000
Miscellaneous metal	26,600
Carpentry	10,400
Waterproofing and dampproofing	2,600
Roofing and flashing	13,400
Metal doors and frames, windows	6,400
Hardware	7,000
Glazing	11,600
Curtain wall	217,600
Lath and plaster	4,600
Drywall	106,000
Tile work	5,600
Accoustical ceiling	10,400
Resilient flooring	2,000
Carpet	2,000
Painting	10,000
Toilet partitions	1,600
Elevators	10,000
Plumbing	388,600
HVAC	331,200
Electrical	129,400
Parking	25,800
Total	$2,000,000

Figure 5.8 *Expected Package Costs*

ing process. We might consider the estimated cost of each work package and the total $2 million as target costs.

The range estimating process does not limit itself to an estimate of a single cost for each work package or phase. Instead, the use of the process states a target cost, a lowest estimated cost, a highest estimated cost, and a confidence limit or likelihood that the actual cost will be equal to or less than the target cost. This is illustrated in Figure 5.9. It should be noted that the lowest and highest costs are unlikely, but should be viewed as possible. Although a certain amount of guessing or subjectivity goes into setting out the lowest and highest cost estimates for the individual work packages or phases, an error made in setting these ranges will have less effect on the estimate's accuracy than an error in another estimating process.

The individual work packages or phases illustrated in Figure 5.9 might be compatible with the manner in which individual work packages are subsequently awarded to contractors. If this is done, the estimator can in effect prepare an estimate that may accommodate the awarding and controlling of individual construction contracts. Once the designer or owner has determined its individual work package cost estimates, including the low/high range, the next step is processing the information. Obviously, thousands of cost combinations can result when one starts combining the possible costs of individual work packages. The calculation and weighting of the possible combinations would be extremely time-consuming, if not impossible, if the process were done manually.

As indicated in Figure 5.9, based on a summation of possible *lowest* costs for the individual packages, the total project cost could be as low as $1,858,300. On the other hand, based on a summation of possible *highest* costs, the total project could cost $2,139,920. Neither of these costs is likely. They represent extremes that have only minute probabilities associated with them. More likely, a total project cost within this range will occur.

Work Package	Target Cost	%	Lowest Cost	Highest Cost
General conditions and fee	$ 165,400	65	$ 154,000	$ 186,000
Sitework	26,200	70	24,000	27,000
Excavation	22,000	80	20,500	23,400
Foundation	47,600	40	45,300	57,000
Formed concrete	75,400	80	72,000	75,600
Interior masonry	21,600	75	21,000	27,720
Structural steel	319,000	60	304,000	342,000
Miscellaneous metal	26,600	80	24,400	29,300
Carpentry	10,400	80	9,800	10,400
Waterproofing and dampproofing	2,600	75	2,000	3,200
Roofing and flashing	13,400	60	12,500	17,400
Metal door and frames, windows	6,400	80	6,400	6,400
Hardware	7,000	80	6,400	7,500
Glazing	11,600	80	10,800	15,400
Curtain wall	217,600	90	210,000	242,000
Lath and plaster	4,600	80	4,400	5,500
Drywall	106,000	70	101,000	121,000
Tile work	5,600	65	5,100	6,300
Accoustical ceiling	10,400	70	9,400	12,200
Resilient flooring	2,000	80	1,900	2,200
Carpet	2,000	70	2,000	2,000
Painting	10,000	70	9,400	11,500
Toilet partitions	1,600	60	1,500	2,100
Elevators	10,000	70	9,500	12,000
Plumbing	388,600	80	328,000	346,000
HVAC	331,200	70	315,000	381,200
Electrical	129,400	70	126,000	134,600
Parking	25,800	80	22,000	33,000
TOTAL	$2,000,000		$1,858,300	$2,139,920

Figure 5.9 *Range Costs*

The knowledge of the range of project costs and the likelihood of overrunning a single cost helps the designer and owner in several ways. For one, it is possible to equate risk. It is also possible to budget for contingencies or to redesign aspects of the proposed project to decrease the potential range of costs.

Additional computer output can also be generated to yield information regarding the criticality of individual work packages or phases. In other words, some work packages can result in a greater overall project cost overrun than others. For example, the analysis may yield information that indicates that the foundation-related work accounts for 20% of a projected possible cost overrun. This information can point the designer or owner toward tight control over this phase of the project.

The uses of the cost profile shown in Figure 5.10 and related range estimating output are left to the creativity of the user. Independent of this, the range estimate provides more potential for management control than the single dollar cost estimate illustrated in Figure 5.8.

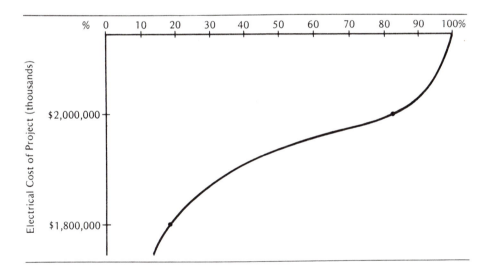

Figure 5.10   *Cost Profile for Range Estimating*

EXERCISE 5.1

The parameter estimating process is dependent on the ability to correlate various components or trade costs for a proposed project to various scope parameters that are known in regard to their quantities early in the design phase. The components or trades and the parameters identified to be part of a parameter estimating process depends in part on the type of building or project being analyzed. Assume that a parameter estimating system is to be designed for estimating the cost of a residential single-family home. The objective is to design a system that correlates the various trades necessary to building a house relative to scope parameters normally defined by a potential homeowner at the time he seeks a contractor. Design a parameter estimating system to include (1) a set of parameters, (2) a list of trades, and (3) an identification of a parameter to which each trade is related in terms of estimating its costs.

EXERCISE 5.2

Assume that a firm has structured a parameter estimating system that utilizes four parameters and ten different work trades or contractor packages. The parameters and trades are as follows:

1. Cross-sectional area of building.
2. Square feet of finished floor area.
3. Number of floors.
4. Linear feet of interior wall.

The trades or contractor packages, along with the identification of the parameter to which each is correlated in the system, and the historical unit cost per parameter trade are as follows:

TRADE OR PACKAGE	PARAMETER CODE	UNIT COST
a. General conditions	2	0.32
b. Excavation	1	0.42
c. Foundation work	1	0.83
d. Exterior walls	2	1.02
e. Interior walls	4	0.60
f. Finishes	4	0.64
g. Roof	1	0.38
h. Electrical	2	0.32
i. Plumbing	3	1.464
j. Mechanical	2	0.38

For a proposed project, the estimator has determined the following approximate quantity for each of the four parameters:

PARAMETER	PROJECTED QUANTITY
1. Cross-sectional area of building	4,200
2. Square feet of finished floor area	34,000
3. Number of floors	9
4. Linear feet of interior wall	28,000

Given these approximate quantities and the historical parameter data, determine a parameter estimate for the proposed project to include the estimated cost per trade or contract and the total project cost.

EXERCISE 5.3

Let us assume that based on collection of past project data, an estimator has determined the following history factor data for an industrial type project. The data tabulated below indicate the percentage cost of each major trade for the building as a percentage of the expected equipment cost.

MAJOR TRADE WORK	PCT. OF EQUIP. COST
Excavation	15%
Foundation	20%
Framing	25%
Finishes	18%
Piping	12%
Electrical	24%
Plumbing	22%
Mechanical	26%
Total Expected Equipment Cost	100%

For a proposed industrial project, the estimator has obtained an equipment quote from the supplier equal to $840,000. Assuming that the proposed project is similar to the projects for which the firm has collected history factor data, determine the estimated cost of the building and the estimated cost for the individual work trades.

6

Overview of the Detailed Contractor Estimate

Components of a Contractor Detailed Estimate/Bid   *
Quantity Take-Off   *   *Pricing Labor and Materials*   *
Job Overhead   *   *Putting all the Components Together to*
Form a Bid

COMPONENTS OF A CONTRACTOR DETAILED ESTIMATE/BID

The previous two chapters discussed the preparation of pre-construction estimates. These types of estimates can be considered as project owner/project designer estimates. While they are vital to the financial success of a project, the term "estimating" in the construction industry also implies the preparation of a detailed contractor estimate. It is usually this estimate that dictates the project owner's construction cost. Similarly, it is the detailed contractor estimate that sets out the potential profit or loss that the contractor will realize for performing the construction work.

An entire construction project is seldom built by a single contractor. Instead, several contractors construct segments of the project. For example, one contractor will perform concrete work, whereas another contractor will do the electrical work on a building construction type of project. A project owner may engage several prime contractors to perform segments of a building project. This process is illustrated in Figure 6.1. But more often, the project owner will engage a single contractor, referred to as the *general contractor*, who will in turn engage several other contractors as subcontractors to perform segments of the overall project responsibilities. This process is illustrated in Figure 6.2. The subcontractors usually perform what is referred to as specialty-type work such as electrical, mechanical, plumbing work, and so forth.

Given the typical project delivery system shown in Figure 6.2, several contractors prepare cost estimates for their segments of the work. The general contractor—who may limit performance to concrete, masonry, and car-

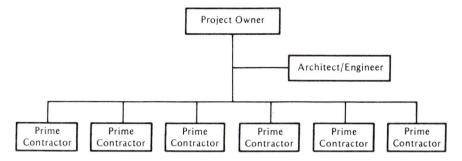

Figure 6.1   *Process of Awarding Separate but Equal Contracts*

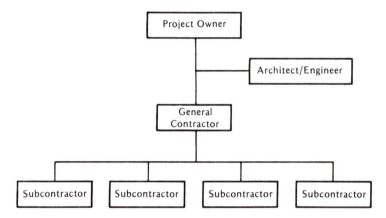

Figure 6.2   *The General Contracting Process*

pentry work—does not prepare cost estimates for the individual subcontractors. Instead each subcontractor prepares his own estimate, which becomes a component of the general contractor's overall project estimate. The general contractor's estimate is therefore a summation of his own plus the subcontractors' cost estimates. The general contractor usually includes a fee for coordinating and administrating of his subcontractors in the overall estimated cost. This "coordinator cost" is sometimes referred to as general contractor *markup* on subcontractors.

The entire book, particularly from this chapter on, focuses on the preparation of the general contractor's detailed estimate. However, one should not lose sight of the fact that individual subcontractors must also prepare their estimates. While similar to the general contractor's detailed estimate in regard to the manner and procedures used to determine cost, subcontractors' estimates are often more technical in regard to the taking off of the quantity of work to be performed and the corresponding cost of performing the work. The general contractor's detailed estimate is referred to as a *contractor estimate* or a *detailed contractor estimate* in this and following chapters.

The detailed contractor estimate includes (1) his determination of his *expected cost* of building a project and (2) his determination of his *desired profit*. Although a project estimate and the profit determination are related in that they sum to yield the contractor's bid proposal, we can view the purpose or objective of the two tasks differently.

The purpose of estimating is to determine the *real cost* of building a construction project. The contractor has to be able to predict all of the costs directly or indirectly associated with building a project. These costs are to be transferred to the project owner, and therefore become a part of the contractor's bid proposal for the project.

Figure 6.3 illustrates the steps involved in the contractor's estimate of his

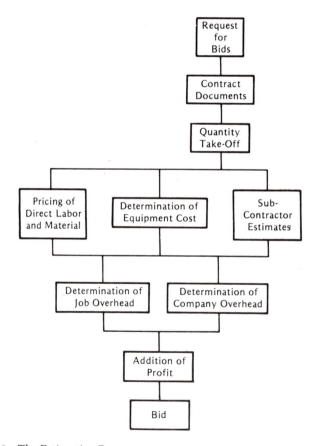

Figure 6.3   *The Estimating Process*

project costs. We give a general overview of these steps in this chapter. Following chapters discuss the individual steps in more detail.

The Contractor's Bid

The contractor's determination of the actual amount of his bid proposal for building an owner's construction project is known as his **bidding task**. The objective of this task is to maximize his rate of return or profitability.

Obviously, a necessary component in bidding is the performance of the project estimating task. However, this is not the only component. In addition, as we have mentioned, the contractor's bid proposal must include the amount of money that constitutes his desired profit. The contractor's cost of building a project is somewhat fixed, since it is a function of the plans and specifications of the project. However, the contractor has complete freedom in determining his bid price.

This does not, however, imply that his bid proposal, and therefore his desired profit, is not subject to constraints. If the contractor includes too large a profit in his proposal, the bid may not qualify as the lowest responsible bid, and he may not receive the contract to build the project in question. On the other hand, if he includes too low a profit in his bid to ensure winning the contract, this may be unprofitable—actual costs may exceed the contractor's estimated costs to the extent that they exceed his profit. Such a situation would result in the contractor losing money on the project. Thus, the contractor is faced with a difficult task when attempting to determine the "best" or most optimal profit to add to his bid proposal.

Owing to the highly dependent nature of the contractor's bid proposals to the success or failure of his company, it is extremely important that estimating and bidding be performed accurately. The contractor should make as few assumptions as possible, and use as much relevant information as is available when determining his bid proposal for a construction project. This bid, which is to be submitted to the owner or his architect/engineer, should represent hours of information seeking, and logical decision making.

QUANTITY TAKE-OFF

Perhaps the most basic as well as the most important aspect of the contractor's estimating and bid functions is the taking off of work quantities from the two-dimensional drawings representing a project. This is sometimes referred to as the quantity take-off or quantity surveying.

The quantity take-off function was briefly discussed in Chapter 1. In particular, the use of a common set of work items or cost objects for the quantity take-off function, pricing work, planning, and control was emphasized.

Small construction firms often tend to discount control in favor of marketing or production; this can result in a lack of controls regarding preparation of an estimate and bid for a project. For example, one individual may prepare extensions with no check on the math, or subbids may not be checked for accuracy. Internal controls such as maintaining footing take-off sheets and having someone other than the person who prepares them check take-off extensions can prevent a significant estimating error that could jeopardize the equity of a contractor.

Internal controls needed are not limited to the quantity take-off function. Controls for the pricing function are also needed.

PRICING LABOR AND MATERIAL

The labor and material costs for constructing a project usually comprise the majority of a contractor's direct and indirect costs in his estimate. This large cost, plus the fact that such costs are often subject to considerable uncertainty, means that these costs involve the greatest risk in the total bid submitted by a contractor.

The contractor must be expected to possess the knowledge necessary to determine as accurately as possible both the labor cost for performing construction work and the cost of the material to be installed.

A contractor establishes his labor cost primarily from knowledge gained from having performed similar work in the past. Often this knowledge is in the form of history data from past projects performed, and structured according to defined work items. (The collection and use of past data is discussed further in a later chapter.) While it is an important ingredient in an accurate estimate, past data alone will not ensure an accurate estimate. Each new project is affected by factors or parameters unique to an individual project, as Figure 6.4 illustrates. A contractor should be aware of each of these factors or characteristics when determining the labor cost for each and every work item in his estimate.

In years past, contractors could consider material costs for performing a construction project as relatively absent of risk. Given an accurate quantity take-off of the project's material, the fact that material could be purchased at a relatively stable price minimized risks. However, in recent years, both the

Worker Morale	Degree of Management
Crew Balance	Leadership
Prior Training	Repetitive Nature of
Prior Experience	Work
Age of Workers	Difficulty of the
Weather Conditions	Work

Figure 6.4 *Some Factors Affecting Labor Cost*

availability of material and its cost have become uncertain. Figure 6.5 illustrates some of the factors or parameters that today make material cost estimations difficult and uncertain. The contractor must be attentive to these factors and parameters in preparing its estimate.

Supply of Material	Location of Material Purchased
Demand for Material	Need of Storage of Material
Quantity Purchased	Quality of Material Needed

Figure 6.5 *Some Factors Affecting Material Costs*

Equipment Costs

Depending on the type of construction project, equipment expenses can comprise a significant portion of the total project cost. Equipment costs are especially characteristic of heavy and highway types of projects. However, given the need for a crane at a building construction project, equipment can also be a significant building construction cost.

Equipment costs incurred at a project site are direct project costs. However, too often a contractor uses an *application process* to estimate the equipment cost component of his estimate. That is, he may determine the labor and material direct cost for a project and then multiply this cost by a percentage to determine the equipment cost. This application process may lead to an under- or overestimate of the actual project equipment cost.

An alternative to treating project equipment costs as an application is to implement an *equipment estimating and control system*. The objective of such a system is to trace equipment expenditures to individual pieces of equipment and to enable the contractor to treat equipment expenditures as *direct costs to projects*. Ultimately, the system should provide the contractor a means of evaluating the profitability of owning an individual piece of equipment, allowing him to better determine the profitability of individual projects. An example equipment cost system is illustrated in a following chapter.

Subcontractor Costs

As discussed earlier in this chapter, the general contractor does not get directly involved in the preparation of his subcontractor bids. But he does have three specific responsibilities in regard to the subcontractor portion of the overall project estimate. He must:

1. Solicit subcontractor bids.
2. Evaluate accuracy of subcontractor bids and select subcontractors.
3. Include subcontractor bids in overall project estimate to include the determination of overhead to be applied to subcontractor work.

The competitiveness of the overall general contractor's estimate is in part dependent on the competitiveness of the subcontractor bids he receives. To ensure that subcontracted work is contracted at an equitable and competitive price, the general contractor usually requests several potential subcontractors (often three or more) to submit a bid for each contract he plans to sub-bid. This process is shown in Figure 6.6. Typically, assuming each of the competing subcontractors is judged to be qualified to do the work, the general contractor awards the contract to whichever one submits the lowest bid. In effect, he engages his subcontractors in the same competitive manner that the project owner uses to engage him, the general contractor.

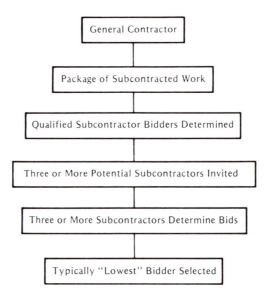

Figure 6.6 *Process of General Contractor Engaging Subcontractor*

Usually the general contractor is free to solicit any potential subcontractors he chooses. In addition to his objective of obtaining competitive prices from subcontractors, the general contractor places emphasis on a subcontractor's willingness to cooperate with the directions of the general contractor and attentiveness to performing the quality of work according to a schedule proposed by the project owner or general contractor. The result is that the general contractor often develops a group of subcontractors he can turn to when he needs to solicit subcontractor bids.

Perhaps the best means the general contractor has for determining the accuracy of subcontractor bids is to compare several competing bids. If he obtains only a single bid for a type of work, he has little assurance that the bid is equitable. However, if three independent bids are made, all of which are within an expected dollar range, the general contractor has a means of comparing the bids and drawing a conclusion regarding their competitiveness. It could happen that all three bids are high and therefore noncompetitive. However, this likelihood is less than the chance that a single submitted bid is high.

A second means the general contractor may use to determine the accuracy of subbids is to compare the sub-bid dollar amounts with the relatively gross estimate that has already been established for the individual work packages to be subcontracted. Even without having previously performed the specialty work to be sub-bid, the general contractor can usually determine a "ballpark" estimate. This approximate amount might be determined as a percentage of all or part of the cost of the work the general contractor is to perform. For example, for a specific type of project, he might estimate the electrical work to be performed to be approximately 10% of the cost of the general contracting work. By comparing this calculated amount with the bid or bids received from potential subcontractors for the electrical work, he might be able to evaluate the competitiveness of the bids. Obviously, owing to the uniqueness of such project and the work to be performed, the actual bids will not likely equal the estimated amount exactly. However, the comparison can detect any obvious inaccuracies or non-competitive bids.

The general contractor can determine the approximate sub-bid dollar amounts by the use of published cost data regarding system or specialty costs as a percentage of total project cost. Several widely circulated cost books can be purchased that contain this type of data. Example data from one of these books is illustrated in Figure 6.7.

Once the general contractor has determined that specific sub-bids are competitive and selects the bids, he then includes them in his overall project estimate. Often the time period in which the sub-bids are received, reviewed, and summarized in the overall bid is short and therefore the process is relatively hectic. Thus, the contractor must be especially conscious of the accuracy of his mathematical calculations and he must take care not to miss or duplicate a subcontract bid in the overall project estimate.

In addition to the dollar amounts of the accepted sub-bids that are included in the overall project estimate, the general contractor usually adds a cost for administering and coordinating its subcontractors. One might refer to this cost as *markup* or a general contractor overhead cost related to "managing" the subcontractors. Typically the competitive nature of the construction industry does not permit the general contractor to mark up the subcontractor work at the same rate he marks up his own work. For example, a general contractor may include a 15% markup on his own cost for supervising or managing its own work. However, he may be limited to marking up subcontracted work to 5 or 6% because each subcontractor has also marked up his own work. In effect, given the competitiveness of the construction

74	SCHOOLS Elementary	S. F.	31.75	37.40	45.25		
002	TOTAL PROJECT COSTS	C. F.	2.11	2.55	3.21		
050	Masonry	S. F.	2.28	4.41	7.05	6.50%	9.10%
073	Miscellaneous Metal		.30	.46	.97	.50%	.90%
112	Water & Dampproffing		.11	.15	.22	.20%	.30%
114	Roofing		1.34	1.95	2.38	2.30%	4.10%
134	Windows		.56	.62	1.73	1.20%	1.30%
135	Glass & Glazing		.22	.38	.71	.40%	.70%
154	Tile & Marble		.12	.16	.52	.20%	.30%
157	Floor Covering		.15	.41	.47	.20%	.90%
158	Painting		.51	.59	.66	.90%	1.20%
180	Equipment		1.45	1.90	2.50	1.80%	4.10%
272	Plumbing		2.22	2.79	3.47	5.80%	7.10%
273	Heating & Ventilating		4.46	6.05	6.70	9.90%	10.70%
290	Electrical		3.15	3.79	4.80	8.50%	10.40%
310	Total Mechanical & Electrical		9.15	11.45	13.80	25.90%	30.40%
900	Per Pupil, Total Cost	Pupil	3,050	5,500	7,250		
950	Total Mechanical & Electrical	Pupil	740	1,300	2,150		

Figure 6.7 *Example Subcontractor Prices*

industry, the more a general contractor subcontracts work, the less he can mark up the subcontract work in the interests of preparing a competitive bid.

The total dollar amount of subcontracted work as a percentage of the total construction work varies. If a general contractor limits the subcontracted work to specialty-type work to include mechanical, electrical, and plumbing work, the subcontracted work may be only about 30% of the total construction cost. However, some general contractors may in effect contract to "manage" the construction work and in effect subcontract all of the work.

JOB OVERHEAD

Costs directly related to a building project in general are often referred to as indirect costs or sometimes as job overhead costs. Examples of job overhead costs might include supervision costs, interest costs, and equipment costs or expenses. Job overhead costs are considered traceable to a project, but are not easily traceable to any one specific segment of a project.

In practice, construction firms have on occasion recognized job-overhead–related costs in the project estimate by using an allocation process; for example, a construction firm might multiply the sum of a project's labor and material cost by 10% to determine the job overhead cost to apply to a specific project.

This method, however, can lead to an inaccurate estimate. The process assumes that job overhead costs are the same percentage of the direct costs for each project. This is usually not the case. For example, assuming supervision is handled as a job overhead cost, it is true that the amount of necessary supervision on projects varies substantially. The amount of supervision required often depends on the complexity or uniqueness of the job.

The correct method of recognizing job overhead types of costs in a project estimate is to *identify each job overhead cost for a specific project and add them for inclusion in the total project estimate.* In other words, job overhead costs should not be allocated. They should be handled as a direct cost to a project in estimating the project's costs.

In practice, certain job overhead costs may be more easily handled if they are allocated to various projects. For example, the construction firm may choose to use an allocation process for equipment rather than identify the equiment to be used on a specific project and the amount of expected use on the specific project. While this procedure may prove to be efficient regarding the time required to recognize a project's equipment costs, it is important to be aware that the potential for accurate estimates diminishes whenever a cost is treated on an allocation basis. Job overhead costs are, as the term itself implies, traceable to projects. The more detailed the construction firm is in identifying specific job overhead costs, the more likely will accurate estimates result.

Company Overhead

Costs that support the production process but are not directly related to the production process also need to be allocated to a project. These costs are often referred to as *general and administrative costs or expenses.* They are also commonly referred to simply as overhead costs, or as company or office overhead. Examples of these costs include the costs of maintaining the home office, entertainment expenses, and marketing costs or expenses.

Two separate questions must be addressed when in the application of general and administrative (G&A) costs or overhead to specific projects. First, the construction firm must decide on the *basis* for the overhead allocation process. Second, once an overhead basis is chosen, there still remains the matter of the *mechanics of determining the amount* to be applied to a specific project. Both of these questions are addressed in Chapter 16.

Profit Strategy

The profit a contractor adds to his bid proposal represents the amount of money in excess of his costs which he desires in return for building a project. Disregarding profit, all contractors bidding on a single project may estimate the same cost for building it. Thus, the profit a contractor adds to his estimated project cost may actually represent the difference between his winning or losing the project contract in the competitive bidding procedure. In reality, various contractors bidding on a single project will probably have different estimates of project cost. This is because of the different structure of cost information, the different construction methods, and the different take-off procedures used by various contractors. However, even in the case of varying contractor project cost estimates, the profit a contractor adds to his bid often determines whether or not he will win the project contract.

Winning the project contract is not the only consideration when one is trying to determine desired profit. Consistent with the overall contracting objective, the contractor wants to maximize his profits or his project rate of return. After submitting a bid proposal in a competitive bidding procedure, the contractor receives the project contract if his bid is judged to be the lowest responsible bid. If he bids too high (not the lowest responsible bid), he receives no work and he must absorb the cost he incurred in making the bid proposal. On the other hand, if the contractor bids low to win the project contract, he risks losing money on the project owing to his small profit margin. Thus, when attempting to determine the optimal profit he can add to his bid, the contractor is confronted with conflicting constraints.

Although the contractor may have a little more freedom in determining his project profit in a negotiated contract environment, his task is not vastly different. To receive work from a project owner, his bid (and therefore his profit) must be competitive with those of other contractors.

Long-term and Short-term Profits: In discussing a contractor's desired profit associated with building construction projects, it is meaningful to consider his desired *long-term profit* and his desired *short-term, or project profit*. Usually, a contractor (or any type of firm) will have a defined desired long-term profit. However, owing to the characteristics of a given construction project, or because of the situation occurring in the contractor's own business at the particular time of bidding a project in question, the contractor may have to adjust his desired long-term profit in favor of using a desired short-term (project) profit—long-term profit objectives may need to be modified for short-term profit considerations. The determination of a strategy for both a short- and long-term profit objective are discussed in more detail in Chapter 17.

PUTTING ALL THE COMPONENTS TOGETHER TO FORM A BID

The total contractor bid fails at its weakest link. If any of the components is inaccurate or for some reason overlooked in compiling the estimate and/or bid amount, the total bid will be inaccurate. Such an inaccuracy will likely lead to one of two unfavorable events: either the contractor will submit a too-high estimate and therefore be non-competitive; or even worse, he may submit too low an estimate and therefore be awarded the contract to do the construction work but subsequently lose money on the project.

A contractor typically uses a standard set of forms to prepare each of the previous components of the overall bid. The forms also provide the contractor the means of integrating each of the components into the summarized contractor bid (this process was illustrated in Figure 6.3).

The specific estimating forms used by a contractor vary depending on the type of work he performs, the size of his operation, and to some degree his preferences. However, as will be discussed in a following chapter, a contractor should use the same types of form for each and every project he estimates in order to maintain and increase efficiency and accuracy.

Direct labor and material costs are often established on quantity take-off forms. Sometimes a single form serves this purpose. On the other hand, some firms use one form for determining *quantities* of work and yet another form for determining the *cost* of the work. (Example forms for determining work quantities and for determining the cost of the work will be illustrated in a following chapter.)

As discussed earlier in this chapter, contractors estimate equipment costs for a project in varying ways. Perhaps the most logical process is to attempt to determine the required hours and cost of equipment for the entire project being estimated. For building construction types of projects, this is likely a more feasible approach than trying to estimate equipment hours and costs for specific phases or work items of the overall project. Either of these two approaches is to be preferred over a process by which equipment costs are applied or allocated to a project as a percentage of another project cost or basis. (An example form that a contractor may utilize to estimate equipment hours and costs for a specific project will be illustrated in a following chapter.)

The gathering and processing of subcontractor bids often occurs within a somewhat hectic timeframe. It tends to be unsophisticated because it lacks controls in regard to written documents and formalized written agreements between the general contractor and the contractors from which he is receiving sub-bids. Often, the general contractor receives the bids orally over the telephone. While a signed commitment letter from a potential subcontractor is preferable, the hectic process and chain of events leading to the final estimate in a competitive environment may prevent this process. At a minimum, the general contractor should formally record any sub-bids received orally from potential subcontractors. (An example form for this purpose will be illustrated in a following chapter.) Use of a standard form somewhat minimizes potential errors that can occur in obtaining and including sub-bids in the overall estimate/bid.

Job overhead costs include several items which, while sometimes relatively small in dollar amount, can and should be estimated as a direct cost to a project. The biggest difficulty related to determining job overhead

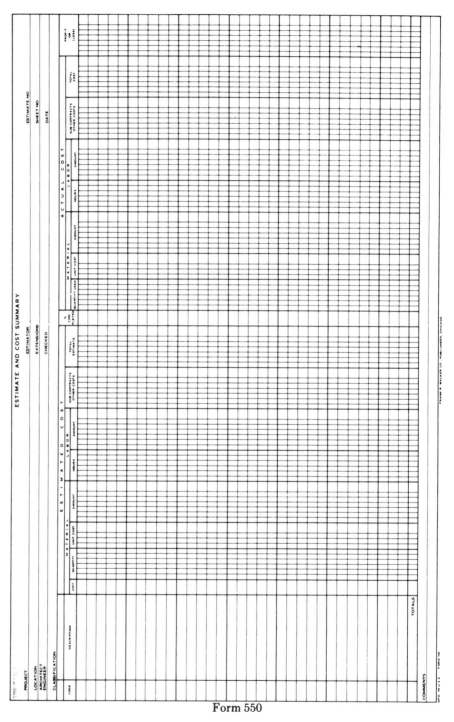

Form 550

Figure 6.8 *Example Summary Sheet*

162

costs can be the mere recognition of each of the items. Often a contractor is well served by the use of a checklist to remind himself of all possible job overhead costs. This checklist can also serve as a form for estimating and summarizing job overhead costs for a specific proposed project.

A contractor determines company overhead for a specific project by using an allocation process. There often is not a single form for estimating the cost for a specific project. However, it is highly recommended that the contractor develop an allocation process and/or formula for determining the company overhead cost to be charged to a project. (This process is discussed in Chapter 17.) The same is true regarding the determination of contractor profit to be included in the overall contractor bid. (The determination of contractor profit is discussed in Chapter 18.)

Perhaps the single most important form for the contractor preparing an estimate/bid is the summary or recapitulation form that integrates each of the component parts of the estimate/bid. An example form for this purpose is illustrated in Figure 6.8. In practice, this form varies significantly. However, in addition to serving as a summarizing form, the form should serve as a reminder to the firm to include each item. A form that has preprinted items is useful for this purpose.

Typically a contractor does not have to disclose the detailed component parts of his bid to a potential project owner client. In fact, on occasion he may have to only submit one single dollar amount—that is, the total bid. In this case, the contractor would not submit a form like that shown in Figure 6.8 to the potential project owner. Instead, he might submit a bid proposal form such as that shown in Figure 6.9. This form summarizes each of the component costs and profits that the contractor has established as part of his preparation of a detailed contractor estimate/bid.

Item Description	Item Bid
General Conditions	$ 1,560,000
Excavation	2,200,450
Concrete	4,500,540
Masonry	2,400,542
Structural Steel	1,850,540
Carpentry	852,000
Miscellaneous Items	547,000
Electrical	2,100,000
Plumbing	1,800,000
Mechanical	1,600,530
Total Bid	$19,411,602

Figure 6.9 *Bid Summary Sheet for Warehouse Building, with Profits Included in Various Item Costs*

EXERCISE 6.1

As part of the preparation of a detailed contractor estimate, the firm may make errors due to varying reasons. For each of the inefficiencies described, describe a procedure the contractor can implement to remove or lessen the inefficiency.

a. During the year, the contractor submitted a bid in which he overlooked the concrete column take-off quantities. As a result of this error, he submitted a bid for less than the cost of the project.

b. The contractor successfully bid a project in February which he anticipated he would complete in August. However, because of overoptimism of the duration and sequence of project events, he later found that estimated project duration was erroneous. The project was still in process on December 31.

c. The contractor submitted an estimate for a project that included a significant amount of concrete placement work. The labor cost for each cubic yard was bid at $32. Your inspection of project records indicates that the actual labor unit cost for the project ws $42 per cubic yard. The result was a $21,000 loss on the project.

d. The contractor incurred a large unexpected cost on a project when a dishonest superintendent purchased many tools from a hardware store in the name of the contractor. The superintendent purchased the tools for his own personal use. The superintendent has since quit his job and the contractor cannot disprove the superintendent's statement that the tools were purchased for the contractor's project and subsequently stolen by the craftsmen.

e. The contractor successfully bid a project during the year in which much of the work was to be subcontracted. After the successful bid was made, one of the subcontractors removed his bid commitment to the contractor because the subcontractor discovered an error in his own take-off procedure. Because the subcontractor's bid had been made orally by telephone just prior to contract letting, the contractor could not prove that the subcontractor even submitted a bid. As a result the contractor, after receiving the project, had to engage another subcontractor at a significantly higher sub-bid to do the work, leading to a loss on the project.

f. In inspecting the cost records for a single project, it appears that the contractor proceeded to do additional work for a project owner outside the original contract. However, the owner disputed the contractor's request for additional payment because the owner now claims he never approved the additional work.

g. The contractor made the embarrassing and costly error of failing to account for all of the masonry work for the fifth floor of a five-story building he had successfully bid. The contractor did use a quantity take-off checklist. Unfortunately, because the masonry was checked off the checklist as the estimator took off the masonry on the other four floors, the checklist did not prevent the costly take-off error.

h. The contractor incurred a significant loss on one project. The actual cost of the project was $1,200,000 versus a $1,100,000 estimate. However, the cost records for the project are such that the contractor cannot pinpoint why the cost overrun occurred.

i.   The contractor successfully bid a project but later detected that he had forgotten to include the cost of public liability insurance required for the project by the project owner. As a result, anticipated profit on the project was reduced.

EXERCISE 6.2

The preparation of a detailed contractor estimate includes the determination of the direct labor cost and material cost. Let us assume the estimated material cost for a project is $400,000 and the estimated direct labor cost for placing the material is $500,000. Let us assume that an estimator has assumed the following range of accuracy can be achieved in determining the direct material and labor costs.

Determination of quantity of material	$\pm 2\%$
Determination of unit cost of material	$\pm 4\%$
Determination of unit labor rates	$\pm 3\%$
Determination of labor productivity	$\pm 10\%$

Given the $900,000 direct-cost estimate, determine the range of possible dollar error based on the above expected accuracies.

EXERCISE 6.3

It is a common occurrence for a contractor to have a project estimate vary significantly from the actual project cost. This variation can prove fatal to the firm.

Inaccuracies in an estimate of direct labor or material costs for a project are compounded by the fact that the contractor may determine the project overhead cost to be applied to an estimate as a function of the estimated labor and/or material cost. Thus the possible compounding of the error.

Assume that an estimator determines the labor and material costs for a proposed $1,000,000 project to be $500,000 and $300,000, respectively. Further assume that the estimator believes either of these cost estimates to be accurate to a range of plus or minus 6%.

Further assume that based on past records, the estimator for the proposed $1,000,000 project applies overhead to the project using an application rate of 15% times the summed estimated labor and material cost for the project.

Based on the above facts, what is the most the estimator could underestimate the total project cost—i.e., his underestimate of labor, material, and overhead costs?

7

Quantity Surveying: Fundamentals for Taking Off Quantities

*Definition of Quantity Surveying * Definition of Work Items * Determining Units of Work for a Defined Work Item * Time and Accuracy: The Two Quantity Take-off Variables * Contract Documents * The Working Drawings * Specifications * Steps for an Accurate and Efficient Quantity Take-Off*

DEFINITION OF QUANTITY SURVEYING

The preparation of a contractor's detailed construction estimate entails the performance of several steps. Perhaps the step that takes the most time and is most clearly identified as part of the estimating function is the determination of the amount of work to be performed through the interpretation of the drawings and specifications for a proposed project. This step is commonly referred to as the quantity take-off or *quantity surveying*.

The completed project take-off or quantity survey serves as the basis for each of the following procedures that comprise the completed contractor estimate: the *direct-cost component* of the estimate, which is based on the calculated quantities; and, to a lesser degree, the *indirect costs and overhead* to be applied to the estimate—these are also based on the amount of work determined in the quantity surveying process.

In varying degrees, each of the construction contract documents prepared either by the project owner or the design team is pertinent to the preparation of a detailed construction project estimate. However, one specific type of contract document, consisting of the *project drawings*, is especially important. In effect, the quantity take-off here is the calculation of work quantities to be performed as set out on the two-dimensional project drawings. Thus the ability to interpret or read a set of drawings becomes critical to the determination of an accurate project estimate. The interpreting or reading of a set of construction drawings (sometimes referred to as prints) is commonly referred to as **blueprint reading**.

The time required to perform the quantity take-off function and the accuracy of the quantities determined vary for a specific project depending on (1) the skills and procedures used by the estimator, (2) the quality of the project drawings and other documents prepared by the project designer, and (3) the degree of the definition of the specific work packages or items to be taken off the drawings. This chapter has the objective of aiding the estimator in taking off the work quantities accurately and efficiently.

DEFINITION OF WORK ITEMS

Basic to the quantity take-off function is the identification of specific packages of work referred to as work items. The estimator has much freedom in his definition of specific work items for a given project. The problem arises as to what is and what is not a work item. Whereas one estimator might choose to define all work related to concrete including forming and stripping as a single work item for the quantity take-off function, another may choose to identify only placing concrete for all types of concrete work as a single work item; and yet another may identify placing concrete in a wall as a work item. Clearly, the problem of identifying specific work items is one of establishing the degree or magnitude involved. For each work item defined, the estimator calculates the amount of work to be performed.

In effect, the definition of work items dictates the manner in which the estimator takes off quantities and collects historical field data for pricing the work. For a given set of drawings, the number of work items the estimator establishes as part of the take-off procedure dictates a specific grouping of items to be identified from the drawings. The broader the scale of a work item, the fewer work items the estimator will single out when taking off quantities for a project. On the other hand, the broader the scale or scope of work items, the less accuracy will be provided for purposes of determining the cost of performing the project's work. That is, if the estimator takes off all the concrete placement as one work item, the fact that it costs the construction firm more to place concrete in an elevated wall than to place a similar amount in a slab on grade, will not be reflected in the estimate. This difference in costs is one major reason why the estimator should delineate separate work items for the two different work processes.

No single rule or principle can be used to determine what is and what is not a work item. However, if the estimator keeps in mind the reason for listing separate work items, this will aid him in determining the specific work items.

There is a relationship between the direct costing or pricing of work and the determining of work packages. If two different types of work require different resources to perform the work, they probably will have different costs for given units of work performed on each of the two types of work. Thus we have an overall guideline: Whenever one or more different resources are required to perform work, the estimator is probably justified in identifying each as a separate work item to reflect the different resources required. However, this general guideline will not be practical for work-item identification in all cases.

In establishing a set of work items that dictate the quantity take-off, the estimator should remember that the work items can prove to be a means of integrating several project management functions. If work items are properly defined, they can serve as a means of interrelating project estimating, project planning, project control, and project accounting and payroll functions. (This relationship is discussed in more detail in a following chapter.)

The difficulty of defining work items for estimating is compounded by the fact that the construction industry is burdened by fragmentation. Work package standards are absent. Units of measure differ for the designer, builder, and material supplier. Using the uniform code as defined by the Construction Specifications Institute (CSI) is a move toward providing a common communication base for the parties involved in the construction process. Definition of work items for an estimating system can be compatible with the CSI division code. Using each division as a broad classification, specific work items can be identified for each. For example, Division 3 of the uniform code is identified as "Concrete." With concrete identified as the base, the estimator may further define "C" work items as follows:

CF1 Forming concrete beams.
CF2 Forming concrete columns, etc.
CP1 Placing concrete in beams.
CP2 Placing concrete in columns, etc.
CR1 Placing reinforcement in beams.
CR2 Placing reinforcement in columns, etc.
where C = designates concrete work
       F = designates forming
       P = designates placing
       R = designates reinforcing

Examples of uniform code work items are presented in Appendix B. Naturally, the number of work items a firm identifies under any one division depends on the type of work it performs.

The benefit in defining work items consistent with the uniform code is that the estimator probably will be better able to relate his data to published cost data averages and systems used by designers. In fact, building construction specifications are increasingly prepared to be compatible with the uniform code.

The uniform code is designed for building construction. A heavy and highway contractor and its estimator would define work items using a data base common to heavy and highway construction. Similarly, a homebuilder would utilize work items relevant to residential construction work. Both of these segments of the construction industry also have published standard sets of work items.

One should not conclude that each and every construction firm and its estimators utilize the published standardized work items as the basis for their respective estimating systems. Right or wrong, in varying degrees, each and every firm seems to define its own unique list of work items for the quantity take-off function.

DETERMINING UNITS OF WORK FOR A DEFINED WORK ITEM

The estimator must also determine the *unit of measure* he will use for the work to be taken off. This is less subjective and thus less difficult than the definition of specific work items to be taken off for a construction estimate, but it remains an important task to be dealt with. For example, should the work be measured by the number of units, its weight, length, area, or volume? Even when a measure of work is determined, the estimator must make further decisions. For example, if the measure is volume, should the volume quantity be measured in cubic inches, cubic feet, or cubic yards?

In deciding what is an appropriate measure of work, one must consider what measure of work is most indicative of how the cost of the work is best measured. The direct material and labor costs for placing concrete is likely most dependent on the volume of concrete to be placed. Concrete is usually purchased in a measure of cubic yards. It follows that the best measure of taking off the concrete for walls is probably cubic yards.

But a volume quantity is surely not the best measure of work for all items. Let us consider the forming of the above-noted concrete walls. Both the direct labor and material costs for forming the walls is apt to be most dependent on the surface area of the walls to be formed. This fact, along with the industry's tradition of measuring the surface area in square feet, has led to the practice of measuring wall forming in square feet of contact area (sfca) or in 100 square feet of contact area.

And a particular type of work—for example, concrete forming—is not always measured in the same units. Beam forming is often measured in linear feet. This practice follows from the fact that the cost of performing this type of work is often more dependent on the length of the beam to be formed than of the area to be formed. For example, consider the two beam cross-sections shown in Figure 7.1. Let us assume that the beams are to be

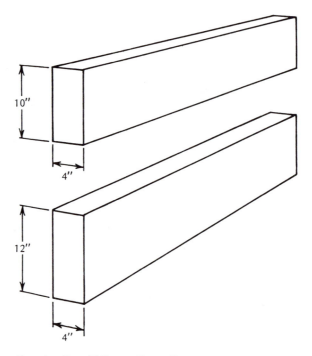

Figure 7.1 *Forming Two Different Beam Sizes*

formed using a job-built system. (*Note:* this term refers to the fact that the forms will be built using lumber/boards fabricated at the job site.) The cost of a 1″ × 10″ board versus a 1″ × 12″ board is not very different. Also, the labor cost to fabricate a 1″ × 10″ board is similar, if not equal, to the labor cost expended to fabricate a 1″ × 12″ board. Given these facts, it follows that the contact area of forming is not indicative of the cost of the forming. Linear feet is a better measure.

For some types of construction work, there may be two different measures of work that match the manner in which each of the labor and material cost components occur equally as well. For example, consider the placement of concrete for a poured-in-place 8-inch concrete slab. The concrete material cost is best measured in cubic yards since concrete is purchased in this measure. On the other hand, the labor cost for placing and finishing the concrete is more dependent on the surface area of the slab than on the volume of concrete. Thus the labor cost per surface area of slab is essentially the same whether the slab is 8 inches thick or 6 inches thick. This is true even though more volume of concrete is poured for the one than for the other.

Given the fact that there are two different equally good measures for the labor and material components of a specified work item, the estimator has one of two choices. He may define two different measures of work for the labor and material components. In effect this is equivalent to defining two separate work items, one for the labor component and one for the material component. Or he can choose the one measure of work that best matches how the combined labor and material occur. While this choice likely results in some inaccuracies in regard to pricing the work, this is the option usually taken by most estimators.

Estimators do not necessarily utilize the same measure of work for a specific work item. For example, some estimators measure screeding (a work process of placing temporary forms for the placement of concrete in a slab) in linear feet of screeds, whereas others may measure screeds based on a calculation of the square feet of slab area to be placed. In other words, the definition of a measure of work for a specific work item is somewhat a matter of personal preference. However, the fact remains that an estimator should attempt to define the measure of work for a work item compatible with the occurrence of the cost of the work performed.

TIME AND ACCURACY: THE TWO QUANTITY TAKE-OFF VARIABLES

The performance of the quantity take-off function is not without its constraints. Because of the very nature of the bidding process and owing to the fact that the quantity take-off time in effect results in contractor overhead costs, the amount of time an estimator can expend taking off quantities is limited. His ultimate objective, it should be remembered, is to prepare as accurate an estimate as possible in a limited amount of time.

In the United States, it is common for the construction firm to take off its own work quantities. However there are some consulting firms in existence that specialize in making quantity surveys. The use of these firms is infrequent by U.S. contractors, but more common in other countries. In Britain, the licensed individuals responsible for taking off quantities are referred to as *quantity surveyors*.

There are probably several reasons why the U.S. contractor takes off his own quantities of work rather than using a consultant. The most obvious reason is that by taking off the quantities, the contractor becomes totally acquainted with the project. Construction method difficulties, possible cost-saving procedures, and information relevant to planning and scheduling the project are among the things learned during the process of taking off quantities.

A more important reason for the contractor taking off his own work quantities is that the process is meaningless unless it is done consistent with the structure of his known cost information. Obviously, it is the contractor who is best able to perform an "optimal" take-off of the project's work quantities.

Another reason the contractor chooses to take off his own work quantities is that he often has more confidence in the accuracy of his take-off than in one made by an outside individual or firm. This confidence is occasionally without merit; however, given the critical nature of the function concerning the financial risk of the contractor, it follows that many prefer not to turn over the process totally to an external entity.

Some Constraints: While taking off the work-unit quantities, the construction contractor is usually subject to two somewhat conflicting constraints: the *required accuracy* of the take-off and the limited amount of *money allocated* for taking off quantities. The money constraint could also be expressed as a time constraint, in that the money spent on taking off of drawings is somewhat dependent upon the time spent on the task. One would suspect that the accuracy of a take-off would decrease as the amount of time spent on it decreases. Faced with these conflicting constraints, the contractor must decide how much accuracy he should try to obtain, and how much time he should spend on taking off quantities. He will often attempt to strive for as much accuracy as possible within an allowable cost constraint, which is usually determined as a percentage of the project's total cost.

Rather than merely striving for accuracy within a time or money constraint, it often assists the contractor to make an analysis of his trade-off of accuracy versus time when deciding on a take-off procedure. One would expect the accuracy of the take-off to continually increase with increasing time spent on the task. However, this is not necessarily so. For example, Figure 7.2 represents a typical accuracy time trade-off curve (graph of accuracy of estimate versus time spent on take-off) for a contractor's take-off procedures. Although such a curve is admittedly difficult for the contractor to produce, he may be able to approximate it by keeping past records of taking-off time versus accuracy (found by comparing the take-off quantities with real quantities determined from actual project records).

The contractor's optimal accuracy–time take-off procedure may be one which corresponds to point X in the figure. It may be seen that point X results in an accuracy of ± 5%. It is true that accuracy is subject to uncertainty; however, let us assume we can at least calculate the expected accuracy. It may be observed from the curve shown in Figure 7.2 that the accuracy increases only marginally with increasing time, *once an accuracy of ± 5% has been accomplished*. This fact, in addition to the contractor's inability to predict with certainty the production of his labor or equipment (which results in his inability to accurately predict the project's costs), may make any additional time beyond point X uneconomical, in terms of taking off quantities. For example, if it is only possible for a contractor to predict production within 20%, it is not really feasible for him to strive for a take-off accuracy of ± 1%. By means of the accuracy–time trade-off model discussed, it becomes possible for the contractor to optimally satisfy his objective, subject to an accuracy and time or money constraint. Obviously, it is difficult for the

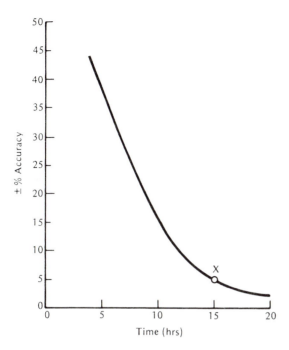

Figure 7.2   *Take-Off Accuracy Time-Cost Trade-Off*

contractor to construct such an accuracy–time trade-off model. However, this does not mean that he should fail to address the problem of trade-off between accuracy and time.

Establishing the desired accuracy of the take-off procedure does not in itself determine the procedure to be used by the contractor. A systematic take-off procedure must be developed that is consistent with the accuracy and time constraint. The contractor requires a procedure that provides for a rapid and accurate calculation of the work quantities. This means that he must be aware of the structure of his cost and production information, know the terminology and meanings of construction drawings, and have a clearly defined systematic approach to the actual taking off of the quantities. Such an approach should include any possible shortcuts that save the contractor time without sacrificing any take-off accuracy.

CONTRACT DOCUMENTS

The 3 parties involved in the construction building process—the owner, the architect/engineer, and the contractor—are brought together by the construction contract documents, which set out the responsibilities of each. Whenever a question arises between two or more of the parties, the contents of the construction documents usually determine the solution.

The contents of the contract documents play an important role in the contractor's decision as to the selection of a project. An evaluation of the contractor's legal responsibilities, the difficulty of the construction methods required, and a measure of project risk may be ascertained from a study of the documents. Although the contractor may not have the time, or be able to allocate the money for a thorough analysis of the documents at the selection decision-making stage, certainly it behooves him to inspect them, looking for characteristics indicative of the overall feasibility of selecting the project.

The documents usually include the drawings, the specifications, the general conditions, the special conditions, the addenda, and the agreement. Occasionally, the owner's advertisement and the instructions to the bidders are also considered part of the construction documents. However, these documents are preliminary to the actual contract and we will therefore not refer to them here.

The Construction Drawings: The construction drawings for the project are undoubtedly the single most important part of the contract documents for the construction contractor. They show the project's structural, architectural, plumbing, mechanical, and electrical work in pictorial form. A typical format for a construction drawing includes an identifying prefix and page number.

The use of standardized drawings is becoming widespread in certain types of construction work. Work such as bridge construction, paving, manholes, and catch basins are fairly standard from one job to the next. As a result, standardized drawings have evolved.

The drawings for the project made by the architect/engineer often do not contain enough detail about the materials or job equipment to enable the contractor to build the project. Detailed shop drawings are used to amplify the drawings. Shop drawings pertain to materials and job equipment prepared by the manufacturers or fabricators of the equipment or materials, and are submitted to the contractor and architect/engineer for their approval. Shop drawings are usually required for work items fabricated away from the building site—for example, structural steel, hollow metal door frames, and cabinets.

The Specifications: The specifications for the construction job describe in words the materials and construction methods required for the project. They vary widely with the type of project and the skills of the specifications writer. The word "specification" may best be defined as a description of a material, process, article, or thing. Construction specifications have several specific objectives: to augment and amplify the plans and the contract by enumerating and describing in detail each and every item or thing within the performance of the work involved in the contract; to furnish a single, definite, and established basis of competition for the awarding of the contract; and to furnish a compact book of instructions for the resident engineer, inspectors, draftsmen, and contractors during the process of the work.

Good construction specifications have certain fundamentals. They should be technically adequate for the purpose intended, clear in phrasing and expression, and just and equitable in effect, purpose, and application. They should also be economical in application, and legally enforceable with a minimum of friction and misunderstanding. Construction specifications that are unclear and indefinite may prove unenforceable.

Special Conditions Document: The working drawings (sometimes referred to as detailed drawings) and specifications for a project are seldom overlooked by the estimator when he is preparing a project estimate. Of the other four contract documents, the one that is the most important to the estimator is the one that outlines special conditions for a project. As noted above, the special conditions sets out factors somewhat unique to the project in question. Included in the special conditions are items that can significantly affect the costs for a contractor constructing a project. For example, the special conditions might state that a contractor is not to be paid for any material for a project until the material is put in place. If a contractor overlooks this fact, he may erroneously assume that he will be entitled to a project draw when he prepurchases some material for a project to store at the job site. This assumption could lead to a significant understatement by the contractor of his finance costs related to the project. The point to be made is that while the drawings and specifications receive the most attention from the estimator (and also in following chapters), the estimator cannot overlook any contract document, especially that containing the special conditions for a project, when he is preparing a detailed project estimate.

THE WORKING DRAWINGS

Previously in this chapter we alluded to the fact that the working drawings for a project are especially critical to the quantity take-off process and the preparation of a detailed contractor estimate. While there is some variation in the style of working drawings prepared by various designers, most present the same type of drawings and in the same order. The typical drawings in the order in which they commonly occur are as follows:

Plot plan	Windows and door schedules
Foundation plans	Interior finish drawings
Floor plans	Specialty system drawings
Elevation drawings	

A typical **plot plan** drawing, illustrated in Figure 7.3, shows the size and shape of the lot and the building on the lot. In addition, the following types of information are often illustrated:

Gas lines	Driveways	Topography
Sewer lines	Sidewalks	Landscaping
Water lines	Fences	

The plot plan usually does not play a significant role in the quantity take-off process. If topography information is included in the plot plan, the drawing does serve as a basis for determining earthwork quantities. However for the most part, the plot plan serves more of a reference purpose than for setting out quantities of work to be performed.

A **foundation plan** for a project shows the location of concrete footings, foundation walls, and any other details concerning the structural aspects of the foundation (sometimes referred to as the building's *substructure*). An example foundation plan is illustrated in Figure 7.4. In effect, a foundation plan is a plan view of a structure projected on a horizontal plane passed through at the level of the top of the foundations. In addition to locating the concrete footings, walls, etc., a typical foundation plan will also show some of the following:

Location of structural steel
Position and types of steel reinforcing
Floor joists

Usually the next drawings that are included in a set of working drawings are the **floor plan** drawings. Often a separate drawing is included showing the plan for each floor in the building. An example floor plan is illustrated in Figure 7.5.

Information on a floor plan includes the lengths, thicknesses, and type of walls on the particular floor, the widths and locations of door and window openings, the lengths and type of partitions, the number and arrangement of rooms, and the types and locations of utility installations. The floor plan may also set out equipment and fixtures, cabinets, fireplaces, and floor and wall finishes.

The working drawings for a building construction project also include **elevation drawings**. These show what the building looks like when finished. They show the front, rear, and sides of a building projected on vertical planes parallel to the planes of the side.

An example elevation drawing is illustrated in Figure 7.6. Usually, there is a separate drawing showing the front, rear, right- and left-side elevations of a building. An elevation drawing usually includes information regarding the exterior finish material, roof materials, roof slopes, and the location of windows and doors.

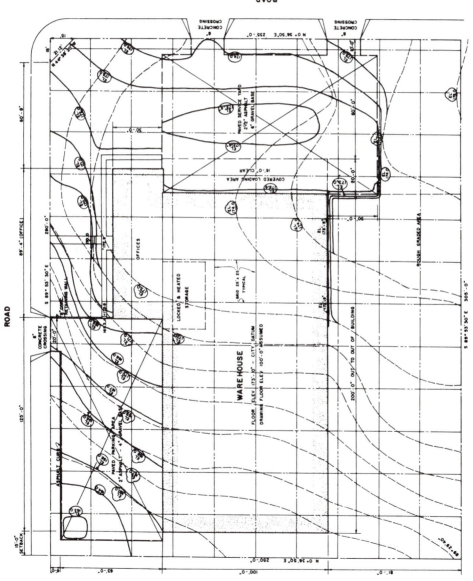

SITE PLAN

Figure 7.3 Example Plot Plan

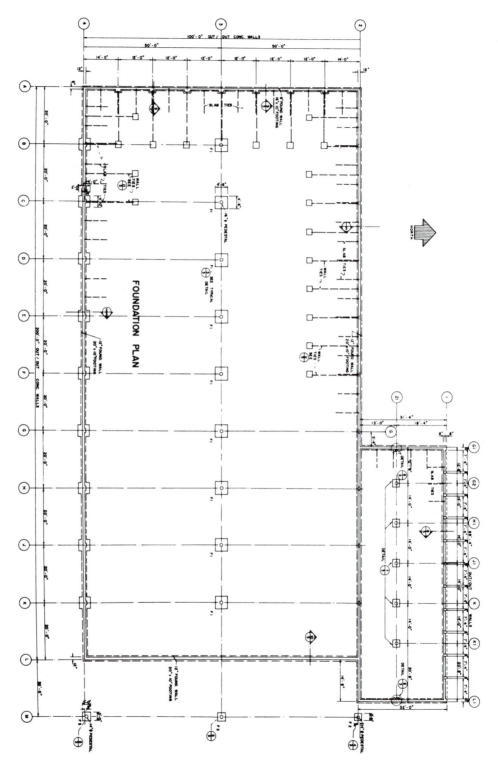

Figure 7.4 *Example Foundation Plan*

178

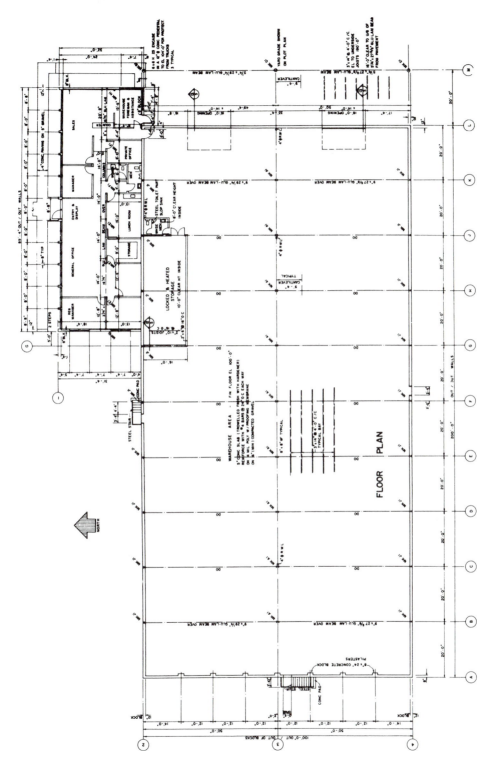

Figure 7.5 *Example Floor Plan*

179

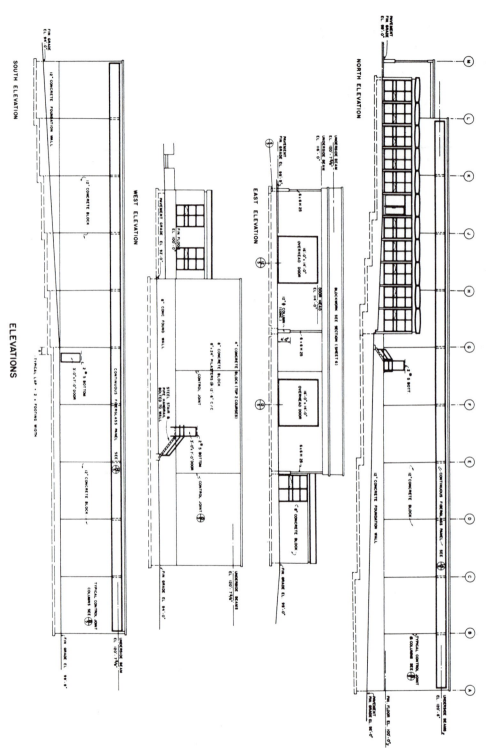

Figure 7.6 *Example of Elevation Drawing*

180

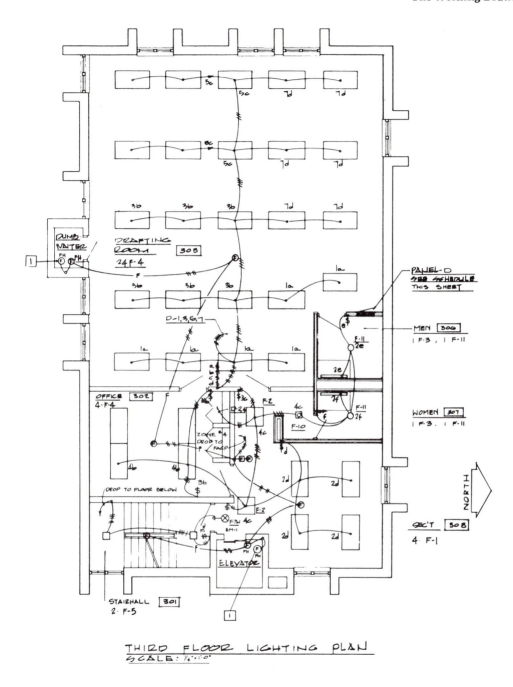

Figure 7.7 *Example of Electrical Drawing*

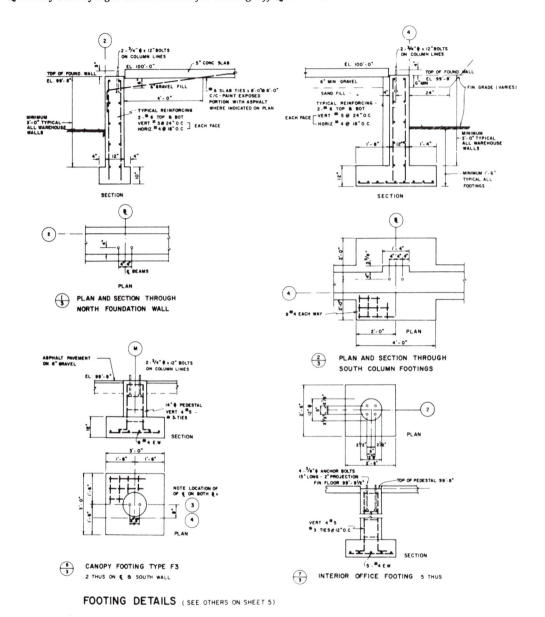

FOOTING DETAILS (SEE OTHERS ON SHEET 5)

Figure 7.8 *Example of Drawing Showing Details*

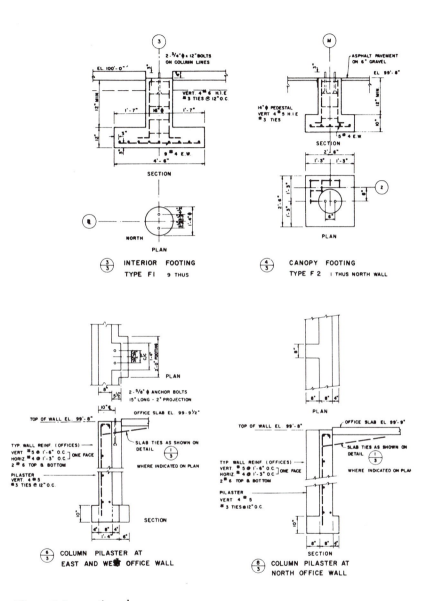

Figure 7.8 continued

In addition to begin located via floor plans and elevation drawings, details concerning windows and doors are often set out by means of **window and door schedules** set out on a separate drawing. Besides describing the size and type of each window and door, the schedules reference the windows and doors to a specific location on floor plan or elevation drawings.

Sometimes the designer numbers or letters each room on the floor plan, and repeats these numbers or letters on a **room finish schedule/drawing**. This drawing is used to specify the finishes used in the room for the floor, wall, and ceiling materials.

The drawings discussed so far can be considered the "building" drawings or sometimes the general condition drawings. They are followed by drawings that set out the specialized work of the building to include the electrical, plumbing, and heating and air conditioning work. Usually there is a set of drawings for each of these three categories of specialized construction work. An example electrical drawing is illustrated in Figure 7.7.

As can be observed in Figure 7.7, **specialized drawings** are very technical. Often the work is performed and estimated by a specialized contractor (e.g., an electrical contractor). These drawings are not as important to a general contractor preparing an estimate as are the building drawings (floor plans, elevation drawings, etc.).

In addition to the types of drawings discussed, a set of working drawings may also include drawings showing **detail and section views**. Sometimes these are included on one or more of the above-noted drawings (e.g., an elevation drawing) rather than a separate drawing. The designer references detail and section views to a location in the building via the use of specific symbols.

The purpose of a detail view is to enlarge a section of the drawing in order to better illustrate the type of construction or work. For example, a detail view might be utilized to enlarge a connection between a beam and column. The enlargement will set out the type of connection to include base plates, type of bolts, etc. An example detail view is illustrated in Figure 7.8.

A section view usually represents a "cut" through a wall, floor, or ceiling section of the building. For example, a section might be taken through a foundation wall (as illustrated on the foundation plan) in order to visually display the reinforcement in the wall. An example section view is illustrated in Figure 7.9. The designer always references a section view to the drawing and place on the drawing from which he has "cut" the section via the use of reference symbols.

Reading Working Drawings

In order to be able to establish quantities of work to be performed, the estimator has to be able to interpret or read construction working drawings. This process is commonly referred to as *blueprint reading*. Fundamental to

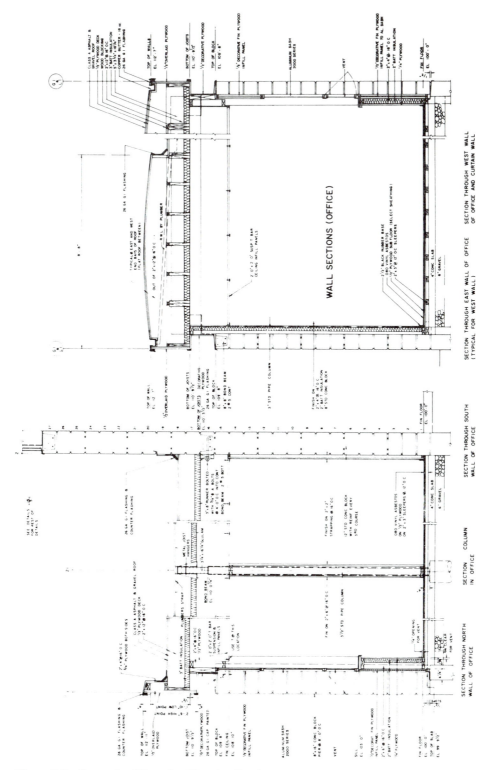

Figure 7.9 *Example of Drawing Showing Sections*

185

blueprint reading is an understanding of the various symbols, lines, notes, and abbreviations commonly illustrated on a set of drawings.

Working drawings for building construction projects are commonly illustrated via an **orthographic projection**. This is a parallel projection to a plane by lines drawn perpendicular to the plane. If the plane is horizontal, the projection is a plan (e.g., a floor plan). If the plane is vertical, the projection is an elevation drawing of the wall of the building, or a sectional elevation of a wall in the building.

In presenting working drawings, the designer uses a **scale**. He commonly uses the same scale reduction in preparing each of the drawings. For example, he might use a scale that equates 1 foot to ⅛ inch on the drawings. Such scales may be in eighths of an inch, tenths of an inch, or in a metric scale. Often an architect uses a scale divided into eighths of an inch, whereas an engineer presents the structural drawings in tenths of an inch. To aid the reader in interpreting the drawings, it is best if all the drawings are prepared using the same scale.

Dimensions are shown on a set of working drawings by means of **dimension lines**. A dimension line is usually a line with an arrow head on

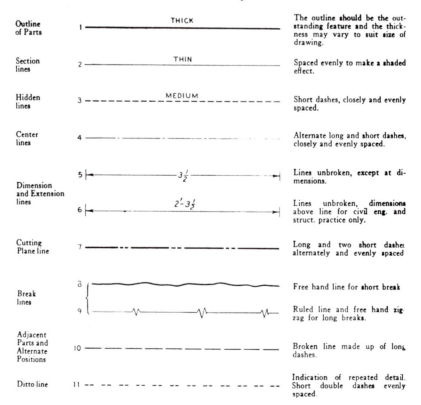

Figure 7.10   *Example of Lines Used on Drawings*

both ends drawn to the point where the referenced distance starts and finishes. Some designers omit the arrow heads and substitute a cross-hatched line. These two different types of dimension lines are illustrated in Figure 7.10. Dimension lines on a set of drawings are usually thin lines.

In addition to dimension lines, the designer will commonly use other various lines in presenting the working drawings. *Thick solid lines* are used to outline the edges of the plans and elevations. Lines consisting of *long and short dashes* denote center lines. Lines with *sequential short dashes* usually indicate hidden lines on the drawing.

Various symbols are commonly used on a set of working drawings to designate construction. Some designers include a symbol schedule in the drawings. Others omit this schedule; this may prove troublesome to the estimator because some designers use different symbols to represent the

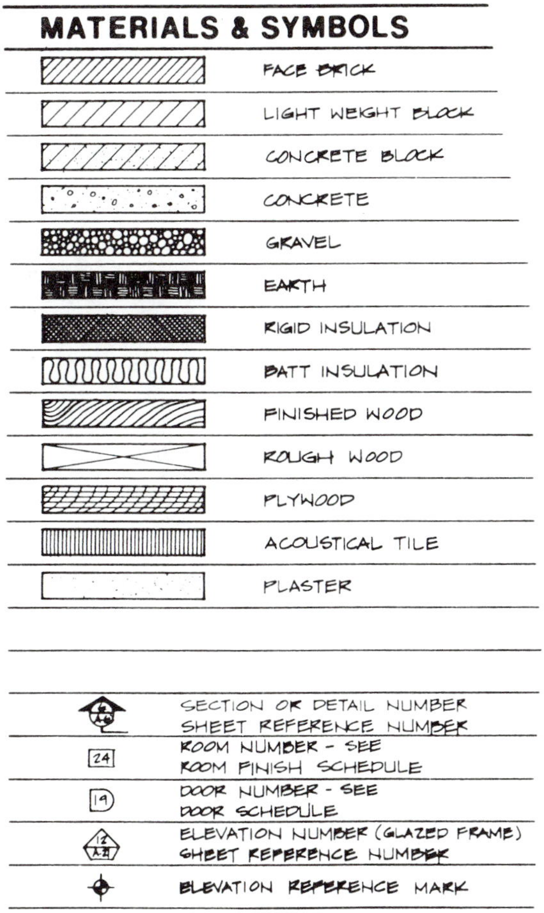

Figure 7.11 *Example Symbols Used on Drawings*

same materials or types of construction. Although there is some variation in practice, the symbols illustrated in Figure 7.11 are somewhat common to working drawings. The estimator should acquaint himself with these symbols in order to expedite the quantity take-off function.

Just as the designer may utilize symbols on the drawings, he may also use a number of abbreviations. The ability of the estimator to interrupt abbreviations is an important part of the preparation of an efficient and accurate take-off. Some of the more commonly used abbreviations that appear on drawings are listed in Figure 7.12.

AB	anchor bolt	EP	expansion bolt	INS	insulation
A/C	air conditioning	EL	elevation	KPL	kickplate
ADJ	adjacent	EGP	equipment	LAB	laboratory
AGG	aggregate	EST	estimate	LAM	laminated
AL	aluminum	FB	face brick	LTL	lintel
BIT	bituminous	FBD	fireboard	MAS	masonry
BLK	block	FBRK	firebrick	MET	metal
BM	bench mark	FGL	fiberglass	MOD	modular
BRK	brick	FLR	floor	MRB	marble
BSMT	basement	FND	foundation	MWK	millwork
CB	catch basin	FP	fireproof	OC	on center
CEM	cement	GD	grade	PAR	parallel
CM	centimeter	GL	glazing	PED	pedestal
COL	column	GLB	glass block	PLAS	plaster
CPR	copper	GPL	gypsum lath	PWD	plywood
CT	ceramic tile	GRN	granite	RAD	radius
CYD	cubic yard	GVL	gravel	RD	roof drain
DIM	dimension	HDW	hardware	ST	steel
DP	dampproofing	HVAC	heating/ventilating/	TZ	terrazzo
			air conditioning	WD	wood

Figure 7.12 *Example Abbreviations on a Drawing*

SPECIFICATIONS

We have already pointed out that both the working drawings and specifications for a project are important contract documents relevant to the preparation of an estimate. Specifications are a written set of instructions that accompany the working drawings for a building. They include details of the building that are not handled on the working drawings. With some exceptions, the specifications address the *quality* of work to be performed, whereas the working drawings focus on setting out the *quantity* of work to be performed.

Specifications for building construction projects are usually in a standardized format that conforms with the Construction Specification Institute (CSI). The 16 divisions in the CSI code are listed in Figure 7.13.

CSI SPECIFICATION DIVISIONS

1. General Requirements	9. Finishes
2. Site Work	10. Specialties
3. Concrete	11. Equipment
4. Masonry	12. Furnishings
5. Metals	13. Special Construction
6. Wood and Plastics	14. Conveying Systems
7. Thermal and Moisture Protection	15. Mechanical
8. Doors and Windows	16. Electrical

Figure 7.13 *Sixteen Divisions for a Building Project*

Each section of the 16 divisions of the specifications is usually further subdivided into the following 3 sections:

1. The scope of work.
2. Materials to include the quality of all the materials to be furnished.
3. A fabrication and erection section that defines the workmanship required.

Included in the scope of work is a listing of the items to be furnished and installed. General requirements including information about shop drawings, samples, tests, and the storage of the materials are included in this section.

The specifications for a project often make reference to various codes or standards set out by technical societies or associations. For example, reference may be made to standards of the American Concrete Institute (ACI), the American Institute of Steel Construction (AISC), or the American Society for Testing Materials (ASTM). The specifications may also reference published specifications of a named manufacturer.

An example excerpt from the specifications for a building project is shown in Figure 7.14. Because specifications are somewhat standardized from one project to the next, and because of the technical nature of the specifications, the estimator sometimes fails to read them. This failure to thoroughly review all the contract documents, including the specifications, can be dangerous. He could, for example, overlook a provision in the specifications indicating that the contractor is not to be paid for on-site stored material until the material is put in place; or a statement that indicates that an unusual amount of soil compaction is required in backfilling earthwork around the building's substructure. Overlooking of a single item in the

CAST-IN-PLACE CONCRETE

.1 Scope

    a. This Contractor shall furnish all labor materials and equipment necessary to the proper execution of the work specified in this Section and/or shown on the drawings.

    b. Work includes, but is not necessarily limited to, the following:

        1. Footings, and pads.

        2. Floor slabs, walks and stoops.

.2 Materials

    a. Cement, for walks, stoops, driveways and other areas exposed to weather — Air entraining Portland Cement, per PCA manufacture, conforming to ASTM C175, Type IA, maximum 5% of entrapped air. Cement for all other areas — conforming to PCA manufacture, ASTM C150, Type 1.

    b. Sand — clean, sharp, hard natural sand, free from scale, flakes, alkali or organic matter.

    c. Coarse Aggregate — hard, durable, uncoated crushed stone or gravel confoming to ASTM C33. Free from organic matter. Sized from No. 4 sieve to 1-1/2 inch, maximum size no larger than 3/4 of clear space between steel and formwork.

    d. Water — clean, free from oils, acids, alkalis and other impurities.

    e. Concrete mix shall be designed for maximum durability, and to meet following compressive strengths and maximum slumps.

        1. Footings and grade beams — 3000 psi, 5 inch.

        2. Column pads and piers — 3000 psi, 5 inch.

        3. Floor slabs, walks and stoops — 3000 psi, 5 inch.

Figure 7.14 *Example Specifications*

specifications can lead to an estimating error that has a negative impact on the contractor greater than a quantity take-off error. The estimator cannot afford to be careless in either of these estimating procedures.

STEPS FOR AN ACCURATE AND EFFICIENT QUANTITY TAKE-OFF

As noted earlier, the estimator has to perform the quantity take-off function (and other estimating procedures) in an efficient as well as accurate manner. Owing to the relatively short period of time available for the preparation of a detailed estimate, the estimator must utilize any procedures that will enhance the preparation of an accurate estimate in a minimum time period.

    Admittedly, "accurate" and "minimal" or "efficient time" are subjective terms. Just how accurate can one expect the estimate to be for a detailed contractor estimate? Surely, given all the variables or uncertainties that characterize the construction of a project, it is unrealistic to believe that an estimate or the determination of quantities (i.e., the quantity take-off) will or should be 100% accurate.

In setting the desired degree of accuracy, it is realistic to assume that an estimator should strive to prepare an estimate that is at least as accurate as the contractor's estimated profit margin. In other words, as a simple statement of sound business theory, *the range of accuracy of an estimate should not exceed the potential for profit.* Published financial data from various financial services indicate that the profit of a contractor as a percentage of the contracts he performs is in the range of 1 to 5% depending on the size and type of contractor. One might conclude that the average profit or revenue or contract value is approximately 3%. It follows that one might make a case for the estimator striving for an accuracy of 3% or better when performing a quantity take-off and estimate.

Even if we assume that a meaningful accuracy range can be established, it is still difficult to determine a set number of estimator hours that should not be exceeded when the quantity take-off or subsequent estimating steps are being performed. Nevertheless, it is a necessary process, because time expended preparing an estimate in effect results in the expenditure of contractor overhead costs; these expenses must be included as a cost of performing work and actually be included in the estimate itself.

Based on the author's experience in preparing detailed estimates for building construction projects, a suggested range of hours for preparing estimates of varying project costs is presented in Figure 7.15. As one would expect given the concept of efficiency as a function of the economics of scale, the estimator hours per dollar value of construction work is shown to decrease as the size of the project increases.

The reader should not conclude that the suggested hours shown in Figure 7.15 are fixed for each and every project. Obviously, the more complex

Cost of Project	Approximate Estimating Time (hours)
$ 200,000	40
500,000	60
1,000,000	80
2,000,000	120
5,000,000	180
10,000,000	300
20,000,000	450
50,000,000	600

Figure 7.15 *Guideline for Estimating Hours Required for Various Size Commercial Construction Project Estimates*

the required construction and the more complex the drawings, the more quantity take-off hours will be required. Similarly, the time that will be required to take off the project quantities is dependent on the experience and skills of the estimator. Independent of the suggested range of hours set out in Figure 7.15, the fact remains that the estimator has to be efficient in his quantity take-off and other estimating functions as well as being accurate. Various means of preparing an estimate in an efficient as well as accurate manner are discussed below.

Order of Quantity Take-Off

The learning-curve principle of industrial engineering indicates that the more times an individual performs a procedure in the same manner, the more efficient he or she becomes in performing the function. Equally important is the fact that as one performs a procedure or function in the same manner over and over again, the *quality or accuracy* of the performance improves.

Given these facts, it follows that the estimator would be advised to develop certain efficient procedures for the performance of the take-off process and to perform the same procedures on each and every estimate. Included in these procedures is the estimator's use of a standard order of material take-off quantities.

There are several orderly ways of taking off quantities. Perhaps the most natural one would be to do so in the order of the 16 *CSI divisions.* The argument for this approach is that the quantity take-off should be performed in the same order as the order of the specifications for the project.

Another approach to determining an order of take-off would be an *efficient calculation process.* For example, some estimators argue that the excavation and sitework should be taken off after the concrete or masonry for the substructure. The reasoning for this is that excavation items depend in part on the below-ground structure (the substructure). By taking off the concrete or masonry substructure first, it can be argued that the estimator determines information and makes calculations that enhance the efficiency of the excavation and sitework take-off. Another argument for taking off the excavation and site work after some other items is that during the take-off process the estimator may get a chance to visit the site. This visit may allow him to determine information relevant to the earthwork and sitework take-off.

Other estimators develop their quantity take-off order to be consistent with the *order that will be constructed.* This aids the estimator in visualizing the quantities and work better than if he merely takes off the items without conceptualizing the construction process.

Independent of the order of take-off selected, the order should not vary from one project to another. The purpose of establishing and keeping a particular order is to follow a methodical pattern for the take-off. In this manner, the take-off performed for all project estimates will go smoothly and efficiently, and the probability of missing an item will be reduced to a minimum.

Use of Checklists

All too often the construction estimator overlooks something when he is preparing an estimate for a project. Given the numerous individual procedures and/or items that he must recognize in his estimate, as well as identification of the numerous types of materials to be included, the estimator must be especially careful not to omit any items. If there are several omissions or any one of them is of a significant dollar amount, the contractor is subject to a potential significant cost overrun on the project, which may in fact result in an unprofitable project.

The use of checklists can reduce or eliminate unintentional omissions of quantity take-off items or the failure to perform various procedures. A checklist is a list of steps or procedures to be part of a process. Some items may be listed that are not appropriate to the preparation of each and every project estimate; however, the very listing of any items, steps, and/or procedures that *may* be relevant to the process may in itself prevent the individual from inadvertently omitting anything.

Various types of checklists are appropriate to the estimating process. A *material checklist* can be used by the estimator to remind himself of all the possible types of material and work that might be part of the project. A sample material checklist is illustrated in Figure 7.16. The estimator can merely check off each item as he locates and takes off the item or determines that it is not relevant to the project in question.

Another useful checklist for the estimator is an *insurance checklist*. Numerous types of insurances are required as part of the construction process. Sometimes certain types of insurance for a project must be furnished by the contractor, whereas on another project, the project owner may furnish the insurance. In other words, the types and amount of insurance the estimator must include in his estimate varies from one project to the next. The types and amounts of insurances required are usually set out in the special conditions and general conditions documents for the project. To enhance the probability that the estimator includes *all* the requested insurances, the use of a checklist is to be recommended. A sample of such a checklist is illustrated in Figure 7.17.

Use of Standard Forms

We have noted earlier that an individual tends to gain efficiency as he or she performs the same process over and over. Given this "learning-curve" characteristic, it is advantageous for the estimator to repeatedly use the same estimating forms when taking off quantities of work and pricing the work.

The use of standardized estimating forms within the firm serves two purposes: (1) as we have seen, a person using the same forms over and over is likely to become more efficient (and possibly more accurate) when using the forms; (2) the use of a standard set of forms promotes communication between the

Basement Excavation	Exterior Doors
Piers	Interior Doors
Footings	Windows
Trenchs	Plywood Walls
Backfilling	Sliding Doors
Grading	Wood Stairs
Shoring	Insulation
Loading and Trucking	Accoustical Tile
Caissons	Wood Lath
Piling	Nails
Dampproofing	Metal Studding
Sidewalks	Channel Iron
Basement Floors	Metal Lath
Concrete Steps	Metal Trim
Concrete Driveways	Plastering
Wall Concrete	Ceiling Panels
Column Concrete	Hollow Metal Doors
Slab Concrete	Hollow Metal Windows
Face Brickwork	Aluminum Windows
Glazed Brickwork	Metal Joists
Fire Brick	Joist Hangers
Common Brick	Metal Gutters
Catch Basins	Hips
Manholes	Moldings
Concrete Blocks	Skylights
Cinder Blocks	Glazing Compound
Cut Stone	Wood Shingles
Scaffolding	Prepared Roll Roofing
Architectural Concrete	Composition Roofing
Tile Floors	Ceramic Tile
Gypsum Block	Linoleum
Structural Tile	Window Glass
Wood Roof Trusses	Painting Interior
Framing Timbers	Painting Exterior
Floor Joists	Structural Steel
Roof Framing	Gratings
Sub flooring	Fireplaces
Stud Walls	Ornamental Iron
Siding	Railings
Asphalt Shingles	Finish Hardware
Corner Boards	Elevators
Shelves	Electrical Rough and Finish Work
Finish Floors	Lighting Fixtures
Hardwood Flooring	Mechanical Equipment
Plank Flooring	Miscellaneous Equipment
Sanding	Door Frames
Window Frames	

Figure 7.16   *An Abbreviated Material Checklist*

firm's various estimators and possibly non-estimating personnel (project managers, superintendents, and so forth). Most if not all project management functions can be performed in a way that is compatible with the project estimate. When a project superintendent or another nonestimator understands the information on the estimating forms, the firm benefits. The ad-

Workmen's Compensation Insurance	Malpractice Coverage
Owner's, Landlords and Tenants Insurance	Automobile Liability Insurance
Products Liability	Physical Damage Insurance
Owner's Protective Insurance	Property Damage Insurance
Contractor's Protective Insurance	Builder's Risk Insurance
Contractural Insurance	Temporary Structure Insurance
Accident Defined Insurance	Installation Floater
Personal Injury Coverage	

Figure 7.17   *Example Insurance Checklist*

vantage is further enhanced if the same set of estimating forms is used throughout the firm for estimating each and every project.

Estimating forms have essentially two sources. A construction firm might design its own estimating forms to serve its specific needs. These **"custom-designed" forms** have the advantage of suiting the practices of the firm, so that the firm does not have to alter its practices to "fit" standardized forms designed for the industry at large.

Custom designing the estimating forms is highly desirable. All construction firms differ somewhat in the type of work they perform, the number of people they have in their estimating department, or the procedures and detail they utilize in preparing their estimates. Given this fact, one might argue that each individual firm needs its own estimating forms. At the very least, it can be argued that if a firm does use standardized forms prepared for the industry at large, it should modify them as necessary to serve its specific needs.

Despite the fact that an argument can be made for preparing and using custom-designed forms, the majority of construction firms utilize **standardized estimating forms** prepared for the industry at large by a publishing company or other company specializing in the design and retailing of such forms.

The justification for the use of these standard forms is that the construction firm can purchase them and save the time it would take to design its own forms. This, coupled with the fact that a standardized form may be very close in appearance to what they might custom-design, may offset the benefits of "starting from scratch" in regard to form design.

Of the companies that sell construction estimating forms, some only sell forms for a certain type of contractor (e.g., a residential contractor), whereas others sell forms for virtually every type of contractor. Two of these later types of companies are NEBS and Frank R. Walker Publishing Company. (Those known as Walker's forms will be utilized in the following chapters.)

Figures 7.18 through 7.20 illustrate three very common standardized estimating forms. The forms are illustrated in the order in which they would be filled out by the estimator.

PRACTICAL
FORM 516 MFD. IN U. S. A.

QUANTITY SHEET

PROJECT		ESTIMATOR	ESTIMATE NO.
LOCATION		EXTENSIONS	SHEET NO.
ARCHITECT ENGINEER		CHECKED	DATE

CLASSIFICATION

DESCRIPTION	NO.	DIMENSIONS									ESTIMATED QUANTITY	UNIT

FRANK R. WALKER CO., PUBLISHERS, CHICAGO

Figure 7.18 *Quantity Take-Off Sheet (Walker's Forms)*

196

PRACTICAL
STANDARDIZED FORMS FOR CONTRACTORS
FORM 314

GENERAL ESTIMATE

BUILDING ______________________ ESTIMATE NO. ______________________

LOCATION ______________________ SHEET NO. ______________________

ARCHITECTS ______________________ ESTIMATOR ______________________

SUBJECT ______________________ CHECKER ______________________

DATE ______________________

DESCRIPTION OF WORK	NO. PIECES	DIMENSIONS			EXTENSIONS	EXTENSIONS	TOTAL ESTIMATED QUANTITY	UNIT PRICE M'T'L	TOTAL ESTIMATED MATERIAL COST	UNIT PRICE LABOR	TOTAL ESTIMATED LABOR COST

MFD. IN U.S.A.

FRANK R. WALKER CO., PUBLISHERS, CHICAGO

Figure 7.19 *General Estimate Sheet (Walker's Forms)*

The Quantity Sheet: The first form shown is referred to as a *quantity sheet* (see Figure 7.18). The purpose of this form is limited to the taking off and determination of work quantities. There is no calculation or determination of pricing performed on this sheet. While it could be utilized by the estimator for calculating all work-item quantities, the form is especially useful for determining quantities that require a significant amount of "subcalculations" — for example, calculations regarding the determination of the number of reinforcing bars or weight of the reinforcing bars.

The General Estimate Sheet: This form, illustrated in Figure 7.19, might be used in one of two different processes by the estimator. He might make his determination of work-item quantities directly on the general estimate sheet; in other words, the estimator may omit the use of the quantity sheet and perform the quantity take-off calculations directly on the general estimate sheet. This might especially be the case if the calculations are not extremely complex and not many are required. (This is in recognition of the fact there is limited space on the general estimate sheet to perform calculations.)

Items to include concrete, masonry, and structural steel are often taken off directly via the general estimate sheet. Or it is possible that in preparing an estimate for a specific project, an estimator will utilize a quantity take-off sheet for determining the quantities of work for some work items and a general estimate sheet for determining the quantities of other work items.

If the estimator performs his quantity take-off calculations on the general estimate sheet, the form in effect serves a dual role of quantity take-off and pricing. As can be observed in Figure 7.19, the form provides the estimator space for multiplying the calculated quantities times unit prices to determine the total number of pieces needed for the work.

If the estimator uses the quantity take-off form for determining all of his work-item quantities, the general estimate becomes a summary sheet for the work-item quantities and price determination. The determined quantity of each work item is transferred from the quantity sheet to the general estimate form. Few, if any, calculations are shown on the general estimate sheet in this case. Once the quantities are transferred and summarized on the general estimate sheet, the pricing is performed on that estimate sheet in the same manner that would be used if the quantity take-off calculations were performed directly on the general estimate sheet. In effect, then, the general estimate sheet is used either for the dual purpose of quantity take-off and pricing, or is limited to just pricing.

The Estimate Summary Form: This form, shown in Figure 7.20, serves to sum the cost of doing the individual work items that are determined on the general estimate sheets and also serves as the worksheet for summarizing various miscellaneous project costs, including job overhead costs such as bonds, insurances, and so forth. In addition to serving a calculation summary purpose, this particular form can be used as a reminder checklist for the estimator in regard to the items to be included in the total bid. As in all checklists, each and every item on the estimate summary sheet will not be included on each and every estimate. However, the use of the form as a checklist enhances the likelihood that the estimator will not omit an item in the estimate.

Standard forms like those illustrated in Figures 7.18 to 7.20 do not match the needs of each and every project or contractor. However, with slight

CLASSIFICATION	TOTAL ESTIMATED MATERIAL COST	TOTAL ESTIMATED LABOR COST	TOTAL SUB-BIDS	TOTAL	ADJUSTMENTS
SUMMARY OF ESTIMATE					
BUILDING	LOCATION			ESTIMATE NO.	
ARCHITECT	OWNER			DATE	
CUBICAL CONTENTS — NO. OF STORIES	COST PER CUBIC FOOT			ESTIMATOR	
FLOOR, AREA, SQUARE FEET	COST PER SQUARE FOOT			CHECKER	
1. GENERAL CONDITIONS AND OFFICE OVERHEAD					
2. JOB CONDITIONS AND JOB OVERHEAD					
3. CONSTRUCTION PLANT, TOOLS AND EQUIPMENT					
4. DEMOLITION AND SITE CLEARANCE					
5. EXCAVATION, GRADING AND DEWATERING					
6. SHEETING, SHORING AND BRACING					
7. PILING AND CAISSON WORK					
8. SITE DEVELOPMENT					
9. CONCRETE FORM WORK					
10. CAST IN PLACE CONCRETE					
11. PRECAST CONCRETE AND CEMENTITIOUS DECK					
12. BRICK, TILE, CONCRETE & GLASS UNIT MASONRY					
13. UNIT MASONRY PARTITIONS & FIREPROOFING					
14. CUT, ROUGH, NATURAL & SIMULATED STONE					
15. STRUCTURAL METALS					
16. OPEN WEB JOISTS AND METAL DECKING					
17. MISCELLANEOUS METALS					
18. ORNAMENTAL METALS					
19. ROUGH CARPENTRY AND ROUGH HARDWARE					
20. FINISH CARPENTRY					
21. CUSTOM MILLWORK					
22. WATERPROOFING AND DAMP-PROOFING					
23. BUILDING INSULATION					
24. SHINGLE AND ROOFING TILE					
25. MEMBRANE ROOFING					
26. PREFORMED ROOFING AND SIDING					
27. ROOFING ACCESSORIES					
28. SHEET METAL FLASHINGS					
29. SEALANTS AND CAULKING					
30. METAL DOORS AND FRAMES					
31. WOOD AND PLASTIC DOORS					
32. SPECIAL DOORS					
33. METAL WINDOWS					
34. WOOD AND PLASTIC WINDOWS					
35. FINISHED HARDWARE & WEATHER STRIPPING					
36. GLASS, GLAZING AND MIRRORS					
37. CURTAIN WALL AND STORE FRONT SYSTEMS					
38. LATHING AND FURRING					
39. PLASTERING AND STUCCO WORK					
40. GYPSUM DRYWALL					
41. CERAMIC, MOSAIC, QUARRY, MARBLE & SLATE TILE					
42. TERRAZZO AND SEAMLESS FLOORING					
43. WOOD FLOORING					
44. RESILIENT FLOORING					
45. ACOUSTICAL TILES AND PANELS					
46. SPRAY ON FIRE PROTECTION					
47. PAINTING AND SPECIAL FINISHES					
48. SPECIAL BUILDING PARTITIONING					
49. BUILDING SPECIALTIES, EQUIP. & FURNISHINGS					
50. ELEVATORS & MECHANICAL TRANSPORT					
51. PLUMBING					
52. FIRE PROTECTION & SPRINKLER SYSTEMS					
53. HEATING, VENTILATION & AIR CONDITIONING					
54. ELECTRICAL WORK					
55. COMMUNICATION SYSTEMS					
56. TOTALS					
57. TOTAL COST					
58. PROFIT					
59. SURETY BOND					
60. AMOUNT OF BID					

PRACTICAL FORM 115

MFD. IN U S A — FRANK R. WALKER CO., PUBLISHERS, CHICAGO

Figure 7.20 *Summary Sheet (Walker's Forms)*

modifications, they often can be used to serve specific estimator needs. The forms we have discussed will be utilized in following chapters to illustrate actual take-offs and pricing.

Estimating Calculations Efficiently

Mathematics plays a significant role in the preparation of an estimate. The adding, subtracting, and multiplying of numbers, the conversion of numbers, and extensions of calculations must be performed as part of every single estimate. The accuracy of an estimate is dependent on the estimator's knowledge and use of efficient and accurate mathematics.

There are certain mathematical shortcuts or procedures that can be used to help the estimator. The word "shortcut" should not be interpreted to mean a mathematical practice that attains efficiency or speed at the expense of accuracy. The mathematical shortcuts we discuss below should be viewed as procedures that enhance the efficiency of preparing an estimate while maintaining or improving the accuracy of the calculation process.

The following list of mathematical shortcuts is not meant to be all-inclusive; it is, however, representative of some of the more important practices:

Preliminary calculations	Calculation of areas and volumes
Use of "super" item or	Use of conversion tables
calculations	Addition/deduction of calculations
Conversion of numbers	Extending and footing of numbers
Rounding off numbers	

Preliminary Calculations: It is common for an estimator to have to utilize the result of a certain calculation several times in determining a quantity of work for several different work items. For example, a calculated area of wall surface may be used in determining the quantity of wall forming required, the amount of concrete in the wall, the number of bricks required for the wall, and the amount of interior finish material to be placed on the wall. Given the fact that the surface area is common to each of these determinations, it may be advantageous to determine the surface area by means of a *preliminary calculation*.

A preliminary calculation can be thought of as a calculation of a length, area, or volume whose resulting calculation will be utilized several times in the estimating process. Given this fact, it is essential that the calculated quantity be correct. It is therefore wise to single it out on the estimating forms and proof it by means of two different mathematical processes.

To illustrate the concept of a preliminary item, let us consider the wall shown in Figure 7.21. Assume that the surface area (i.e., the total area minus the window and door area) will be used in determining the quantity of work for several different work items. Assume also that the surface area is to be determined via a preliminary calculation.

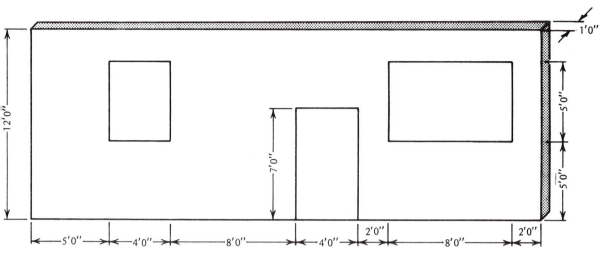

Figure 7.21   *Example Wall with Blockouts*

One means of determining the surface area of the wall is by breaking the wall into separate rectangles and summing the areas. The rectangles are shown in Figure 7.22 and the calculation of the area is shown as method 1 on the estimating form illustrated in Figure 7.23. Given the importance of the calculation, it is advantageous for the estimator to check it out carefully. This can be done by calculating the area of the total wall and deducting the area of the windows and doors. (The areas are shown in Figure 7.24 and the calculation process is shown as method 2 in Figure 7.23.) Given an equivalent calculated answer via the two methods, the estimator can then be assured that the quantity is correct and proceed to use it for several work items without recalculating the number for each individual work item.

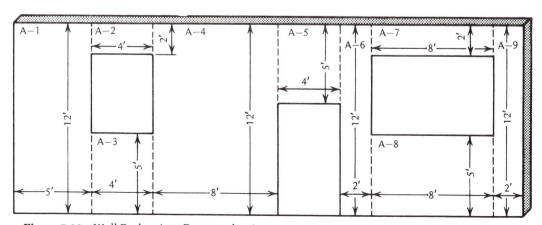

Figure 7.22   *Wall Broken into Rectangular Areas*

QUANTITY SHEET

PROJECT	ADD/DEDUCT EXAMPLE	ESTIMATOR		ESTIMATE NO. EXAMPLE
LOCATION	ANYWHERE	EXTENSIONS		SHEET NO. 1
ARCHITECT ENGINEER	JJA	CHECKED		DATE 9-12-X0

CLASSIFICATION EXAMPLE CALCULATION

DESCRIPTION	NO.	DIMENSIONS Height	width		Sq. ft					ESTIMATED QUANTITY	UNIT
CALCULATION OF WALL AREA											
METHOD 1:											
AREA A-1		12.0'	5.0'		60.0						
A-2		2.0'	4.0'		8.0						
A-3		5.0'	4.0'		20.0						
A-4		12.0'	8.0'		96.0						
A-5		5.0'	4.0'		20.0						
A-6		12.0'	2.0'		24.0						
A-7		2.0'	8.0'		16.0						
A-8		5.0'	8.0'		40.0						
A-9		2.0'	12.0'		24.0						
TOTAL AREA										308	ft²
METHOD 2:											
TOTAL WALL AREA		12.0'	33.0'		396.0						
DEDUCT B-2		5.0'	4.0'		- 20.0						
DEDUCT B-3		2.0'	4.0'		- 28.0						
DEDUCT B-4		5.0'	8.0'		- 40.0					308	ft²
TOTAL AREA											

Figure 7.23 Use of Add/Deduct Process

202

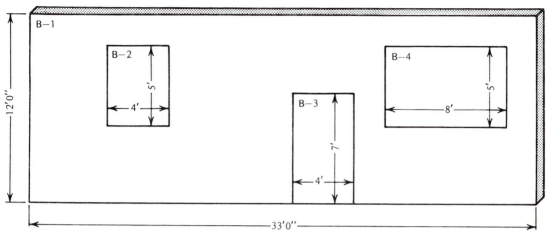

Figure 7.24 *Wall Broken into Deduct Areas*

Use of "Super" Item Calculations: Somewhat similar to this use of a preliminary calculation is the estimator's use of a *"super" item or calculation.* This is a number that is utilized several times by the estimator. Rather than recalculating the number several times, the first time the calculation is made on an estimating form, the calculated quantity is set out, underlined, or simply identified as a "super" item. This "setting out" of the resulting quantity assures that the estimator will not forget that he has made the calculation and waste time making it several more times. Instead, the estimator merely makes reference to the "super item" calculation when confronted with the calculation subsequent to its first use, and transfers the calculated quantity to the determination of the new work item.

As an example of a super item, let us consider the calculation of an area of a slab. Assume that the first time the estimator is confronted with the need to determine the area is when he is determining the concrete quantity. While this quantity is to be determined in cubic yards, the estimator might recognize that the area of the slab is a quantity that will be needed for determining the quantity of other work items (required gravel fill, excavation, and area of finish material to be laid on the concrete slab). Given this fact, the estimator sets out the calculated area as a super item. This might be done as shown on the estimating form in Figure 7.25. Once this area is determined it can be used for other work items, as is also illustrated in Figure 7.25. The use of a super item aids the estimator in the efficiency of his quantity take-off.

Conversion of Numbers: In taking off quantities, an estimator may often need to *convert the numbers* shown on the drawings to another unit of measure. This task is made more complicated by the fact that the units on the drawings may themselves be mixed. For example, it is common for the designer to express a length in a measure of both feet and inches—e.g., 2'-10".

QUANTITY SHEET

PROJECT _Example of Using A Common Calculation_ ESTIMATOR   ESTIMATE NO. _Example_

LOCATION _Anywhere_   EXTENSIONS   SHEET NO.

ARCHITECT ENGINEER _JJA_   CHECKED   DATE _9-12-X0_

CLASSIFICATION _Example_

DESCRIPTION	NO.	Length	Width	Depth	ft²	ft²	x Depth	ft³	ESTIMATED QUANTITY	UNIT
03201										
Concrete Slab	1	22.0'	28.0'	1.5'	616.0					
Northeast Sect.	1	18.5'	12.0'	1.5'	222.0					
Southeast Sect.	1	24.5'	22.0'	1.5'	539.0					
Southwest Sect.	1	32.0'	16.5'	1.5'	528.0					
						1905.0 (Super)	1.5'	2857.5 ÷ 27 ft³/yd³	105.9 yd³	
03202										
Gravel Fill (Super-Conc. Slab)	1								1905	ft²
03203										
Conc. Finishing Slab - Super Area	1								1905	ft²

Figure 7.25 *Example Use of a Common Calculation*

204

To enhance the efficiency of the take-off process, it is advantageous for the estimator to transfer a "mixed-unit" number to a single unit when the number is transferred from the drawing to the estimating sheet. Usually it is a good idea to convert a length set out in feet and inches to feet alone, because unit prices are often given in measures of feet or yards in construction. In this manner, the 2'-10" length measure would be converted to 2.87 feet on the estimating form. To facilitate the conversion of inches to feet, the estimator should make use of conversion tables such as that shown in Figure 7.26.

INCHES TO THE DECIMAL PART OF A FOOT												
$\frac{1}{4}''$	$\frac{1}{2}''$	1"	2"	3"	4"	5"	6"	7"	8"	9"	10"	11"
0.02	0.04	0.08	0.17	0.25	0.33	0.42	0.5	0.58	0.67	0.75	0.83	0.92

Figure 7.26 *Table for Converting Inches into Feet*

To improve the efficiency and accuracy of the estimating process, the estimator should try to minimize the number of mathematical conversions required. To illustrate this point, consider the calculations shown in Figure 7.27. Assume that the work item is to be measured in square yards. However, the units shown on the drawings are indicated in feet and therefore the estimator has taken off the entries shown in feet (see Figure 7.27). Having established the two relevant lengths for each entry, the estimator could then convert each length to yards and multiply the two lengths to determine the square yards for each entry. For example the 9.0' times 12.0' initial entry could be converted to 3 yards times 4 yards and multiplied to yield 12 square yards. Each entry *could* be similarly converted to square yards and the total square yards be determined by summing the individual entries. However this process would be inefficient as well as potentially inaccurate.

As an alternative to converting each length to yards, a more efficient process would be to determine the square feet for each entry via the multiplication of the two lengths expressed in feet. Once this is done for each work item, the total square feet can be determined by summing the individual entries, as shown in Figure 7.27. After the sum is determined, the total square feet can be converted to square yards via a *single conversion* (also illustrated in Figure 7.27). In addition to cutting down the time required for the mathematics by reducing the number of conversions, the accuracy of the final answer is enhanced because the smaller the number of mathematical conversions performed, the less likely the estimator will be to make a mathematical error (an incorrect multiplication, inaccurate rounding off of a number, or

PRACTICAL
FORM 516 MFD. IN U.S.A.

QUANTITY SHEET

PROJECT	*EXAMPLE OF MINIMIZING CALCULATIONS*		ESTIMATOR		ESTIMATE NO. *EXAMPLE*
LOCATION	*ANYWHERE*		EXTENSIONS		SHEET NO. *1*
ARCHITECT ENGINEER	*JJA*		CHECKED		DATE *9-12-X0*
CLASSIFICATION	*EXAMPLE*				

DESCRIPTION	NO.	DIMENSIONS Length	Width			ft²	÷9 ft³/yd		ESTIMATED QUANTITY	UNIT
02801 LANDSCAPING										
AREA 1	1	9.0'	12.0'			108.0				
" 2	1	24.0'	18.5'			444.0				
" 3	1	13.5'	14.0'			189.0				
" 4	1	18.5'	16.0'			296.0				
" 5	1	12.0'	9.5'			114.0				
" 6	1	28.0'	16.0'			448.0				
" 7	1	48.5'	94.5'			4583.0				
" 8	1	112.0'	24.0'			2688.0				
" 9	1	18.5'	19.0'			351.0				
						9221.0	÷9		1,025 yd²	

FRANK R. WALKER CO., PUBLISHERS, CHICAGO

Figure 7.27 Example of Minimizing Calculations

incorrect placement of a decimal place). The fewer the number of mathematical calculations, the greater the potential for an accurate result.

Rounding Off Numbers: Multiplication and division are two mathematical procedures common to the determination of work-item quantities. Given these mathematical calculations, there often is a question of *rounding-off of numbers*. The rounding off of numbers is often referred to as the determination of "significant" figures. The estimator has to be aware that a careless rounding off of numbers can lead to significant errors in a calculated total. For example, consider the multiplication of the following series of numbers:

.01344 × 2.54 × 1584.25

Let us assume that each of these numbers represents a dimension of a rectangular expressed in feet. Assume that the estimator is attempting to determine the cubic feet of volume. Further assume that he starts the multiplication process by multiplying the first two numbers. However, in performing this calculation, he carelessly rounds off the number .01344 to .01 to obtain the following result:

.01 × 2.54 = .0254 sq. ft.

Assume further that the estimator then rounds off the number .0254 to .02 and proceeds to multiply this times the number 1584.25. The result is as follows:

.02 × 1584.25 = 31.685 cu. ft.

Let us compare this result to the number that would have resulted if the estimator had done no rounding-off of numbers in performing the calculation. The result would be as follows:

.01344 × 2.54 × 1584.25 = 54.0825 cu. ft.

The careless rounding off of the numbers has resulted in an approximately 41 percent error in the calculated number! This error would carry through to any determination of the calculated cost of doing the work item. It is obvious that it is essential that the estimator know when and how to deal with significant figures and the rounding-off of numbers.

The following two principles should be followed by the estimator in regard to rounding off numbers.

1. Never round off any numbers that are part of a multiplication or division process *before* the multiplication or division process is made. This is especially true when making a series of calculations and when multiplying or dividing a large number by a small number.
2. The result of a multiplication or a division process is only accurate to the number of significant figures of the smallest number. For example, when multiplying 28.42 × 38.564 the result is only accurate to two

decimal places. The resulting number (i.e., 28.42 × 38.564 = 1095.9888) should be rounded to 1095.99.

The question of rounding off numbers also occurs when the estimator is adding numbers or summarizing the final calculated number for a work item before he prices the work item. For example, in taking off individual earth-work quantities, the estimator might determine the quantities in cubic yards. Let us assume that, based on the calculation process, each entry is deter-mined to two decimal places—for example, 26.14 cubic yards. Should the estimator round this number off to 26.1 cubic yards before he adds a series of similar numbers? The answer to this is likely no. However, it is true that the degree of rounding that is desirable is somewhat dependent on the type of work item being estimated. Given all the assumptions that are usually part of the excavation take-off, and given the relatively low unit cost that will subse-quently be applied to the work item, one might argue that taking off the ex-cavation work to the nearest cubic yard is accurate enough. While this is true, the estimator is better served by his rounding off the *sum* of the in-dividual entries rather than rounding off each individual entry.

The question as to what number of significant figures each calculated work item quantity should be rounded is dependent on the criticality of the item in regard to cost. This issue will be discussed for individual work items in the following chapters that address the take-off and pricing of individual types of work items.

Calculation of Areas and Volumes: The preparation of an estimate occa-sionally requires the estimator to be able to make *area or volume calculations* for shapes or figures that are frequent in occurrence. The estimator's knowl-edge of formulas for calculating areas or volumes is essential to the accuracy and efficiency of the overall estimate. Some of the more common shapes and figures along with formulas for determining their areas or volumes are shown in Figure 7.28.

Use of Conversion Tables: Similar to the use of standard formulas for cal-culating areas and volumes, the efficiency of preparing an estimate can be enhanced through the use of conversion tables to transform standard combi-nation of numbers into units compatible with the estimator's pricing objec-tives. For example, the tables shown in Figures 7.29 and 7.30 convert various sizes of boards into board-foot measure and typical structural steel rectan-gles into a weight measure. Various estimators have designed conversions tables such as those shown in Figures 7.29 and 7.30 to serve their specific needs. For example, an estimator that is often confronted with taking off wall forming may develop tables that convert various wall sections to a sur-face area of required forming.

Add and Deduct Calculations: Yet another means of improving the effi-ciency of the estimating process is through the estimator's use of *common di-mensions* and the use of *add and deduct items* in the calculation process.

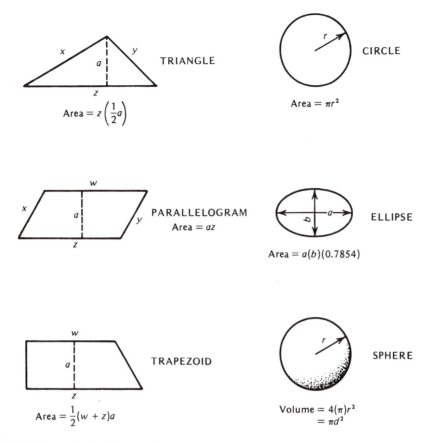

Figure 7.28 *Formulas for Common Shapes*

Often, when he is making several intermediate entries as part of the determination of the quantity for a work item, the estimator observes that several of the individual entries have a common dimension. For example, for several isolated column footings he may determine that many of them are 1'-0" deep. However, many may have varying section areas such as 4' × 6' or 5' × 5'. In performing the volume calculations on the estimating forms, it would be advantageous for the estimator to merely multiply the sectional area for each entry, add the areas for each entry, and then multiply by the depth (i.e., 1 foot) to convert the quantity to a volume. This volume can then be converted from cubic feet to cubic yards.

The process described above is shown in Figure 7.31. This process of singling out the common dimension (i.e., the 1-foot depth) is more efficient than the process whereby the estimator multiplies each and every entry by the common number. This process not only adds estimating time, it might lead to decreased accuracy because of the possibility of more multiplication and round-off errors.

Lineal Foot Table of Board Measure
Number of Feet of Lumber, B.M., Per Lineal Foot of any Size

2″ × 4″ = 0.667	4″ × 4″ = 1.333	8″ × 14″ = 9.333
2″ × 6″ = 1.	4″ × 6″ = 2.	8″ × 16″ = 10.667
2″ × 8″ = 1.333	4″ × 8″ = 2.667	10″ × 10″ = 8.333
2″ × 10″ = 1.667	4″ × 10″ = 3.333	10″ × 12″ = 10.
2″ × 12″ = 2.	4″ × 12″ = 4.	10″ × 14″ = 11.667
2″ × 14″ = 2.333	4″ × 14″ = 4.667	10″ × 16″ = 13.333
2″ × 16″ = 2.667	4″ × 16″ = 5.333	10″ × 18″ = 15.
2½″ × 12″ = 2.5	6″ × 6″ = 3.	12″ × 12″ = 12.
2½″ × 14″ = 2.917	6″ × 8″ = 4.	12″ × 14″ = 14.
2½″ × 16″ = 3.333	6″ × 10″ = 5.	12″ × 16″ = 16.
3″ × 6″ = 1.5	6″ × 12″ = 6.	12″ × 18″ = 18.
3″ × 8″ = 2.	6″ × 14″ = 7.	14″ × 14″ = 16.333
3″ × 10″ = 2.5	6″ × 16″ = 8.	14″ × 16″ = 18.667
3″ × 12″ = 3.	8″ × 8″ = 5.333	14″ × 18″ = 21.
3″ × 14″ = 3.5	8″ × 10″ = 6.667	16″ × 16″ = 21.333
3″ × 16″ = 4.	8″ × 12″ = 8.	16″ × 18″ = 24.

Figure 7.29 *Table for Converting Board Sizes into Board Feet*

Pounds per Linear Foot for Various Structural Steel Sections

Width (inches)	Thickness (inches)						
	1/4	3/8	1/2	5/8	3/4	7/8	1
1	0.85	1.27	1.70	2.12	2.55	2.97	3.40
2	1.70	2.54	3.40	4.25	5.11	5.94	6.80
5	4.26	6.36	7.83	10.63	12.76	14.86	17.00
10	8.53	12.73	15.66	21.26	25.53	29.73	34.00
13	11.10	16.60	22.10	27.60	33.20	38.7	44.20
15	12.8	19.1	25.5	31.90	38.3	44.6	51.00

Figure 7.30 *Table for Converting Structural Steel Rectangles to Weights*

The concept of an add/deduct item was illustrated in Figure 7.24. In calculating the area of the wall shown, the total wall area was determined and the area of the windows and doors was then subtracted. In effect, the area of the windows and doors was a "deduct" item from the total wall area. This type of calculation process may prove more efficient than breaking the wall area into several rectangles and calculating the surface areas.

The use of "deduct item" calculation is common when one must take off areas that are regular in shape except for a few obstructions or "blockouts."

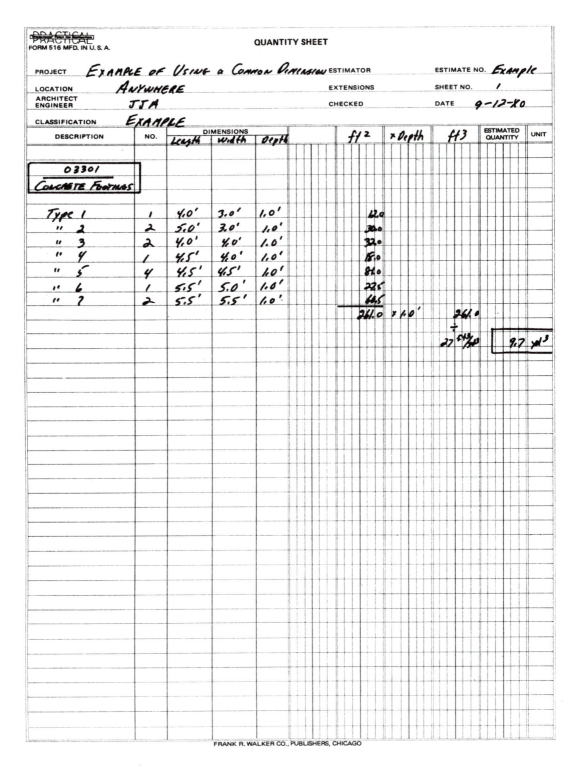

QUANTITY SHEET

FORM 516 MFD. IN U.S.A.

PROJECT	Example of Using a Common Dimension	ESTIMATOR		ESTIMATE NO.	Example
LOCATION	Anywhere	EXTENSIONS		SHEET NO.	1
ARCHITECT ENGINEER	JJA	CHECKED		DATE	9-12-X0
CLASSIFICATION	Example				

| DESCRIPTION | NO. | DIMENSIONS | | | | ft² | x Depth | ft³ | ESTIMATED QUANTITY | UNIT |
		Length	Width	Depth						
03301										
Concrete Footings										
Type 1	1	4.0'	3.0'	1.0'		12.0				
" 2	2	5.0'	3.0'	1.0'		30.0				
" 3	2	4.0'	4.0'	1.0'		32.0				
" 4	1	4.5'	4.0'	1.0'		18.0				
" 5	4	4.5'	4.5'	1.0'		81.0				
" 6	1	5.5'	5.0'	1.0'		27.5				
" 7	2	5.5'	5.5'	1.0'		60.5				
						261.0	x 1.0'	261.0		
								÷ 27 cu. ft./yd³	9.7 yd³	

FRANK R. WALKER CO., PUBLISHERS, CHICAGO

Figure 7.31 *Example of Using a Common Dimension*

211

For example, a foundation wall for a building may be uniform in thickness and height except for a few "blockouts" for pipes, vents, or perhaps windows. The calculation of the concrete volume or wall-forming area can be efficiently calculated by first considering the wall without blockouts and then subtracting the blockouts. This process and the means of labeling a deduct item is illustrated in Figure 7.32.

The use of "add" item calculation is less frequent. An example of an instance in which the estimator may use an "add" item in his calculation process is when he is faced with the need to determine an area for a straight surface that has a few protrusions. Rather than again break the wall into sections, the estimator might calculate the wall area independent of the protrusions, and then include the protrusions by means of an add item.

Extending and Footing of Numbers: An essential part of working with the addition and extension of numbers is the "proofing" of calculations. One of the more common mistakes an estimator makes when he is doing mathematics on the estimating forms is the erroneous extension of numbers from one column to the next. Because of this potential it is essential that he *foot* and *cross-foot* all columns of numbers on the various estimates. The term "foot" refers to the adding of the individual columns and showing the result at the bottom of the page. Usually the symbol "f" is placed alongside the individual totals. This is illustrated in Figure 7.33. The accuracy of the extensions is checked by footing the last column (i.e., the one that is the result of the extending of the individual entries). If this last column sums to the total of the "footed" columns shown in Figure 7.33, the estimator has in effect proofed the mathematics and the extensions. He signifies that the mathematics and extensions have been checked by placing the symbol "cf" (the symbol for "cross-footed") alongside the summed right-hand column. This is also illustrated in Figure 7.33. Footing and cross-footing, while adding time to the take-off process, is essential to making sure the calculations and estimate are accurate.

Similar to the need to foot and crossfoot numbers as part of the estimating process is the *subtotaling of each page* of an estimate when a calculation process extends beyond a single page. For example, let us assume that calculations for concrete footings extend several pages. If each page is footed and in effect subtotaled, the possibility of a math error diminishes, compared to a process whereby the estimator sums all the entries on the several pages via one calculation process. This process often also aids the efficiency of the estimating process because, if the estimator does make an erroneous entry in totaling numbers, he may only have to recalculate one page of numbers rather than *all* of the numbers.

Some of the details regarding the use of mathematics discussed in this section may not be appropriate to each and every estimate. Also, the items or procedures discussed are not meant to be all-inclusive. However, the fact remains that the estimator should be able to use mathematics correctly and

PRACTICAL	QUANTITY SHEET									
FORM H6										

PROJECT *EXAMPLE USING DEDUCTS* ESTIMATOR _____ ESTIMATE NO *EXAMPLE*

LOCATION *ANYWHERE* EXTENSIONS _____ SHEET NO _____

ARCHITECT ENGINEER *JJA* CHECKED _____ DATE *9-12-X0*

CLASSIFICATION *EXAMPE*

DESCRIPTION	NO.	DIMENSIONS						ft^3	$\div 27 \frac{ft3}{yd3}$	ESTIMATED QUANTITY	UNIT
		Length	Height	Depth							
03301											
CONC.-FDN WALL											
NORTHEAST WALL	1	52.0'	8.0'	1.5'				624.0			
SOUTHEAST WALL	1	34.0'	8.0'	1.0'				272.0			
SOUTHWEST WALL	1	54.0'	8.0'	1.5'				648.0			
NORTHWEST WALL	1	36.0'	8.0'	1.0'				288.0			
MINUS								1832.0			
Deduct (Windows)	4	5.0'	4.0'	1.5'			-	120.0			
Deduct (Doors)	2	4.0'	7.0'	1.0'			-	56.0			
Deduct (Pipe Areas)	9	1.0'	1.0'	1.0			-	9.0			
								1647.0 $\div 27$		61	cu yd

Figure 7.32 Use of a Deduct Item

JOB TITLE *EXAMPLE Project - Footing Numbers*   NAME   *JJA*

Date *9-12-X0*   Approved by   *SLL*

			QUAN.	MATERIAL	INST.	UNIT COST	TOTAL
Concrete - Footings			215.0 yd	10750.00	7525.00	85.00	18,275.00
Concrete - Walls			1400.	6720.00	3920.00	76.00	10,640.00
Concrete - Slab			3150	16380.00	11500.00	88.51	27,880.00
Concrete - Beams			400.0	18000.00	14300.00	80.75	32,300.00
Concrete - Columns			45.0	2340.00	1850.00	93.11	4,190.00
Concrete - Piers			120.0	6240.00	3840.00	84.00	10,080.00
Masonry - Brick			2200.br	780.00	1460.00	1.02	2,240.00
Masonry - Block			2400.bl	1050.00	1280.00	0.97	2,330.00
Structural Framing			25.0 ton	7500.00	11400.00	756.00	18,900.00
				69760.00	57075.00		126,835.00

Figure 7.33   *Footing Number as a Check*

efficiently in order to minimize estimating time and maximize his potential to make accurate calculations.

EXERCISE 7.1

When taking off the quantity of work for a proposed project, the estimator must evaluate the need for accuracy versus the time required to take off the work quantities. At some point in time, the marginal increase in accuracy is not justified by the added time required to take off the quantities in more detail. Assume a contractor has kept records regarding the amount of take-off time required to prepare a $500,000 project estimate and has also kept records regarding the expected range of accuracy of the prepared estimate relative to the actual cost of the project. The history data is as follows:

Estimating Time Required (in hours)	60	70	80	90	100	120	140	250	300
Range of Accuracy (+)(−)	40%	30%	20%	10%	8%	7%	6%	5.5%	5%

Assume that an hour of estimating time results in a contractor incurring a cost of $25. Given the history data illustrated and the estimator's hourly rate, what range of estimating accuracy would you recommend for preparing a $500,000 project estimate (i.e., estimating hours that should be expended)?

EXERCISE 7.2

A contractor expends a significant amount of time taking off quantities of work from the project drawings. He often expends in excess of 100 hours to take off quantities for a $1,000,000 project. Given the fact that a multiple number of contractors may be invited to bid a project, and given the fact that several subcontractors in turn prepare estimates for bids, a significant amount of take-off time and cost is expended.

Two alternatives to the general contractor taking off the project quantities would be the following:

1. The designer could take off the work quantities and stipulate these on the drawings which in turn would be distributed to all contractors bidding the work. The contractors would merely price the work.
2. The contractor could engage a consultant to take off the work quantities. The contractor would then price the taken-off quantities.

Discuss these two alternatives relative to their cost effectiveness compared to the process in which the general contractor takes the quantities off; discuss the potential accuracy of each of these two alternatives; and discuss the impact the use of either of these alternatives would have on the structure of the construction industry.

Quantity Surveying: Fundamentals for Taking Off Quantities

EXERCISE 7.3

The amount of estimating take-off time an estimator can expend is limited by the fact that the time results in costs; moreover, the estimator is usually under a time constraint to complete the estimate.

The amount of time required to take off any one type of work relative to the amount of time required to take off another type of work should be based on the unit cost of the two different types of work. Assume that an estimator needs to take off 3 different types of work: items X0, X1, and X2. Assume that the estimator has determined that the amount of time required to take off a unit of work of each item and the unit cost of each item to be as follows:

Work Item	Unit Cost	Estimating Time	Accuracy ±
X0	2.80	0.5 hrs.	10%
X0	2.80	1.0 hrs.	4%
X0	2.80	2.0 hrs.	2%
X1	36.00	3.0 hrs.	8%
X1	36.00	4.0 hrs.	6%
X1	36.00	5.0 hrs.	5%
X2	120.00	6.0 hrs.	8%
X2	120.00	7.0 hrs.	7%
X2	120.00	7.5 hrs.	3%

Assuming that the estimator is allotted a total of 11 hours of time, how many hours should the estimator spend on each of the 3 work items to minimize the potential for estimating dollar errors? Assume 100 units of each item are estimated.

8

How To Determine
Direct Costs for the Project

Types of Direct Costs * *Establishing Direct Costs* *
Accumulation of Direct Costs * *Materials Costs* *
Labor Costs * *Equipment Estimating Systems* *
Collecting and Processing Job-Site Data * *Factors That
Distort Historical Data* * *Correlation: A Means of
Forecasting*

TYPES OF DIRECT COSTS

A detailed contractor estimate includes 3 types of components: direct costs, overhead costs, and desired profit. Usually the largest cost component and the one that is most subject to uncertainty and risk involves the direct costs for the project. As noted in Chapter 3, a direct cost can be considered as a cost that is easily traceable to a defined cost object. In regard to the preparation of a detailed contractor estimate, the job-site labor cost, and temporary (e.g., concrete forms) or put-in-place material are obvious direct costs to a project. Equipment costs at the job site, while usually not directly traceable to phases of the job, can be traced to the job itself. Therefore, equipment costs should also be considered part of the direct costs of the project.

In addition to direct labor, material, and equipment costs, other costs—including bond costs, job trailors, supervision, and finance costs—can be easily traceable to a project. These types of direct project costs are often referred to as *job overhead costs*. The determination of direct labor, material, and equipment costs are discussed in this chapter. Job overhead costs are discussed in Chapter 16.

ESTABLISHING DIRECT COSTS

Unfortunately, the prediction of future direct costs incurred to build a project will never be a science. As long as the construction project is subject to uncertainties, the detailed contractor estimate will be subject to risk. It will never be possible to determine a "magic" formula to estimate future costs with certainty. The contractor should be willing to accept this fact as a characteristic of his project.

While risk is a necessary part of the estimated direct cost, the contractor strives for a means of minimizing the risk. His ability to minimize his risk is subject to the amount of feasible time he will need to prepare the estimate, as well as to uncertain factors that will become evident during the construction of the project. In other words, the contractor cannot expend unlimited time in preparing his estimate. Too much estimating time results in a job or company overhead cost that is excessive, and thus can result in the contractor being non-competitive.

There are essentially two complementary means of determining the direct costs for a defined amount of work to be performed for a construction project. For one, the contractor can and should utilize cost data collected from past projects to estimate costs for a proposed project. The science of mathematics indicates that the more data that is collected regarding the performance of a specific event, the more representative and accurate the data is in predicting the result of a future event.

While historical data significantly aids the contractor in estimating future direct costs, the uniqueness of each new work activity and the conditions in which it is performed cause variations in predicted results. A variable environment including weather considerations, different degrees of supervision, a varying labor morale at the job site, and other factors are some of the conditions that are somewhat unique to each particular construction situation. While not totally a scientific process, there is only one way of predicting these variable conditions, and that is **forecasting**. The ability to forecast accurately is somewhat dependent on the contractor's experience—his familiarity with the construction industry and with the impact of each possible variable. However, forecasting is also dependent to a degree on "educated guesswork." It is this element of forecasting that cannot be taught or quantified.

A contractor can estimate direct costs for a project using only *one* of the two means of estimating direct costs—i.e., he can use historical data independent of forecasting, or vice versa. However, it is the contractor who combines these methods that has the best potential for an accurate, minimum-risk estimate. Perhaps the best approach is for the contractor to utilize his historical data for establishing a direct cost, and then to modify this predicted cost for factors or conditions he *expects* to occur (forecasts) on the project being estimated. In this way, historical costs become the basis for the estimate, and the contractor's experience and forecasting ability are used to modify the somewhat mechanical process of using only past data.

ACCUMULATION OF DIRECT COSTS

Bookkeeping is a necessary ingredient in a construction firm's direct-cost estimating and accounting system. Bookkeeping entails the procedures, forms, and distribution channels for the collection and formulation of production cost data. The bookkeeping system is aimed at providing the firm a means of project cost control and a means of determining project costs. In addition, the documentation and classification of costs by means of a bookkeeping procedure are necessary in order to produce reports that are required by the multiplicity of tax requirements that have been placed on all forms of business.

Bookkeeping is a rather cumbersome process. It requires many hours of clerical work in addition to the time and cooperation of those initiating the production cost reports (i.e., timekeeper, supervisor, and so forth). As such, bookkeeping is often a significant business cost in itself. However, this cost is necessary in order to prohibit much greater costs owing to a lack of control.

The bookkeeping process in the construction industry is made difficult by the fact that project costs are incurred at a distance from the location at which the accounting for them is done (i.e., the home office). Normally, the construction project is not large enough in regard to cost or long enough in regard to duration to justify setting up an entire bookkeeping personnel team at the project location. This is not to say that it has not been done occasionally. In the case of large power plants or similar projects, the construction firm may in fact set up a "home office–branch office" at the project site. However, it is much more common to periodically send the required bookkeeping cost data from the field office to the home office. This distance between the location of the occurrence of costs and the location of formulating them creates several problems in regard to control. In addition to making necessary more forms and procedures, the distance creates problems concerning the time at which the cost data is received, formulated, and analyzed for control purposes. The following discussions will assume that bookkeeping data are sent from the field to the home office.

MATERIALS COSTS

The bookkeeping for construction materials has to recognize some of the unique characteristics of the industry. Most construction material is delivered directly to the project site. In other types of manufacturing industries, material is first received at a central receiving location and then is distributed by means of a purchasing department. This latter process facilitates better control of the materials costs.

Material buying for a construction project is normally planned at the time of estimating a project's cost. This often differs from other manufacturing industries in which material buying is planned and carried out throughout the manufacturing process.

Regardless of when construction material is purchased or to where it is initially delivered, the objectives of bookkeeping for material are the following:

1. Control of purchasing.
2. Control of receiving.
3. Identification of materials to projects and work items.
4. Protection against theft, misplacement, and damage.

Purchasing

Construction materials are normally purchased through one of two procedures: they may be purchased as needed by means of purchase orders; or they may be purchased for the entire lump-sum contract. Purchasing material as needed (i.e., as indicated by individual purchase orders) is advantageous because it eliminates a great number of problems associated with theft, misplacement, and damage.

If the material is to be secured by purchase orders, the initial request comes from the field. Only authorized personnel should be permitted to initiate purchase orders. More than likely, the project superintendent will be responsible for initiating such requests. Purchase requisition forms should be used for initiating this type of material request. Such a form is shown in Figure 8.1. A good system is to have triplicate forms, so that a copy of the purchase requisition can be kept by the field office and a copy each delivered to the accounting and estimating or planning departments. This will provide the accounting department a means of controlling purchase requests against invoices and receiving statements. In addition, a form will alert the estimating or planning department as to the progress of the project and material usage versus the predetermined project plan.

Let us assume that the material requested in the field is not kept in inventory by the firm. In this case, the home office will send a purchase order to the material supplier who has been selected. The material supplier will be requested to deliver the material to the job site. He might send the invoice for material ordered to either the home (central office), the field office, or both. Because the field office is removed from the actual purchase or paying of the purchase price, the invoice must eventually be received by the central office—thus it is preferable to have the invoice sent directly there.

PURCHASE ORDER

TO

PLEASE DELIVER THE FOLLOWING ORDER TO:

SHIP TO

SHIP VIA

ORDER
NUMBER

DATE

JOB

JOB NO.

F.O.B.

TERMS

DELIVERY TO BE MADE ON OR BEFORE   OR RIGHT IS RESERVED TO CANCEL ORDER

In Accepting Verbal Orders, Purchase Order Number Must be Obtained Before Making Delivery.

QUANTITY	DESCRIPTION	PRICE	AMOUNT

INVOICES Must State Order Number and Point of Delivery

PRICES On This Order Not Subject to Change

BY _____

Figure 8.1 *Example Purchase Order*

221

Receiving

The next step in the process is the potential bad link in the control process. Because the material is sent directly to the field, it is the field personnel who are responsible for checking to see that the material delivered corresponds to that requested and eventually paid for by the home office. That is, field personnel are responsible for counting or measuring, inspecting, and signing for each incoming delivery. The key to performing this vital step in the control process is the assignment of the duty to dependable individuals who make themselves available when in fact the material is delivered.

On large projects, this usually creates little difficulty. The project is large enough to justify the employment of a warehouseman or receiving clerk who is accountable for incoming materials. On the smaller job, it is common to allocate this responsibility to the timekeeper. On even smaller jobs, the responsibility is sometimes given to the foreman. These latter two practices, especially the last, are generally not as desirable as employing a full-time receiving clerk. The fact that the timekeeper or foreman is responsible for other tasks often results in his viewing the materials-receiving responsibility as somewhat secondary. This can result in a haphazard control procedure. Subcontractors often arrange with the general contractor to have that individual's personnel receive the subcontractor's material. The potential for adequate control is increased by this practice.

Whoever is responsible for acknowledging the receipt of materials will normally be requested by the supplier to sign for the material. Since his signature has legal implications (i.e., title passes), the individual signing should first assure himself as to the quantity and quality of the material. If it is inferior, he should not accept it. His failure to inspect the material in a responsible manner removes his right to later reject the material.

Identification of Materials

The field office, by way of the individual responsible for acknowledging receipt of construction materials, should account for all materials received for the project in a weekly production report, or a separate report such as the one shown in Figure 8.2. Here again, the form should be sent as well to both accounting and the estimating or planning departments. The departments use these reports in a continuing evaluation of the progress of the project.

In addition to documenting direct material for a specific project, it is advantageous for the construction firm to document materials costs related to work items or phases of its projects. Assuming that the contractor has a materials budget for each of the defined work items as part of his estimating function, the ongoing documentation of material expended on work items provides the contractor with a means of monitoring and controlling materials expenditures throughout the project, and establishing material stan-

			Direct Material				Indirect Material				
Date	Req. No.	Material Units	Total $	Job 101	Job 102	Job 103	Total $	Maintenance	Job 101	Job 102	Job 103
5-17-74	101	100	125.00		125.00						
5-21-74	108	215	135.00	135.00							
5-22-74	109	106					35.00	35.00			
5-30-74	112	2	193.00		193.00						

ABC COMPANY
Stores Requisition Analysis   **Week ended**

Figure 8.2 *Example Stores Requisition Form*

dards and expected material wastage factors for estimating the material costs on future projects.

The calculation of material put in place on a work-item basis relative to the previously prepared estimate/budget provides the contractor a means of determining the percentage of completion of a project as it progresses. An example form for this purpose is illustrated in Figure 8.3. Once a project is completed, the material put in place for a work item relevant to the budget can be analyzed and summarized for future estimating. A sample form for this purpose is shown in Figure 8.4. In effect, it is the documentation of material used on previous projects that becomes the basis of estimating future projects.

If it is feasible for the firm to maintain a central supply of materials that are needed for building its various projects, such a system offers better control over purchasing and receiving materials. In addition, through centralization and bulk purchasing the firm may be able to purchase materials more economically.

The bookkeeping and control process differs when a centralized inventory of materials is kept by the construction firm. Purchase requisitions are still initiated in the field; these are sometimes referred to as *stores requisitions*. If an adequate supply of the material is not on hand, the central office may initiate a purchase order to a supplier. However, not every stores requisition necessarily initiates a purchase order. Control of material ordered, received, and sent to projects is now centralized. That is done by individual material accounts referred to as *stores ledgers*. Each type of material is accounted for in subsidiary ledgers referred to as *stores cards* which are in turn summarized in the control stores ledger. Purchases are entered into a company register and when received are entered in the stores card for the material. Purchases are listed with their unit price, and by means of one of several available inventory flow methods, cost levels of inventory are maintained and adjusted for materials issues.

The issues of materials as reported by the stores requisition forms are periodically analyzed by the home office as a way of controlling the usage of materials on the various projects. This analysis serves as a means of evaluating project progress.

Protection of Materials

The fourth objective of bookkeeping for material cost—that of protecting against theft, misplacement, or damage—may seem removed from the cost accounting process. However, the analysis of purchases, deliveries, and issues can be used to alert the firm to any loss of material. Without an effective cost accounting and bookkeeping system, such losses may never be apparent to the firm and they will continue to be accounted for as ordinary project expenses. Determination of acceptable material wastage factors is part of an effective cost accounting system. In addition, protection of material through the setting out of authorization and security procedures are important elements of the cost accounting and bookkeeping system. If it is worthwhile to keep an inventory of construction materials, it is worthwhile to maintain adequate protection and control over it. Usually, it is financially rewarding to employ a specific individual on a project site to receive, protect, and to do general housekeeping for project materials. This is especially true when the project is large enough to dictate a large buildup of construction materials on the project site.

LABOR COSTS

Whereas the lack of proper documentation and poor control of construction materials seldom leads to overall lack of profit for the project or to serious difficulty in estimating material quantities, such practices in regard to *labor* productivity and costs are among the prime causes for project and company financial loss. Labor is the most variable of all construction costs.

MATERIAL LEDGER							
Material							
Date	√	Purchase Quantity	Purchase Price	Quantity Removed	Quantity Returned	Balance Quantity	Balance $

Figure 8.3 *Analysis of Material Put in Place*

CONTRACTOR NAME
PROJECT
DATE

CONTRACT $ AMOUNT
PROJECT MANAGER

Work Item	Planned Quantity	Actual Quantity	Planned Quantity	Actual Quantity	Planned Quantity	Actual Quantity	Planned Quantity	Actual Quantity	Planned Quantity	Actual Quantity
Resources Required to be at Project										
Expected $ Payout										
Working Day										
Calendar Day										

Figure 8.4 *Analysis of Material from Use on Projects*

Construction project labor bookkeeping is commonly referred to as **time-keeping**. The actual timekeeping process has two objectives. First, it determines the total amount of time for which each worker is to be paid. Thus, it is part of the firm's payroll task. Second, and of more importance in regard to control and cost accounting is the contribution of timekeeping to efficient allocation of labor hours and costs to the various project work items. It is also the timekeeping function that serves to collect labor productivity data that are subsequently available for estimating the labor cost of a proposed project.

Monitoring Work Hours

The actual timekeeping on a construction project is normally done by a project foreman or a timekeeper. The employment of a timekeeper is usually not justified on small construction projects. Instead, the foreman is given the responsibility of daily documentation of worker hours for both the number of hours and the allocation of the hours to the performed work items. He does this by recording the hours in a small book kept in his pocket or by filling out a *timecard* (see Figure 8.5) at the end of the day. Because such timecards are stored at the field office there is little chance that they will be lost, so this procedure is preferred.

If a project is large enough to justify the employment of a timekeeper (who can also be responsible for receiving and protection of project materials), there is better potential for control than if a foreman is assigned the timekeeping responsibility. Since this function is the timekeeper's primary job, this person generally has more time and is more committed to the timekeeping task than are foremen. In addition, less potential exists for collusion between the timekeeper and the workers than between a foreman and his workers.

When a project is spread out over a relatively large area, the timekeeper may have difficulty documenting late starting or early quitting of individual workers. In addition, there may be so many workers on the project site that he may in fact fail to observe and document the absence of a given worker. Such a failure will lead to documenting and payment of wages for which no work is performed. There are several well-known procedures for documenting the starting and finishing time of workers. Probably the most frequently used timekeeping device used in the manufacturing system is a *time clock*. Workers insert a card into a punch-type device which records the time of the day on the card. Time clocks, however, have not proved popular in the construction industry. The mobility of the construction project, workers' general dislike of them, and the difficulty associated with monitoring such a system are some of the reasons for the nonpopularity of time clocks.

An alternative method is the use of a *badge procedure*. Small metal badges with numbers on them are kept at a central location maintained by the project timekeeper. Upon arriving for work, a worker picks up his badge

DAILY TIMECARD						
Date				Preparer		
Worker Name	Job No.	Work Item	Work Item	Work Item	Work Item	Work Item

Figure 8.5 *Example Timecard for Cost System*

with his assigned number and pins it on his work uniform. At the end of the day he returns the badge. Such a procedure ensures documentation of a worker's presence, his starting time, and his quitting time. In addition, it brings attention to long absences of a given worker. The main difficulty of the badge system is that workers have to report to a central location to pick up and return their badges. Such a procedure may prove time-consuming when workers are spread out over a project.

A related alternative is to have workers continuously wear a badge number. This badge is not turned in, but is kept by the worker. The timekeeper, in making his rounds, documents the presence of workers by noting their number. Such a procedure is especially advantageous when many workers are employed and the timekeeper does not know each one by face. This procedure, while useful for documenting the presence of a worker, does not help record whether a worker starts or quits on time.

Regardless of whether a foreman or a timekeeper is responsible for documenting labor time, occasionally someone from the home office should visit the construction project to check the foreman's or timekeeper's reports. This provides a means of control in regard to potential collusion between a worker and the foreman or timekeeper.

No matter who makes out daily timecards, they have to be sent to the home office. Preferably, they should be sent in daily or, at the minimum, weekly. The weekly reports lack the control potential that daily reports offer. The sooner labor inefficiencies or overruns are recognized, the better the chance for corrective management. In addition, weekly reports encourage fraudulent reporting procedures. A worker might be laid off on Tuesday and carried on the payroll through Friday due to fraudulent timekeeping.

Allocation of Hours to Work Items

The daily or weekly labor cost reports that are turned in by the foreman or timekeeper should indicate both total worker hours and the allocation of the hours to project work items. However, the project foreman cannot be expected to document a very detailed analysis of his workers' time. That is, it is awkward to require him to account for coffee breaks, inefficient production, etc. His effectiveness as a foreman might be weakened if he is asked to be a watchdog. This is yet another reason for justifying the employment of a timekeeper on a project.

To effectively monitor a project's direct labor cost and to collect and structure labor productivity data for estimating the cost of future projects, it is necessary for the contractor to collect labor-hour time on a work-item basis. For example, the labor time of a carpenter on a given day might include work on concrete wall forming, beam forming, and placement of concrete. The timecard shown in Figure 8.5 accommodates the recording of direct labor time on a work-item basis.

The daily or weekly cost reports that are sent to the home office are used by the accounting, payroll, and estimating or planning department. The payroll department uses the reports to initiate worker paychecks. The individual in charge of this payroll function is referred to as the *paymaster*. The accounting and estimating/planning departments record the labor costs and analyze them in terms of individual project work items. This analysis, an important part of the cost accounting system, serves to bring attention to cost overruns, work improvement, and potentially costly delays. This can be done by comparing labor hours expended per unit of work-item production performed with planned labor hours per unit of work item. Productivity trends can be singled out and analyzed. More and more firms are starting to use documented work-item labor costs as a means of estimating future projects. Several of these firms are transforming the work-item labor costs into manhours for their future estimating. These manhour records are less sensitive to changes in labor rates in regard to future estimates.

Once the timecards, such as the one shown in Figure 8.5, are processed, it is possible to process the labor hours recorded to a *project job-cost ledger*. An example of this type of form is shown in Figure 8.6. The job-cost ledger form enables the contractor to identify potential direct labor cost overruns during the process of a project. This is done by comparing the percentage of work put in place for a specific time period to the percentage of labor hours expended at the same point in time.

The timecard data that are subsequently summarized on the job-cost ledger can eventually serve another purpose—that of developing standards for estimating a future project. The direct labor hours and work performed for a completed project are merely summarized from the completed job-cost ledger to yield a direct labor productivity record. As jobs are completed, this data is accumulated to yield work-item direct-labor productivity that potentially improves in accuracy as more data are collected. The productivity data for future estimating might be collected on a form similar to that shown in Figure 8.7. These data will complement the forecasting function and determining future direct-labor costs.

In addition to the timekeeping aspect of accounting for labor, other related bookkeeping procedures include forms and authorization procedures for hiring and firing employees, handling of payroll fringe benefits, and payroll deductions. These procedures have a less significant role in the control and estimating process and vary substantially from firm to firm.

EQUIPMENT ESTIMATING SYSTEMS

Depending on the type of construction project, equipment costs can comprise a significant portion of the total project cost. Equipment costs are especially important to heavy and highway types of projects.

Equipment costs incurred at a project site are direct project costs. However, too often a contractor uses an application process to estimate the equipment-cost component of his estimate. For example, a contractor may determine the labor and material direct cost for a project and subsequently multiply this cost by a percentage to determine the estimated equipment cost. This application process may lead to an under- or overestimate of the actual project equipment cost.

JOB COST LEDGER

Job Name_____

Job Code _____

Job Address _____

Job Description_____

Contact Individual_____

Contract Price $_____

Progress Billings and Receipts			
Date	Amount	Date	Amount

	Labor Hours																		Total	MAT	EQ	Total
Date	GSU	GCU	EXC	EGR	ECR	CFU	CPF	CPP	CGF	CGP	CGR	BPP	CVR	CVP				Hrs	$	$	$	$
Est Quan																						
Est Hour																						
Tot																						

Figure 8.6 *Example Job Cost Ledger*

WORK ITEM FILE

Page ____________ Units____________ Work Item Name _______________ Code_______

Job	Comp.	Est. Matl.	Total Hrs.	Total Prod.	Prod Cprd	✓ Ent	Cum. Matl	Total Chrs	Total Cprd

Figure 8.7 *History Data from Past Projects*

An alternative to treating project equipment costs as an application is to implement an *equipment estimating and control system.* The objective of such a system is to trace equipment expenditures to individual pieces of equipment and to enable the contractor to treat equipment expenditures as direct costs to projects. Ultimately, the system should provide the contractor a means of evaluating the profitability of owning an individual piece of equipment and also enable the contractor to better determine the profitability of individual projects. The purpose of this section is to illustrate the basics of developing an equipment system that serves both an estimating and control objective. There are 6 steps in this process.

Establishing Hourly Rates per Piece

Step 1 in an equipment estimating/control system is to establish an hourly *rate* per piece of equipment to be included in the system. Not every piece of equipment should be included. As a guideline, only equipment that is considered to be manageable in regard to time or cost by field personnel should be included. Small or minor pieces of equipment like small saws, compressors, and so forth should likely be handled on a project application process. These types of equipment are often budgeted for an entire project and the equipment is left at the project for its entire duration.

Once the equipment to be included in the system is identified, an hourly rate for each piece of equipment must be established. This calculation is illustrated in Figure 8.8.

The initial hourly rate used in the system for a piece of equipment is subject to uncertainty. Lacking actual usage hours and maintenance and repair costs, the initial rate has to be estimated based on only a subjective evaluation. However, as data are collected regarding actual use and expenditures, the intent of the system is to *determine an accurate hourly equipment rate based on past data.*

The calculation of the hourly rate for any piece of equipment must recognize the depreciation of the equipment and its expected annual maintenance and repair. The depreciation charge for a year may be calculated by dividing

---

1. Establish hourly rate per piece of equipment

$$\text{Hour rate} = \frac{\text{Depreciation} + \text{Maintenance \& Repair}}{\text{Annual Estimated Hours}}$$

Example:
Cat 977 K Loader, Account 023
$$\text{Rate} = \frac{3555 + 2745}{900} = \frac{6300}{900} = \$7/\text{hr}$$

Figure 8.8 *Determination of Hourly Rate for Equipment*

the initial cost of the equipment by its expected life, whether or not the equipment is fully depreciated. Independent of the calculation of an hourly rate that recognizes depreciation and expected maintenance and repair, an hourly rate *not* related to the calculation can be used. For example, while a rate based on depreciation and maintenance and repair costs may indicate $20 per hour for a piece of equipment, the company may override this with a $30 hourly rate to reflect market conditions.

Establishing Hours of Use per Piece

Step 2 in developing an equipment estimating and control system concerns estimation of the equipment hours of use by individual piece of equipment for a project. In other words, the contractor estimates usage of a piece of equipment in the same way that he estimates the direct labor hours for a project. This step is illustrated in Figure 8.9.

---

2.   New Project estimated — Job 124

Estimated equipment hours for Cat 977 K for Job 124 = 200
Estimated equipment budget for Cat 977 K for Job 124 = $1,400

Figure 8.9   *Estimating Equipment Cost for a Project*

In establishing required equipment hours, there is a question of whether the contractor should base the system on productive hours of usage or hours during which the equipment is at a project site. Given the fact that equipment may be idle for a significant amount of time when it is at a job site, the difference between productive and total equipment hours at a project can be significant.

The calculation of the estimated hours illustrated in Figure 8.9 assumes *productive* hours. The use of productive hours, and the calculation of an hourly rate in Step 1 based on productive hours will likely give the contractors more control potential than a system built on total hours.

Accounting for Use of Equipment

Let us now assume that the contractor initiates a project. The accounting for the use of equipment at the project is illustrated in Figure 8.10. In effect, the accounting entries made in Figure 8.10 have the effect of charging the project for the use of the equipment and in effect recognizing revenue for the equipment account. One might think of this process as one of renting the equipment from the general ledger account to the project.

3. Job starts. Productive hours for Cat 977 K input. Assume 180 hours in total. Entry made as follows. *Note: (180 × 7 = $1,260)

Equipment 023 (Expense)		Job 124 (Expense)	
D	C	D	C
	1,260		1,260

Figure 8.10 *Accounting for Use of Equipment*

Recording Actual Cost per Piece

Step 4 in the process, illustrated in Figure 8.11, recognizes the actual expenditures for a piece of equipment as an equipment account expense and not a specific project cost. The theory behind this process is that an equipment expenditure, such as for the repair of tires, is one that will benefit future projects as well as the project on which the actual maintenance was performed. Equally important, the project manager or superintendent can usually control the use of the equipment, whereas he likely cannot control the actual maintenance or repair expenditure. The system is consistent with the principle that a manager should only be accountable for expenditures he can control.

4. Maintenance and repair cost for Cat 977 K incurred during year equal $800 in total. Individual expenses recognized as occur. In total entry would be as follows.

Cash		Equipment 023 (Expense)	
D	C	D	C
	800	800	

Figure 8.11 *Accounting for Equipment Expenditures*

Assignment of Total Equipment Expenses

In Step 5, all equipment expenses should be assigned to the equipment account; this includes periodic recognition of depreciation entries such as shown in Figure 8.12.

5. End of year (or quarterly) depreciation entry made for each piece of equipment. For example, for Cat 977 K, depreciation entry would be made as follows:

Equipment Asset Acct.		Equipment 023 (Expense)	
D	C	D	C
	3,555	3,555	

Figure 8.12   *Depreciation Entry for Equipment*

Analyzing Equipment Accounts

Step 6 involves analysis of the accuracy of equipment rates and the profitability of each piece of equipment. This should be done at periodic points in time, perhaps quarterly, or at least annually. This step is illustrated in Figure 8.13. This analysis may result in the contractor deciding to rent a specific piece of equipment rather than buying it, or adjusting the hourly rate for a specific piece of equipment to more properly reflect its ownership cost and accurately determine project costs.

The process illustrated in Steps 1 through 6 is an ongoing process. As the process continues, the system is refined to yield a more scientific means of estimating and controlling equipment costs.

6. Year end summary entries made. Prior to adjustment equipment and job expense accounts would be as follows:

Equipment 023 (Expense)		Job 124 (Expense)	
D	C	D	C
800	1,260	1,260	
3,555	2,800	2,800	
4,355	4,060		
295			

Jobs other than 124

The net equipment 23 debit balance of $295 would be adjusted to zero by crediting equipment 023 in amount of $295 and debiting Underapplied Equipment Job Expense $295. In the subsequent year, the budgeted hourly rate for the Cat 977 K would be adjusted upwards.

Figure 8.13   *Evaluating Equipment Profit or Loss*

COLLECTING AND PROCESSING JOB SITE DATA

A job-cost system is only as good as the data that are entered into the system. Most, if not all, of the data that are needed to prepare a job-cost control report is initiated at the job site. This includes timecards, daily reports, and progress reports. All too often we hear complaints from the owner of the construction firm that he cannot get good data from the job site. For example, data on collected timecards may be erroneous or be misclassified. This is somewhat commonplace in the construction industry. If the contractor believes it is impossible to obtain accurate data for a job-cost control system from the job site, he is accepting the conclusion that he cannot implement an effective job-cost control system.

Perhaps the potential for inaccurate job-site data stems from the project decentralization—in other words, the request for job-cost reporting is decentralized. Independent of whether the collected data are processed at the individual job sites or is sent to the contractor's home office for processing, the source of the initial data is the field personnel, including a foreman, superintendent, or perhaps on larger projects, a project timekeeper.

Data Collection by Field Personnel

It is not realistic to believe that the potential difficulties of collecting accurate job-site data can be alleviated by merely substituting accounting personnel for production personnel. Most contractors cannot afford this luxury. A more realistic approach is to attempt to get field or job-site personnel to be more attentive to the collection and use of accurate job-site data.

One might conclude that personnel at a job site are averse to the collection of accounting data. In reality, some contractors have found that employees at a job site, if approached correctly, may actually enjoy and take pride in being part of the information collecting and reporting associated with effective project control. In fact, involving field personnel in the project reporting may prove an effective way of bridging the apparent gap between the goals of the production personnel and the owners or managers of the contracting firm.

Contractors often complain about their inability to obtain accurate data from field personnel, but in fact the contractor himself can promote inaccurate data through the use of ineffective forms or ineffective practices regarding the use of the forms.

In general, no individual in the reporting process, including a foreman or superintendent, should be asked to fill out any forms unless:

1. He has been instructed on the purpose and importance of the form and accompanying data.
2. He is assured, by means of example, that the data are in fact used.
3. He is given some feedback via a simple but meaningful report.

All too often one or more of these steps is missing in a contractor's job reporting process. For example, a foreman may be asked to fill out a daily report without being told what purpose the form serves; he may never be given any feedback from the data input; and consequently the foreman may get the idea that the data he provides isn't used anyway. Given these conditions, it is not surprising if the contractor cannot obtain accurate data.

Effective Data Collection Forms

Even assuming that the contractor does fulfill the 3 steps listed above, accurate data are still not assured. The very design of the forms that are part of the information reporting system affects the ability to collect accurate data.

| NAME _____ | | | | | | | |
| CRAFT _____ WEEK ENDING _____ | | | | | | | |
WORK ITEM		WED	THR	FRI	SAT	MON	TUE	TOTAL
	ST							
	ST							
	ST							
	ST							
	ST							
	ST							
	ST							
TOTAL								

Figure 8.14 *Example of a Timecard That May Prove Non-Functional*

Forms should be simple to use, non-redundant in regard to the data requested and processed, and designed to be compatible with the characteristics of the construction industry.

Consider the timecard shown in Figure 8.14. On the surface the timecard appears functional. However, upon further analysis we can detect the fact that it is a weekly timecard and requires a breakout of daily worker time to individual work items or phases. Given the fact that workers and a foreman are busy performing construction work, in all likelihood the timecard will be filled out weekly instead of daily. Thus labor-hour distribution for individual work items will probably be inaccurate because a worker or foreman will not be able to remember exactly what a worker did two or three days ago. A daily timecard would enhance the possibility for collecting accurate job-site data.

Attention to small details such as this can make the difference between accurate and inaccurate input data for a job-cost system. One must remember that if inaccurate data are fed into a job-cost system, the resulting reports will be meaningless.

The contractor should also be attentive to the need to use forms and procedures that minimize the time needed to process data. Data should not be collected if they are not used. Similarly, whenever possible, the same data should be processed for multiple purposes whenever possible. For example, labor time cards should be processed to serve both the payroll function and job-cost reporting simultaneously to the degree possible. There is no substitute for getting good job-cost data in the field in regard to the potential to control a project.

FACTORS THAT DISTORT HISTORICAL DATA

Estimating direct costs for a proposed project from past data will not in itself always result in a totally accurate prediction of the costs. To some degree each project and the conditions that surround each work process that makes up the total project are unique. Thus direct costs for a project will never be completely predictable. Moreover, variables such as weather conditions, varying degrees of supervision, and worker morale can result in inconsistent labor productivity and direct costs even if a work process is "identical" to a previously performed process (bearing in mind that in reality, two work processes are seldom *completely* identical).

Given the uniqueness of working conditions and construction methods in the construction industry, the preparation of an estimate of the direct cost for a project is partially dependent on the estimator's *forecasting ability*.

Some individuals have a knack for accurate forecasting, while others do not. One individual may be very good at forecasting the weather, another may be lucky in horse races, some may "sense" economic indicators—all independent of specific knowledge or previous training regarding the matter or event being forecast.

While some might argue that an individual's ability to forecast is a non-teachable trait or based on luck, there is little question that an individual's ability to forecast an event is also dependent on the experience he has had related to the event and the factors that weigh on the occurrence or non-occurrence of the event. Assuming equal intelligence, a construction estimator with years of experience in estimating and observing construction work processes is likely to be more accurate in forecasting the performance of a future construction work process than a construction estimator who is preparing his first estimate. While there are exceptions to this statement, time and again this has usually proven to be the case.

The forecasting aspect of the preparation of a construction estimate can also be viewed as the one characteristic of the construction industry that will always retain a certain element of risk and uncertainty. Associated with uncertainty and risk, however, is the potential for profits.

If there were no risk or need to forecast, in theory, all contractors could collect past project data and over a long period of time prepare estimates for a project that would be equal or near equal in regard to the estimated cost. In other words, without variable conditions and the need to forecast, contractors would "tie" when submitting their respective estimates to a project owner!

Variables to be Forecast in Estimating Direct Costs

The performance of construction methods is a very complex process. As discussed in Chapter 2, the performance of a construction work process is affected by numerous factors and/or variables. Figure 8.15 illustrates the dependence of the performance (or non-performance) of a work process on labor, equipment, material, capital, environmental, managerial, and external factors. Various characteristics or states of each of these factors has a cause and effect relationship to a construction work process and also on the cost of the work process. This is true in regard to both the quantity of the factor and the status or quality of the factor. For example, the labor productivity for the performance of a construction method, such as the forming of a wall, is dependent on both the number of workers used and the "quality" of the workers, including their morale, degree of required supervision, prior experience, and so forth.

In effect, each of the factors shown in Figure 8.15 is in turn characterized as being independent within a more detailed or minute set of conditions. For example, let us consider the environmental considerations shown in Figure 8.15. Within this broad category are several others that affect the cost of a construction method. Temperature, humidity, wind, noise, and precipitation are all environmental factors that have an impact on the performance of a construction work process or method. The construction estimator is faced with predicting the status of each of these factors along with the

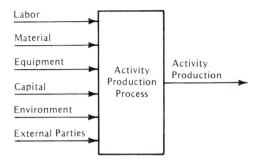

Figure 8.15   *Factors Affecting Productivity*

relationship between their predicted status and the performance and cost of a construction method.

When we recognize that each of the factors shown in Figure 8.15 is in fact dependent on other factors, the list of variable factors that can influence the performance of a construction method is indeed long. Some of these more detailed factors or conditions are illustrated in Figure 8.16. One might argue that the status of each of these factors or conditions along with its impact on productivity must be forecast in order to enhance the accuracy of a

Labor	*Capital*
Worker Morale	Availability of Cash of Contractor
Skills of Workers	Project Owner's Payment Policy
Crew Size	Interruptions in Cash Flow
Management of Crew	Cost of Money
Material	*Environment*
Quality of Material	Temperature
Quantity of Material	Humidity
Type of Material	Rainfall
Material Flow to Project	Working Conditions
	Physical Surroundings
Equipment	*External Parties*
Type of Equipment	Safety Regulations
Age of Equipment	Set Aside Programs
Skill of Operator	Wage Laws
Maintenance Requirements	Equal Opportunity Requirements
Amount of Equipment Available	

Figure 8.16   *Example Factors Affecting Work Performance*

construction estimate. Surely this would be a very tedious and difficult, if not impossible, task.

Simplification of the Forecasting Process

We have seen that there are numerous factors that must be forecast in predicting the performance of a construction method. We also noted that it is unrealistic to believe that an estimator can, even if time would permit, estimate each of these numerous factors. This is true whether or not the estimator has the availability of a computer.

Several attempts have been made to simplify the forecasting problem faced by the estimator. These attempts have usually involved such methods as limiting the number of factors to be forecast, or grouping several of the factors into a broader category and forecasting the status of the grouped factors as a whole. For example, let us consider a construction forecasting process proposed by Dallavia.

Dallavia's Estimating System: Using Dallavia's estimating system, a construction estimate first utilizes published tables that indicate ideal productivity based on past data. One of these tables for a sample work process is illustrated in Figure 8.17. It is assumed that the productivities (i.e., the work performance) published represent work performed under ideal work conditions.

The second step in Dallavia's estimating process recognizes that there are factors or conditions surrounding a work process that will cause its performance to vary. Dallavia somewhat oversimplifies the problem of forecasting numerous factors by itemizing 8 factors that affect the productivity or performance of a work process.

The 8 factors that Dallavia utilizes as a forecasting process to adjust the historical data illustrated in Figure 8.17 are part of what he refers to as a *Production Range Index*. The Production Range Index is shown near the top of the illustration in Figure 8.17. They are shown as percentages from 25 to 100.

The Production Range Index has been designed to facilitate accurate estimates for projects governed by a wide variety of working conditions and local restrictions. Production is arbitrarily classified into 3 basic ranges: low, average, and high. Low production indicates a job efficiency of 25 to 55%; average production, 55 to 85%; and high production, 85 to 100%. These percentages are based on the amount of work a typical shift crew can perform under local conditions. Given ideal circumstances, such a crew might achieve an approximation of 100% efficiency, but for estimating purposes, one should not count on this degree of production unless the operations in a project are highly mechanized. In a manual operation such as bricklaying, for example, the scheduling of high-range production may very well mean that the contractor will have to push his workers every inch of the way.

Production elements	Production efficiency, per cent		
	25 35 45 55 65 75 80 85 90 95 100		
	Low	Average	High
1 *General economy*	*prosperous*	*normal*	*hard times*
local business trend	stimulated	normal	depressed
construction volume	high	normal	low
unemployment	low	normal	high
2 *Amount of work*	*limited*	*average*	*extensive*
design areas	unfavorable	average	favorable
manual operations	limited	average	extensive
mechanized operations	limited	average	extensive
3 *Labor*	*poor*	*average*	*good*
training	poor	average	good
pay	low	average	good
supply	scarce	normal	surplus
4 *Supervision*	*poor*	*average*	*good*
training	poor	average	good
pay	low	average	good
supply	scarce	normal	surplus
5 *Job conditions*	*poor*	*average*	*good*
management	poor	average	good
site and materials	unfavorable	average	favorable
workmanship required	first rate	regular	passable
length of operations	short	average	long
6 *Weather*	*bad*	*fair*	*good*
precipitation	much	some	occasional
cold	bitter	moderate	occasional
heat	oppressive	moderate	occasional
7 *Equipment*	*poor*	*normal*	*good*
applicability	poor	normal	good
condition	poor	fair	good
maintenance, repairs	slow	average	quick
8 *Delay*	*numerous*	*some*	*minimum*
job flexibility	poor	average	good
delivery	slow	normal	prompt
expediting	poor	average	good

Figure 8.17 *Dallavia Model*

Dallavia assumes that in general, crew productivity is dependent on two major factors: (1) the present economy, and (2) the specific local circumstances under which the work is to be accomplished. No matter what his trade may be, the individual construction worker is directly influenced by the general economy. In good times, when jobs are plentiful and labor scarce, his productivity usually decreases and costs rise. In normal times, average productivity and costs become the rule. During depressions or slumps, labor becomes plentiful and more productive, and consequently costs decline. The Production Range Index is designed to reflect all these conditions.

Dallavia assumes that the second factor affecting productivity relates more directly to the job itself and is compounded by the many variables that influence all construction activity, such as the character of the job site, the volume of work to be performed, and the quality of available labor. Each of these variables, called *production elements*, is listed in the Index. The estimator must study them in light of both the existing and foreseeable conditions that will affect the proposed work. At the same time, he must make an ordered evaluation of the construction firm as a productive unit in order to determine its potential performance under a given set of circumstances.

To calculate the overall efficiency of a project, the estimator must develop a percentage value for each of the 8 production elements listed in the Index. The overall efficiency is then obtained by averaging the 8 values. This percentage will be subject to further analysis by the estimator, as will be explained later.

Let us consider the following example. After studying a project on which he is bidding, a contractor makes the following evaluations of the production elements involved:

PRODUCTION ELEMENT	PERCENT EFFICIENCY
1. Present economy	75
2. Amount of work	90
3. Labor	70
4. Supervision	80
5. Job conditions	95
6. Weather	85
7. Methods and equipment	55
8. Delays	75
Total	625

The total of the 8 elements is 625, so the average value will be 625/8, or 78%. This figure represents the overall efficiency rating that the contractor will give his firm for the execution of the proposed project. It will be subject to revision, as has been noted, in the detailed estimates that must be made for the individual operations required by that project. Each construction method must be evaluated in a similar manner and adjusted against the overall rating before the production of the typical shift crew can be determined in actual units of scheduled work. In making such evaluations, those production elements which do not seem directly applicable to a particular operation may be ignored.

Once an efficiency index, for example 78%, is determined, the percentage is multiplied by the historical data work standard to yield a forecast of work performance.

Pages Estimating System: Dallavia is not alone in his attempt to quantify and simplify the forecasting problem faced by the estimator. A similar process has been proposed by Page, who simplifies the process even more than Dallavia. He considers only 5 factors in determining the adjustment or productivity index. While these processes are workable and add structure to the process of forecasting, the construction estimator should recognize that at best they are an approximation of the more challenging process of forecasting a more detailed and numerous set of factors.

CORRELATION: A MEANS OF FORECASTING

In a previous section, attention has been drawn to the fact that an estimator's ability to accurately forecast is dependent in part on his prior experience in observing a work process and the factors that affect that work process. The reason for this is because the estimator somewhat subconsciously correlates past events and factors in forecasting a future occurrence. This correlation takes place even if the individual does not go through the process of formally documenting the past event in writing.

The conscious or subconscious analysis of factors and their impact on the occurrence of an event might be referred to as *correlation*. One "correlates" a cause and effect relationship between an event and the status of factors that have varying effects on the event.

Correlation in effect allows an estimator to somewhat reduce the risk of forecasting. Because any one individual is only able to retain a limited amount of information in his memory, documentation of data provides an improved means of correlating the occurrence of an event as a function of external factors; thus the documentation and correlation of data provides a basis for accurate forecasting.

In order to illustrate the use of correlation as an aid in forecasting the performance of a construction method as a function of external factors, let us assume that a contractor has collected the following information regarding the past performance of a mason work process:

EVENT NO.	MASON OUTPUT (BLOCKS/DAY)	WEATHER F°	CITY	SUPERINTENDENT
1	150	40	Chicago	Joe
2	170	40	Chicago	Ed
3	140	40	St. Louis	Ed
4	180	80	St. Louis	Ed

While the data are limited in quantity, they can be used to illustrate the correlation process. In particular, analysis of the data can yield information about the impact of weather, location, and supervision on the amount of work performed by a mason for a specified time period.

The use of correlation is based on the study of an event as a function of external factors. For example, if one can focus on the occurrence of an event when all other factors, other than a single varying factor, are constant or in the same state, it is possible to measure the cause and effect of the varying factor for the event being analyzed.

Let us focus on the performance of the mason process noted above as a function of the weather. Inspection of Events 3 and 4 indicate that the location and superintendent for these two occurrences of the work process were identical, except for the temperature. The mason placed 140 blocks on the day the temperature was 40°, whereas he placed 180 blocks when the temperature was 80°. It appears that the mason places approximately 28.5% more blocks when it is 80° than when it is 40°. One might further assume that the mason would place approximately 160 blocks per day when it is 60°. Such a conclusion is obtained by assuming a linear relationship between the output and the weather. Without obtaining more data, especially, say, for a day in which the temperature is 60° and the work is performed in St. Louis and Ed is the superintendent, such a conclusion might be in error. However, the point here is that by isolating the event being analyzed—i.e., the mason process—as a function only of the weather or any other factor, it is possible to correlate a cause and effect relationship between the event and an external event.

Analysis of the data regarding the mason process also indicates that only the superintendent varied during the performance of the work process for Events 1 and 2. It would appear that when the work process is supervised by

Ed, the productivity is 13.3% higher (i.e., 170 blocks/150 blocks times 100) than when Joe is the superintendent. The estimator, faced with estimating a new project, should adjust to past data regarding mason productivity to reflect the impact of the superintendent to be utilized.

Finally, inspection of the mason work process indicates that no other factors were constant other than the location of the work for the performance of the work for Events 2 and 3. Observation of the data indicates that an additional 30 blocks (i.e., 21.4%) were placed during a day when the work was performed in Chicago relative to St. Louis. Knowledge of this would again enable the construction estimator to improve his forecast of a work process as a function of the *location* of the work to be performed.

Obviously the data given and the analysis illustrated above represent an oversimplification of correlation analysis. In practice, the estimator would have to collect a much larger set of data and analyze the data as a function of a more detailed list of factors in order to rely on the data and the correlation analysis. This task imposes a certain degree of paperwork on a firm. The amount of data and subsequent analysis required may result in a decision that the process is unfeasible because of the time required, especially if the process must be performed manually. The increasing use of computers by construction firms is alleviating the time problem, however. In addition to being able to store a significant quantity of past project information, the computer can be used to actually determine the correlation of an event with several factors. This is done through highly scientific mathematically based curve fitting and data searching algorithms. The use of computers has added a scientific dimension to forecasting. (The use of computers for construction estimating is discussed in Chapter 18.)

EXERCISE 8.1

Assume that a contractor uses a forming crew consisting of 4 carpenters and 2 laborers. Assume that the hourly rate for a carpenter is $16 per hour and for a laborer is $14 per hour. Further assume that at year end the carpenter's hourly rate increases 15% and the laborer's hourly rate increases 12%.

Calculate the hourly crew cost rate at the end of the year.

What is the percentage increase in the hourly crew cost at the end of the year relative to the beginning of the year?

EXERCISE 8.2

During the year 19X0, a contractor has incurred the following expenses for his company-owned crane:

Gasoline	$3,000
Oil and lubrication	1,200
New tires and tire repair	800
Repair of crane boom	1,500
Overhaul motor	3,000
Repair clutch	2,000

The equipment was purchased as a new piece of equipment in the prior year. The purchase cost was $160,000. The company has decided to depreciate the equipment in 8 years using a straight-line depreciation and no salvage value. During the year 19X0, the equipment was used for productive work a total of 1,000 hours. The company considers a normal year to include 2,000 work hours.

The company is now estimating a new project to start at the beginning of 19X1. The company has decided to use the 19X0 data regarding the crane to establish an estimate of the equipment costs for the new project. The new project is estimated to take 50 working days or 400 hours. The equipment will be at the project site about one-half of this time period (i.e., the equipment will be at other project sites during the other time periods). While the equipment is at the project site, the contractor estimates that it will be used for productive work approximately 80 hours.

Calculate the equipment cost to be estimated as part of the project cost if

a. Productive hours are to be used to establish the equipment rate.
b. Total hours are to be used to establish the equipment rate.

Discuss the practicality of implementing the alternative systems at a job site in regard to the recording of equipment hours.

EXERCISE 8.3

Given rapidly increasing material costs, some contractors have pre-purchased material to include lumber and masonry units and have inventoried the material for future use on projects. The benefits of pre-purchasing material must be weighed against the loss of use of cash and the decreasing of the firm's liquidity.

Let us assume that a contractor has determined that it will need the following quantities of concrete block during each of the following months in the coming year:

January	10,000	May	24,000	September	18,000
February	12,000	June	28,000	October	22,000
March	15,000	July	22,000	November	14,000
April	10,000	August	18,000	December	10,000

The contractor can purchase all of the blocks for $.65 per block on January 1. As an alternative, the firm can purchase the amount of blocks the firm needs in any one month on the first day of the month—e.g., the contractor can purchase on February 1 the 12,000 blocks needed in February.

Assume that the price of blocks is expected to escalate in cost at a rate of 1.5% increase per month from the January 1 price. Assuming that the interest rate a contractor can invest its money at is 10% compounded annually, determine whether the contractor should pre-purchase the blocks, or purchase the blocks on a month-by-month basis.

9

Integrating Estimating With Project Planning and Control: A Project Management System

Performing Project Management Functions as a System *
Quantity Take-Off * *Activity Planning* * *Estimating*
Direct Cost * *Project Planning and Scheduling* *
Resource Scheduling * *Cash Budgeting* * *Payroll* *
Cost Control * *Accounting Reports* * *Example of a*
Manual Project Management * *Information System*

PERFORMING PROJECT MANAGEMENT FUNCTIONS AS A SYSTEM

In performing project management functions systematically, the contractor performs several more or less separate steps. Essentially, he plans work and subsequently controls the work. Both the planning and the control function consist of more definitive steps. For example, planning entails estimating, activity planning, material scheduling, cash budgeting, and so forth.

While it is possible for a contractor to perform these functions independently, this may prove too time-consuming and costly for the small- or medium-sized contractor. It is, however, possible to perform them as *interrelated functions*. Many of the individual project management functions can be interrelated and performed as a system via a common data base. This concept of an integrated project management estimating/planning/control system is illustrated in Figure 9.1.

The key to an effective system—one that facilitates planning and controlling—is the *definition of work items or segments that are to be the cost objects of the system*. This definition can provide the base for collecting labor, material, and overhead costs whereby historical work-item data can be collected from ongoing and past projects. The data serve as the basis for planning, estimating, and budgeting future projects. These projects are in turn controlled through the monitoring and comparing of work-item costs and progress to the plan and the estimate.

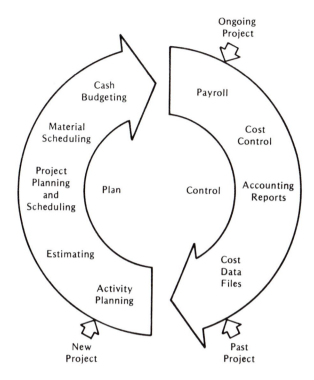

Figure 9.1 *Overview of a Project Management System*

Let us now illustrate the concept of an integrated project management estimating/planning/control system by walking through the cycle shown in Figure 9.1.

Assume that the company has already defined its work items and has collected labor productivity and material usage data for the work items from the performance of past projects. We will enter the cycle shown in Figure 9.1 at the starting point of a new project.

QUANTITY TAKE-OFF

Faced with a new project, a contractor starts the project management process by taking off quantities from the project drawings. Let us assume that one of the work items or cost objects in a specific contractor's system is defined as the placement of formwork for concrete walls. The contractor would systematically review construction drawings such as those illustrated in Figure 9.2 and calculate the amount of work to be performed. The calculations and calculated quantity of work are usually placed on an estimating sheet such as that shown on Figure 9.3. Assume that for the project being estimated,

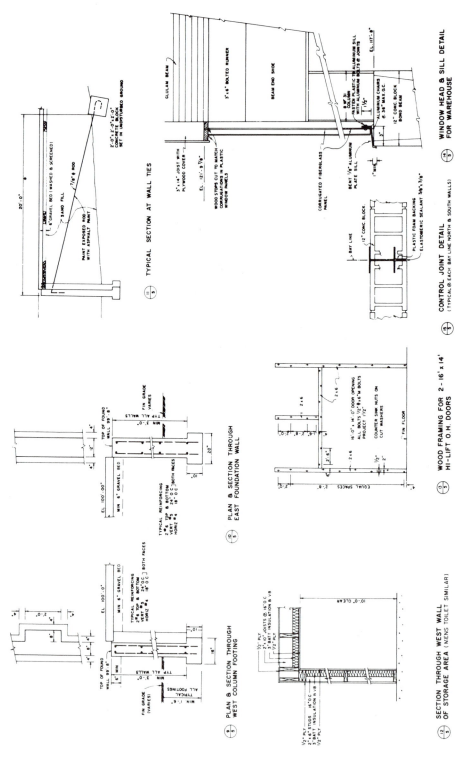

Figure 9.2 *Example Construction Drawing*

252

School 106, the contractor estimates that he will have to perform 50,000 square feet of contact area of formwork for concrete walls. This quantity is shown in Figure 9.3.

ACTIVITY PLANNING

The objective of activity planning is to determine the optimal combination of resources to be used in performing the required quantity and quality of work. Activity planning means determining the best crew size, combination of labor crafts, and types of equipment to be used to perform the projected work.

Too often in the construction industry the traditional methods of performing a job have been accepted, and the lack of innovation has weighed heavily on the industry, causing low productivity. By comparing different combinations of labor and equipment, a company can spot inefficiencies and in effect optimize performance. The goal of the activity planning function should be to establish the number of employees by type of labor craft to be employed along with the type of equipment to be utilized for the method. Given the combination of resources available, the work-item data from past projects could then be used to effectively predict or plan each new project. Productivity can be established as units of work item per manhour.

Productivity records are structured from the data collected from the performance of the work in question on past projects. For example, let us assume a contractor has performed forming for concrete walls on 5 previous projects. Through the collection of data at the project sites, let us assume that the contractor has collected the data shown in Figure 9.4. Based on the performance of 5 projects, the contractor's cumulative average productivity for placing formwork (expressed in manhours per 100 square feet of contact area of forms placed) is 10.0 manhours per 100 square feet of contact area of forming.

Let us further assume that the project being considered by the contractor, School 106, is judged to have similar characteristics represented by the average of the 5 projects summarized in Figure 9.4. In other words, the cumulative productivity shown in Figure 9.4 for forming is judged to be representative of the productivity expected on School 106.

Based on the calculated quantity of work to be performed for formwork (50,000 SFCA, as shown in Figure 9.3) and the calculated cumulative productivity for formwork (10.0 manhours/100 SFCA) as shown in Figure 9.4, the estimated duration of manhours for the formwork to be performed on School 106 can be calculated. This calculation is illustrated in Figure 9.5.

Given the calculated 5,000 manhours of duration, the contractor can now decide upon crew size and then estimate the resulting required workdays. For example, he may consider the 3 different crew sizes indicated in Figure 9.6 to be feasible for School 106. Assume that the contractor judges

PRACTICAL FORM NO				QUANTITY SHEET						
PROJECT *SCHOOL 106*				ESTIMATOR *JJA*		ESTIMATE NO *EXAMPLE*				
LOCATION *ANYWHERE*				EXTENSIONS		SHEET NO				
ARCHITECT ENGINEER *JKA*				CHECKED *SLL*		DATE *9-12-X6*				

CLASSIFICATION DESCRIPTION	NO.	DIMENSIONS Length	Height				ft²		ESTIMATED QUANTITY	UNIT
03101										
FORM-CONC WALL										
Northeast Wall	1	120.0'	10.0'				1200			
" "	1	140.0'	10.0'				1400			
" "	1	180.0'	8.0'				1440			
Southeast Wall	1	420.0'	12.0'				5040			
Northwest Wall	1	220.0'	12.0'				2640			
" "	1	260.0'	12.0'				3120			
" "	1	260.0'	10.0'				2600			
Southwest Wall	1	540.0'	12.0'				6480			
Annex - North W	2	180.0'	10.0'				3600			
Annex - East W	2	140.0'	12.0'				3360			
Annex - South W	2	380.0'	12.0'				9120			
Annex - West W	2	500.0'	10.0'				10000			
									50000	sfca

MFG. IN U.S.A.   FRANK R. WALKER CO., PUBLISHERS, CHICAGO

Figure 9.3 *Quantity Take-Off*

HISTORY DATA
FORMING-CONCRETE WALLS

Unit of Measure:
  Square Feet of Contract Area–SFCA

Productivity:
  Manhours/100 SFCA

Project	Date Completed	Quantity of Work Performed	Crew Size	Duration (Hours)	Manhours	Productivity Manhours/ 100 SFCA	Cum. Productivity Cum. Manhours/ 100 SFCA
School 101	7/5/X0	20,000	10	200	2000	10.0	10.0
Hospital 102	8/8/X0	24,000	15	140	2100	8.75	9.32
School 103	9/6/X0	40,000	8	400	3200	8.00	8.69
Office Building 104	11/5/X0	18,000	8	200	1600	8.88	8.72
School 105	2/3/X0	41,000	20	270	5400	13.2	10.00

Figure 9.4 *History Data from Past Project*

that based on considerations of manpower availability and productivity, he
will use 10 men to perform the work. Based on this crew-size selection, a 500-
hour duration or 12.5 work weeks (assuming a 40-hour workweek) is selected, as illustrated in Figure 9.6. This calculated duration will now be used
to calculate a cost estimate for the forming work, and an overall project
schedule.

CALCULATION OF MANHOURS FOR SCHOOL 106
FORMWORK-CONCRETE WALLS

1. Quantity of Work to be Performed:
    Quantity Take-Off   =   50,000 square feet of contract area (SFCA)

2. Assumed Productivity:
    Cumulative Productivity   =   10.0 manhours/100 SFCA

3. Calculated Duration (Manhours)
    (Quantity of Work to Perform) $\times$ Productivity = (50,000 SFCA) $\times$ 10 $\dfrac{\text{Manhours}}{100\,\text{SFCA}}$ = 5,000 manhours

Figure 9.5 *Estimating a New Project*

The point here is that the common work item—the formwork for concrete walls—is being utilized to integrate the quantity take-off function and activity planning. Other work items would be taken off and planned in the same manner.

We will now proceed to see how the common work-item concept will facilitate other contractor project management functions to include estimating, project planning, and control.

ESTIMATING DIRECT COST

No single management function plays a more important role in the financial success of the company than estimating. The secret competitive bidding process that characterizes the industry emphasizes the need for accurate estimates. Too often in the construction industry a company establishes part of its estimate from "the seat of its pants." With little other than a vague recollection of past performances of the work in question, the estimator is faced with predicting future costs based on hunches and unreliable word-of-mouth information. Admittedly, each time a company undertakes a new project, the project will likely be somewhat unique as to factors that dictate cost. However, knowledge of structured past project data, along with recognition of varying and unique project conditions, can result in reduced uncertainty in the estimator's project bid.

The estimating function can make great use of past project data regarding the performance of individual work items. For example, let us consider the forming of concrete walls work item illustrated in Figures 9.4 to 9.6. In Figure 9.6 we calculated an expected work-item duration of 500 hours or 12.5 workweeks for the forming of concrete walls for the project identified as School 106. The direct-labor cost component of the cost estimate for the

---

DETERMINATION OF CREW SIZE AND DURATION
FORMWORK-CONCRETE WALLS

Work to be Performed: 50,000 square feet of contact area (SFCA)
Required Manhours: 5,000

Possible Crews:	Size of Crew	Resulting Duration–Hours	Resulting Duration–Weeks
	8	625	15.6
	10	500	12.5
	12	416	10.4
Selected Crew:	Size of Crew	Resulting Duration–Hours	Resulting Duration–Weeks
	10	500	12.5

Figure 9.6 *Activity Planning for a New Project*

DETERMINATION OF DIRECT LABOR COST
FORMWORK-CONCRETE WALLS

Estimated Manhours Duration: 500 hours

Cost per Craftsman: $18.00/hour

Estimated Direct Labor Cost = Duration (Hours) $\times$ Craftsman rate/hour $\times$ number of Craftsmen

Estimated Direct Labor Cost = 500 hours $\times$ 18.00 $/hour $\times$ 10

Estimated Direct Labor Cost = $90,000.00

Figure 9.7 *Estimating Cost for a New Project*

forming of concrete walls can be determined by multiplying the work-item duration in hours by the crew cost per hour.

We assumed in Figure 9.6 that the contractor decided to use 10 craftsmen to do the forming operation. Let us assume that the average wage rate plus labor fringes for each craftsman is $18. Based on this per-craftsman cost per hour and the duration calculated and illustrated in Figure 9.6, the estimated direct labor cost for the concrete work is calculated as illustrated in Figure 9.7. The direct material cost can be established through identification of quantities of material for individual work items (referred to as the quantity take-off), material wastage factors as indicated from historical data, and the relevant material prices.

In addition to labor and material cost components of a contractor's estimate, equipment and overhead costs must be determined. A construction company often spends many hours attempting to be very accurate in its estimates of project labor and material costs. However, too often this effort is followed by a less than detailed approach to equipment costs and overhead cost allocation. For example, after determining direct labor and material costs for a project, a construction company may add 40% of the summed direct costs to cover equipment costs and overhead cost and profit. Such a procedure often leads to substantial inaccuracy in the total project bid price. Less than accurate overhead allocation usually can be traced to a company's lack of a cost accounting system. Inability to develop an accurate application rate and base along with an analysis of the over- and underapplied overhead leads to unprofitable estimates.

PROJECT PLANNING AND SCHEDULING

Closely related to the management functions of activity planning and estimating is that of overall project planning and scheduling. The purpose of a contractor's overall project planning and scheduling function is to integrate all of the planned project activities into an overall project schedule that is compatible with project time and cost objectives. A project plan and schedule for a project is often prepared via the use of a bar chart or a CPM diagram. (Both bar charts and CPM are discussed in a following section of the text.)

A project plan and schedule is prepared by constructing the technological and resource logic between the various required construction activities that are part of the overall project. It is possible for the contractor to define project activities for the project planning and scheduling function to be compatible with the same work items/activities that are defined for the quantity take-off, activity planning, and estimating functions.

For example, the previously discussed work item/activity identified as forming concrete walls can be defined as one of the activities on a bar chart or CPM diagram for an overall project plan and schedule. This is illustrated in Figure 9.8. The activity duration used for calculations performed via the project plan and schedule is the duration previously calculated in Figure 9.6.

The point here is that the overall project plan and schedule can be prepared in conjunction with other project management functions if a common data base (i.e., work items) are utilized. This use of a common data base for several functions enhances the benefit cost ratio for any one of the functions relative to performing the functions independently.

RESOURCE SCHEDULING

Closely related to the management functions of planning and scheduling is project resource scheduling, including the scheduling of labor, material, purchases, and equipment. The quantity take-off segment of estimating establishes what resources are required. Overall project planning and scheduling dictates when the resources are needed.

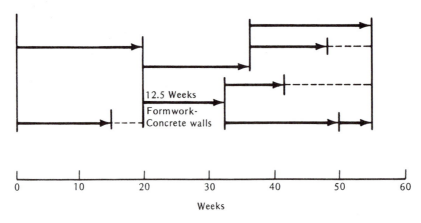

Figure 9.8 *Integrating Planned Activities into an Overall Project Plan*

By establishing the resource requirements for the individual project activities and subsequently summing them via the prepared overall project plan and schedule, it becomes possible to determine the cumulative demand for any project resource as a function of time.

Let us consider the labor resource for the project plan and schedule illustrated in Figure 9.8. This simplified illustration is shown again in Figure 9.9 with the labor requirement for each activity shown in the block alongside the activity arrow. Note that the labor resource requirement for forming concrete walls is the 10 laborers selected as the crew in Figure 9.6.

By summing the cumulative laborers required for activities occurring simultaneously, the overall labor resource requirement at any one point in time can be established as shown in Figure 9.9.

While laborers may not be a critical resource to a specific project, the process illustrated in Figure 9.9 could be performed for any resource we might consider. Once again, the common work item/activity is utilized to perform another project management function—i.e., resource scheduling.

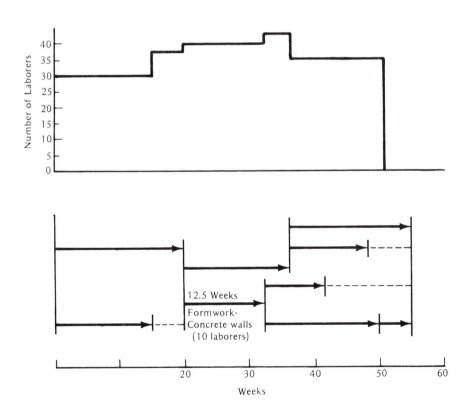

Figure 9.9 *Integrating Activity Planning, Project Planning, and Resource Scheduling*

CASH BUDGETING

The benefits of preparing cash budgets have been well documented. However, the fact remains that construction companies are often negligent in their preparation of cash budgets. As a result, many of the problems these companies have can be traced to the lack of adequate preparation and updating of project cash budgets. The cash budget must include the identification of expected dollar amounts of cash receipts and disbursements, and the time at which the various receipts and disbursements occur. The amounts of the receipts are determined by the contractor's total project bid. The disbursements are, for the most part, identified by the estimating function.

The activities plan establishes the time at which disbursements will have to be made. The timing of receipts of cash from the project owner are determined in part by the activities plan and in part by the owner–contractor project payment agreement.

The cash budget for a project is essential for determining financing needs as a function of time. Because deferred payment for work performed is very common, and because of low working-capital ratios, the construction company is often faced with the need to secure financing at various times during a construction project. The amount of this financing and its cost should be determined before the project is bid. This is necessary so that the interest cost can be reflected in the estimate of the project costs, and subsequently in the project bid. Financing costs are as much a part of project cost as are the labor or material costs. Their identification in the project estimate is essential to a profitable estimate and project bid.

The very ability to obtain financing at a critical time in the project schedule is often dependent on how soon the company initiates the search for funds. The availability and cost of loan money from financial institutions varies as a function of many economic factors. Unless a plan exists for obtaining funds in advance, the construction company may find itself without a source of funds when they are needed. The preparation of these supporting schedules and the cash budget is facilitated by the use of a common set of work items. Just as the work items can serve as the base for establishing work segment durations, costs, and resource schedules, the cash budget and its supporting schedules can be produced as a function of the defined work items.

Let us assume that the direct labor cost for each of the project activities illustrated in Figures 9.9 are determined as was done for the forming of concrete walls in Figure 9.5. By plotting the direct labor cost for each activity as is done in Figure 9.10, it is possible to determine the cumulative cash requirement for the direct labor requirement as a function of time. Similarly, the total project cash requirement for the project as a function of time can be determined by performing a similar process for all of the component costs of the project.

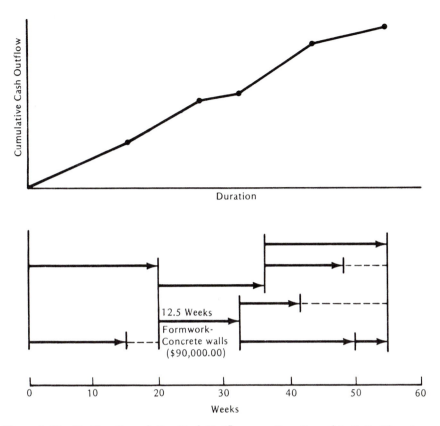

Figure 9.10 *Plotting Cumulative Cash Outflow as a Function of Activity Planning and Project Planning*

When a proposed project becomes reality, the contractor's project management function does not terminate. Planning, which includes estimating and the preparation of budgets, provides the potential for a profitable project. But it is the control function that brings these potential profits to reality.

PAYROLL

The labor component of the control function is initiated with the performance of the payroll obligation. The payroll function is fundamental to the daily operation of any firm that is dependent on the use of labor. The preparation of employee paychecks, the establishing of withholding liabilities, and the filing of required forms to taxing bodies requires the company to document individual employee hours worked. Timecards are often limited to documenting total time worked by a given employee. With only a slight modification of the timecard, it is impossible to document labor time to specific segments of work performed by the employee.

The work-item cost objects that are used for documenting labor hours can be the same work items that are used for project planning, scheduling, and estimating. In this manner, historical work-item files are established from past projects and in turn used to aid the planning and estimating of future projects.

As is true of the definition of work items that serve planning and estimating, practical considerations must guide the defining of work items for data collection. Considerations such as the ability to differentiate time to specific work items, the cost of collecting and processing the data, and the benefits to be gained from the processed data all play a part in work-item definition.

As with labor, procedures can trace *material* to specific work items. From this collected data, material standards and wastage factors can be established for planning and estimating future projects. Equally important is the role material placement work-item standards can have in the control of both material and labor costs.

COST CONTROL

In order to have an effective cost control program, the trend of the cost must be determined as soon as possible and compared against the progress of the plan so that required corrective action can be taken. The monitoring of production costs cannot be overemphasized in regard to the operation of the construction company and its projects.

The need to have a means of quickly documenting and monitoring project production and costs is emphasized by the fact that the construction company's planned production and costs are often less than accurate estimates. Regardless of the soundness of the company's planning and estimating practices, numerous factors prevent the plan and estimate from being totally deterministic. The ever-changing environment surrounding construction work, its variable nature, and the industry's high dependence on labor productivity all inject uncertainty into the project plan and estimate. The result is that plan and performance differences are partly due to lack of control and partly due to a less than accurate plan. Independent of the cause, early recognition of inefficiencies and deviations from planned progress are essential to corrective management decisions.

Effective project control of labor and material costs can be gained through documenting material quantities and labor hours for previously defined work items. Let us assume that for a work item such as the previously defined forming of concrete walls, the company estimates 50,000 square feet of contact area (sfca) of work to be performed. On the basis of historical work-item data collected from previous projects, the company estimates that it will need 10 carpenter manhours per 100 sfca of work. The total estimated manhours for the work item is therefore 5,000. Let us now assume that the company receives the contract to do the work. Further, assume that inspection of the work after a month of placing forms indicates that 5,000 sfca of work has been put in place. This determination of work performed is necessary for progress billings. In fact, the documentation of work performed in work items can result in more accurate billings.

On the basis of payroll records structured to report labor hours to work items, 1,250 carpenter hours have been charged to the wall-forming–related work. Analysis of the material put in place indicates that 10% of the work is completed. However, based on an estimate of 5,000 manhours, the cost accounting system indicates that 25% of the estimated labor hours and cost have already been expended. This calculation is shown in Figure 9.11. Assuming that labor hours should be proportional to the wall forming performed, it appears that the company is faced with a substantial labor cost overrun.

The variation in the percentages of labor hours versus material placed should receive attention immediately. It may be that it can be explained and is not reason for alarm. However, it may also be indicative of inefficiencies. The early detection of these inefficiencies can bring corrective action before cost overruns become excessive.

Merely comparing actual costs with estimated costs does not always give a good measure of performance. It can be that excessive overtime labor costs

---

Estimate

Quantity of Work to be Performed = 50,000 square feet of contact area (SFCA)

History Productivity Data: $\dfrac{10 \text{ carpenter hours}}{100 \text{ SFCA}}$

Estimated Carpenter Hours $= \dfrac{50,000 \text{ SFCA (10 hrs)}}{100 \text{ SFCA}} = 5,000$ carpenter hours

Accumulated Project Data

1,250 Carpenter Hours to Date = 25% of estimated hours
10 Percent of Work Performed, i.e., 5,000 SFCA

Projected Cost Overrun

10 Percent of Work Equates to 1,250 Hours Expended
therefore 100% of Work Equates to 12,500 Hours

Projected Hours	=	12,500
Estimated Hours	=	5,000
Estimated Over-		
Run in Hours	=	7,500

Figure 9.11 *Analysis of Labor Report*

are hidden by high labor productivity. Looked at another way, the benefit of high employee productivity may be hidden because of the inefficiencies of management in scheduling work. The point to be made is that effective cost control is achieved by analyzing the factors that make up the total cost. In the case of material costs, it is necessary to analyze the purchase price as well as the budget price, and the amount of material used as well as the estimated amount from the quantity take-off. Similarly, the analysis of the components of overhead costs must be analyzed in order to effectively control the total overhead cost.

ACCOUNTING REPORTS

All too often the construction company finds out about a project's profitability or loss only after all the bills are paid and the final payment is received from the owner. The lack of cost accounting and in particular the inability to generate interim financial reports regarding a project keep the contractor in the dark and with a sense of uncertainty.

Collected material, labor, and overhead data from an ongoing project can serve the control function while at the same time providing the basis of interim progress and financial reports. These reports aid the company in its future bidding activities, in improving relations with creditors, and perhaps more importantly, in measuring its own performance. Using a uniform set of work items for planning, estimating, and controlling the preparation of the interim reports is a natural outgrowth of data use that serves these management functions.

Collected data that aid control and establish standards for planning and estimating can also satisfy the needs of financial accounting. Documented material, labor, and overhead costs can be summarized and categorized through ledger accounts so that financial statements are prepared directly from the collected data. The financial accounting function can be carried out continuously by the utilization of a cost accounting system which isolates project costs. Modern-day computer systems can automatically post to the general ledger while structuring cost data for project control.

The strength and internal controls of the cost accounting system improve the validity of the financial statements. Better record-keeping, cost accounting procedures, and control can reduce the rather widely held view by auditors, bankers, sureties, and investors of the unacceptable nature of the construction company's financial information and the related financial risk.

The cycle illustrated in Figure 9.1 is repetitive. Planned projects become ongoing projects. In turn, the ongoing projects become past projects. Throughout these steps historical data structured through the company's cost accounting procedures continue to be updated. Using cost accounting methods, each and every one of the management functions can be performed with increased benefits and in a manner that complements the performance of each of the other functions. The concept discussed in this section will now be illustrated via the use of an example set of forms in the next section.

EXAMPLE OF A MANUAL PROJECT MANAGEMENT INFORMATION SYSTEM

An effective job-cost control system should integrate the project estimating, planning, and control functions by means of a common set of data. By integrating these and other project management functions to include project accounting, the benefit/cost ratio for performing any single one of these functions can be improved.

The typical contractor seeking an effective project management information system that includes project control cannot always afford the luxury of automatic data-processing equipment. However, this does not mean that an effective system of project planning and controlling work item cost is beyond the capabilities of the firm. Through in-depth analysis of the defined work items, data collection forms, and structured procedures, it is possible to obtain a manual work-item planning and controlling cost system.

A manual management system for planning and controlling project labor and material costs will now be illustrated. The system emphasizes management data for project planning, estimating, and controlling. In addition, the system will aid the function of payroll, owner billings, and financial accounting. While cash budgeting and material scheduling are not dealt with by the system, it can be extended to serve these functions.

A previous section discussed the critical task of defining work items. The importance of selecting work items remains the same regardless of whether the system used is automated or non-automated.

In order to illustrate a typical system that traces labor and material costs to project work items, we will assume that a set of work items has been defined. In particular, we will assume the set of 44 work items shown in Figure 9.12. The list corresponds to what might be used by a general contractor and is not unique. Another general contractor with a different type of work or data collecting and processing capabilities might develop a list that varies from that shown in Figure 9.12. Nevertheless, a list of about this number of work items is manageable as to the time required to process the project data.

Example System Based on Productivity

The system to be illustrated is based on productivity rather than unit costs; thus the collected labor data are somewhat insensitive to time. An overview flowchart of the system is shown in Figure 9.13. The system is manual except for payroll, which is processed through a computer service bureau, although this too could be carried out manually. However, manual

WORK ITEM CODES

GENERAL

GS1 SUPERVISION
GC1 CLEAN UP
G01 OTHER

EXCAVATION

EM1 MISCELLANEOUS EXCAVATION
E01 OTHER

CONCRETE

CF1 FORM FOOTINGS
CF2 FORM WALLS–GRADE AND 1st FLOOR
CF3 FORM WALLS–ABOVE 1st FLOOR
CF4 FORM BEAMS
CF5 FORM RECTANGULAR COLUMNS
CF7 FORM SLABS
CF8 FORM STEPS
CF9 FORM DRIVES AND WALKS
CR1 REINFORCING RODS
CR2 REINFORCING MESH
CC1 CONCRETE PLACING–FOOTINGS
CC2 CONCRETE PLACING–WALLS
      GRADE AND 1st FLOOR
CC3 CONCRETE PLACING–WALLS
      ABOVE 1st FLOOR
CC4 CONCRETE PLACING–BEAMS
CC5 CONCRETE PLACING–COLUMNS
CC6 CONCRETE PLACING–SLABS
      GRADE AND 1st FLOOR
CC7 CONCRETE PLACING–SLABS
      ABOVE 1st FLOOR
CC8 CONCRETE PLACING–STEPS
CC9 CONCRETE PLACING–DRIVES AND WALKS
CP1 CONCRETE PRECAST
C01 OTHER

MASONRY

MB1 BRICK
MB2 BLOCK AND BRICK WALL
MC1 CONCRETE BLOCK–4"
MC2 CONCRETE BLOCK–8"
MC3 CONCRETE BLOCK–12"
MS1 SCAFFOLDING
MT1 TILE, GLAZED
MT2 TILE, CLAY
M01 OTHER

STRUCTURAL STEEL

SS1 STRUCTURAL STEEL
SS2 STAIRS AND RAILING
S01 OTHER

WOOD

WS1 SHEATHING
WF1 FRAMING
WD1 DOORS, JAMBS AND HARDWARE
WT1 TRIM
WM1 MILLWORK
W01 OTHER

Figure 9.12 *Example List of Work Items*

payroll calculations often prove tedious and costly. Computer service bureaus for payroll processing are becoming readily accessible to construction firms, often provided by their banks at little charge.

The integrated management system illustrated in Figure 9.13 is designed to be compatible with the possibility of modification to a fully automated system at a future date. While a computer system may not be affordable at a given point in time, it may in the future. A manual system designed with a view toward the future can help minimize conversion costs.

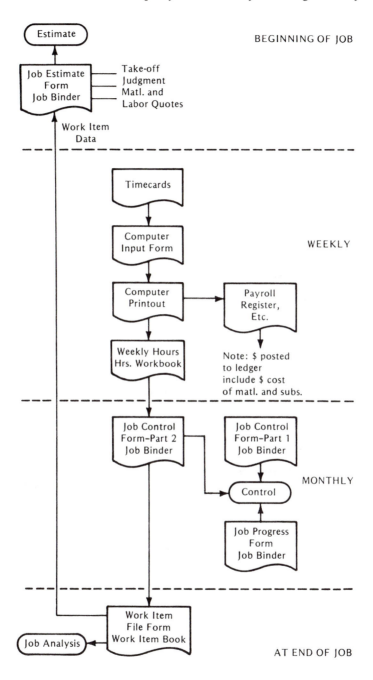

BEGINNING OF JOB

WEEKLY

MONTHLY

AT END OF JOB

Figure 9.13 *Overview of Job Cost System*

The time at which each of the forms comprising the integrated system is filled out and utilized is indicated in the overview flowchart in Figure 9.13. As will be illustrated, various forms may be used on a daily, weekly, yearly, or end-of-project basis.

The project management system is initiated with the planning of a potential project. The plan and estimate are the initial management project functions to be performed. If this is the first project for which the system is being implemented, no historical work files would be available from past projects. Therefore, the estimate of manhours for each work item for which there is work would have to come from a source external to the system. However, as the system is applied to projects, these projects would provide the source data for estimating.

Job Estimate Sheet

The first form that is part of the project management system, the job estimate sheet, is illustrated in Figure 9.14. The purpose of this form is to estimate the material and labor cost for the project in terms of required material, unit material prices, required labor hours, and labor rates for each of the individual work items. The rows on the job estimate sheet correspond to work items. An abbreviation for a work item is shown in each row. (The abbreviations are those shown in Figure 9.12.)

After the first column, which identifies the individual work items, the next three columns are used for calculating the work item material costs. The first of these columns, "Est. Matl.," is for listing the quantity of material work to be performed for each work item. These quantities are determined from the take-off of quantities from the project drawings and from recognition of material wastage. For example, for the work item labeled CC1 (the Concrete Placing-Footings), the quantity take-off might indicate that 200 cubic yards of concrete is to be placed. Assuming a 2% waste factor, or 4 cubic yards, there are 204 cubic yards of concrete work to be estimated. This number is placed in the second column in the CC1 row.

The material price per unit of material is placed in the next column. For concrete work it would be the purchase price per cubic yard. For structural steel it would be the purchase price per ton of steel. The unit of material used for pricing should be the same as that used for the quantity take-off. If the carpentry work is measured in board feet, the material unit price should be entered as the price per board foot.

The material estimate for each work item is determined by multiplying the "Est. Matl." column and "Matl. $ Unit" for each of the work items. The result is placed in the "Tot. $ Matl." column. The total estimated material cost for the project is found by summing all the work-item entries in the "Tot. $ Matl." column. It is possible and even likely that for any one project, there will be several defined work items for which no work is performed. The work-item list is designed for the firm and not for any one project; so,

JOB ESTIMATE SHEET

ESTIMATOR _____ DATE _____ JOB NAME _____ CODE _____

WORK ITEM	EST. MATL.	MAT. $ UNIT	TOT. $ MATL.	SUPER-VISION HRS.	CARPENTER			MASON			LABORER			FINISHER			IRON-WORKER			OPER-ENG.			TOTAL LAB. $
					HRS.	$/H	$	HRS.	$/H	$	HRS.	$/H	$	HRS.	$/H	$	HRS.	$/H	$	HRS.	$/H	$	
GS1																							
GC1																							
G01																							
EM1																							
E01																							
CF1																							
CF2																							
CF3																							
CF4																							
CF5																							
CF7																							
CF8																							
CF9																							
CR1																							
CR2																							
CC1																							
CC2																							
CC3																							
CC4																							
CC5																							
CC6																							
CC7																							
CC8																							
CC9																							
CP1																							
C01																							
MB1																							
MB2																							
MC1																							
MC2																							
MC3																							
MS1																							
MT1																							
MT2																							
M01																							
SS1																							
SS2																							
S01																							
WS1																							
WF1																							
WD1																							
WT1																							
WM1																							
W01																							

Figure 9.14 *Job Estimate Sheet*

whereas one project may entail all the work items, yet another may contain half or even fewer of the work items listed.

A number of columns are used for calculating the estimated labor costs. It is assumed that the firm utilizes 7 different types of labor: supervision, carpenters, masons, laborers, finishers, ironworkers, and operating engineers. It is unlikely that all 7 types of labor would be utilized for any one work item. In fact, analysis of the work items indicate that no work item would require more than 3 crafts of labor.

Each of the 7 types of labor is listed on the job estimate form. The first column in each of the 7 labor categories is for documenting the labor hours required for each work item. These hours can be determined from experience, past records, or external sources such as published cost books.

The estimated hours per work item are multiplied by the dollar rate per hour for the labor in question, to yield the labor cost estimate per craft for each work item. The supervision column uses a different approach for handling the entry of the hourly rate. The supervision hourly rate is entered, but it is assumed that this rate is constant for each of the work items and thus is not entered separately for each work item. On the other hand, under each other labor type is a place for multiplying the estimated hours by various labor rates. This is in recognition of the fact that a given labor craft's rate may depend in part on the type of work being performed.

Timecard

Ideally, estimated projects become ongoing projects. Assuming this to be the case, the next form that is part of the project management system is the timecard. The purpose of the timecard is to allocate labor hours to work items and to tabulate total labor hours for each worker. The timecard is an essential form in the management system—it serves as the means of collecting essential input data.

Timecards should be designed with attention to ease of use and accuracy of data input. The very design of the timecard can have a positive or negative effect on worker productivity and motivation.

A timecard designed to be compatible with the sytem is illustrated in Figure 9.15. The work items and codes shown in Figure 9.12 are printed on the reverse side of the timecard shown in Figure 9.15. The timecard is to be filled out daily and returned to the office weekly for processing. The labor craft foremen probably would have the responsiblity for filling out the timecard. It should be noted that the firm's week starts on a Wednesday and ends on a Tuesday. This provides the potential to process the timecards on Wednesday and Thursday and generate payroll checks for Friday.

The front of the timecard provides space for the individual worker's name, labor craft, and the ending date of the timecard period. The front part of the timecard also provides space for documenting labor hours on a given date with regard to individual work items. There is room for documenting work on 7 work items in any one week. In addition, there is space for documenting standard, or straight-time hours, and overtime hours. Finally, the form provides 2 ways of calculating and checking the total hours charged for the individual worker in the time period.

The back side of the timecard has 2 functions: (1) a reprint of the system's work items is duplicated there to aid in on-site identification of labor hours to specific work items; (2) there is also space for calculating net pay—hours worked can be transferred to this side of the form, multiplied by the relevant rate, and total earnings calculated. From these earnings, deductions such as

| NAME _____ |||||||| |
| CRAFT _____ WEEK ENDING _____ |||||||| |
WORK ITEM		WED	THR	FRI	SAT	MON	TUE	TOTAL
	ST							
	ST							
	ST							
	ST							
	ST							
	ST							
	ST							
TOTAL								

Figure 9.15   *Timecard*

FICA are subtracted to yield net pay. Using the form to calculate net pay is optional depending on the way payroll is processed.

The construction firm is normally required to generate payroll checks weekly. If the firm is of even modest size, the number of weekly payroll checks may exceed 100. The necessary payroll calculations include adding individual worker hours from the timecard and calculating and subtracting withholding taxes and insurances. Performed manually, these calculations can be very time-consuming. The clerical time required becomes even greater if the clerk is called upon to add labor-craft hours worked for specific work items. This is not to imply that such a procedure is unfeasible. If efficient clerical procedures are developed, manual payroll calculations, including distribution and summing of labor hours to specific work items, can be accomplished.

However, using a computer service bureau is a more economical method. Such service bureaus are available to the construction firm through independent operators, time sharing service companies, or in some cases, as has been mentioned, the firm's bank. In many cases these computer services have the capability of generating a distribution summary of labor hours into specific categories (work items) while producing paychecks and a payroll register.

Let us assume that the system we are describing for project planning and control utilizes a computer printout of labor distribution hours. This distribution sheet for labor hours per work item would be generated weekly as the payroll was produced. If such a service is not utilized, the same information could be produced manually. However, undoubtedly the total cost for performing the manual calculations would exceed the cost of a computer service. The choice of which method is best for a particular firm in terms of cost obviously depends on the number of employees and timecards that are processed. As the number of timecards increases, the economics favor using automated data-processing equipment for generating payroll and work-hour distribution.

Weekly Hours Form

The next form used as part of this system is a weekly hours form. This serves as a worksheet for compiling labor cost hours related to specific work items for a given month. As noted earlier, the payroll is normally generated every week. In this case, the labor cost hours per work item would appear weekly on the labor distribution form. What is needed is a form to compile the weekly hours into a monthly total. The form, which is designed to be compatible with the work and labor-craft items previously defined, is shown in Figure 9.16. As can be seen, the 7 different labor types identified with the general contractor in question are listed. Each work item is listed vertically. The first column labeled "BG" under each labor craft is for the beginning total of labor hours from previous months. The following 5 columns are for transferring the labor hours by work item for each of the weeks in the month. These are taken from the labor distribution forms that were printed as part of the payroll calculations. Finally, the last column, labeled "TT," is for adding the beginning hours and hours for each of the weeks in the month in question.

The weekly hours form represents data for a month. Each new month a new weekly hours form is initiated. The monthly weekly hours form is

MONTH	WEEKLY HOURS						JOB NAME							CODE						
WK ITM	FINISHER							IRON WORKER							OPER. ENGR.							TOTALS						

WK ITM	BG	W1	W2	W3	W4	W5	TT	BG	W1	W2	W3	W4	W5	TT	BG	W1	W2	W3	W4	W5	TT	BG	W1	W2	W3	W4	W5	TT
GS1																												
GC1																												
G01																												
EM1																												
E01																												
CF1																												
CF2																												
CF3																												
CF4																												
CF5																												
CF7																												
CF8																												
CF9																												
CR1																												
CR2																												
CC1																												
CC2																												
CC3																												
CC4																												
CC5																												
CC6																												
CC7																												
CC8																												
CC9																												
CP1																												
C01																												
MB1																												
MB2																												
MC1																												
MC2																												
MC3																												
MS1																												
MT1																												
MT2																												
M01																												
SS1																												
SS2																												
S01																												
WS1																												
WF1																												
WD1																												
WT1																												
WM1																												
W01																												
TOTAL																												

Figure 9.16 *Weekly Hours Form*

Integrating Estimating With Project Planning and Control: A Project Management System

MONTH_____   WEEKLY HOURS_____   JOB NAME_____   CODE_____

WK ITM	SUPERVISION							CARPENTER							MASON							LABORER						
	BG	W1	W2	W3	W4	W5	TT	BG	W1	W2	W3	W4	W5	TT	BG	W1	W2	W3	W4	W5	TT	BG	W1	W2	W3	W4	W5	TT
GS1																												
GC1																												
G01																												
EM1																												
E01																												
CF1																												
CF2																												
CF3																												
CF4																												
CF5																												
CF7																												
CF8																												
CF9																												
CR1																												
CR2																												
CC1																												
CC2																												
CC3																												
CC4																												
CC5																												
CC6																												
CC7																												
CC8																												
CC9																												
CP1																												
C01																												
MB1																												
MB2																												
MC1																												
MC2																												
MC3																												
MS1																												
MT1																												
MT2																												
M01																												
SS1																												
SS2																												
S01																												
WS1																												
WF1																												
WD1																												
WT1																												
WM1																												
W01																												
TOTAL																												

Figure 9.16 *Continued*

basically a worksheet that compiles the monthly hours. These monthly hours are transferred to yet another form for the control function. The responsibility for filling out the weekly hours form lies with the clerical help. The monthly forms can be filed in a job project binder for storing the estimate sheets.

A weekly check of the distribution of hours can be made by adding the column for the week in question and comparing this to the sum of the weekly craft hours as printed in the payroll distribution. Errors found in this comparison should be corrected. More than likely, an error indicates that the clerical help erred in translating labor hours per work item to the weekly hours form.

Job Control Sheet — Part One

The next form that is part of the project management system is the job control sheet, shown in Figure 9.17. This two-part form is used to control ongoing construction projects. The first part is for the *estimating* function; it is used to indicate the estimate of total material and labor hours for each required individual work item. Initially, the entries on this part of the form are those that have been made on the job estimate form (shown in Figure 9.14). If there are no change orders, this part of the job control sheet will be identical for each of the months. If there are change orders during a project, the entries of the estimated additional material and additional labor hours are placed in the columns labeled "CHG" and "EST," respectively, on the estimate part of the job control sheet.

Job Progress Form

Before we discuss how the second part of the job control sheet is used, we ought to divert our attention to yet another form that is part of the system. This form will provide us the data for the second part of the job control sheet. Every month the construction firm is usually responsible for making a progress report to justify its monthly billings. In some cases this estimate of monthly progress is made in very crude fashion which may result in either overbillings or underbillings. A more accurate estimate of job progress is a benefit of the project management system being considered. In particular, as part of the process of determining how much work has been done so that the work can be controlled, the firm can obtain an accurate estimate of progress for billing purposes.

The form used to determine job progress is the job progress form, shown in Figure 9.18. The purpose of this form is to determine the progress of each work item as determined by job inspection. The estimated progress will be input to the job control function and can also aid in billing. It is filled out at the end of each month, typically by the job superintendent.

The columns in the job progress form provide space for updating the monthly progress on one individual form for a year's period of time. The work items are listed vertically in the first column. (These are the same work items that are used for the estimating and job control.) Next, the job material estimate is shown for each of the work items. This estimate of materials consists of what was determined in the initial estimate and updated for any change orders. The next column "BEGIN BAL.," is for carrying over the

Integrating Estimating With Project Planning and Control: A Project Management System

JOB CONTROL SHEET (PART 1) ESTIMATE

MONTH   JOB NAME   CODE

WORK ITEM	MAT'L. EST.	SUPERV.		CARPENTER		MASON		LABORER		FINISHER		IRONWORKER		OPER. ENG.		TOTAL	
		EST	CHG	EST	CHG	EST	CHG	EST	CHG	EST	CHG	EST	CHG	EST	CHG	EST	CHG
GS1																	
GC1																	
G01																	
EM1																	
E01																	
CF1																	
CF2																	
CF3																	
CF4																	
CF5																	
CF7																	
CF8																	
CF9																	
CR1																	
CR2																	
CC1																	
CC2																	
CC3																	
CC4																	
CC5																	
CC6																	
CC7																	
CC8																	
CC9																	
CP																	
C01																	
MB1																	
MB2																	
MC1																	
MC2																	
MC3																	
MS1																	
MT1																	
MT2																	
M01																	
SS1																	
SS2																	
S01																	
WS1																	
WF1																	
WD1																	
WT1																	
WM1																	
W01																	
TOTAL																	

Figure 9.17 *Job Control Sheet—Part 1*

WRK ITM CODE	JOB MATL. EST.	BEGIN. BAL.	JAN	FEB	MAR	APR	MAY	JUN	JUL	AUG	SEP	OCT	NOV	DEC	END. BAL.
GS1															
GC2															
G01															
EM1															
E01															
CF1															
CF2															
CF3															
CF4															
CF5															
CF7															
CF8															
CF9															
CR1															
CR2															
CC1															
CC2															
CC3															
CC4															
CC5															
CC6															
CC7															
CC8															
CC9															
CP															
C01															
MB1															
MB2															
MC1															
MC2															
MC3															
MS1															
MT1															
MT2															
M01															
SS1															
SS2															
S01															
WS1															
WF1															
WD1															
WT1															
WM1															
W01															

JOB PROGRESS FORM

YEAR _____   JOB NAME _____

SUPERINTENDENT_____   JOB CODE _____

Figure 9.18   Job Progress Form

progress to date when we overlap a year. For example, if in 19X0 the balance of work done for an item was 225 cubic yards, this number would appear in the BEGIN BAL. for the first month of 19X1. As the job superintendent reviews the job at the end of the month, he identifies the material put in place in that month. This is done for each work item for which work was done.

There are two different ways of documenting job progress. Both are keyed to the amount of material put in place. One procedure is to document the actual quantity of material put in place at a given time. For example, for concrete-related work items this would mean that the job superintendent would document the actual amount of concrete put in place in a given month. Normally the unit of measure would be cubic yards.

As an alternative to documenting the actual quantity of material put in place in a period of time, the individual making the job progress report could merely document the percentage of work performed in a given period of time. For example, if the entire estimate indicates 100 cubic yards of concrete for a given work item, the individual making the job progress report might establish that approximately 25% of the work was completed. Naturally, given this estimate we could determine the cubic yards of concrete put in place. It may be easier for the individual making the job progress report to think of progress in terms of percentages rather than quantities. Either way, we should get an accurate estimate of job progress as a function of the individual work item.

The information from the job progress form will be used for job control. Each month after it is used, the job progress form can be stored in the individual job binder where it can serve as evidence to back up job progress billing. There is a job progress form for each job and each serves only one job. Naturally, if the project exceeds a year's duration, the design of the form shown results in the need to use more than one job progress form.

Job Control Sheet — Part Two

Let us now return to the job control sheet. We previously discussed part of the form used for the estimating function; let us now discuss the *control* segment of the form, shown in Figure 9.19. On this part there is space for documenting the materials put in place for each work item on a project to the specified date. In addition, there is space for documenting cumulative hours for each identified craft for each individual work item. The first column, labeled "ACT. MAT'L. TO DATE," is used for documenting the actual material identified with specific work items that have been used to date. The source for this information may be the invoices received from material vendors. This information is independent of the monthly job progress report and, in some cases, will not be available.

The second column, labeled "EST. MAT'L. TO DATE," is used for documenting the materials put in place for each of the work items as indicated in the monthly progress report. This information comes from totaling the

WORK ITEM	ACT. MAT'L. TO DATE	EST. MAT'L. TO DATE	ACT. EST.	EST. MAT'L. TO DATE TOTAL EST.	SUPER-VISOR HRS	%	CAR-PENTER HRS	%	MASON HRS	%	LABORER HRS	%	FINISHER HRS	%	IRON-WORKER HRS	%	OPER. ENGR. HRS	%	TOTAL HRS	%	VR.
					JOB CONTROL SHEET (PART 2) PROGRESS																	

JOB CONTROL SHEET (PART 2) PROGRESS

MONTH_____ YEAR_____ JOB NAME_____ CODE_____

WORK ITEM
GS1
GC1
G01
EM1
E01
CF1
CF2
CF3
CF4
CF5
CF7
CF8
CF9
CR1
CR2
CC1
CC2
CC3
CC4
CC5
CC6
CC7
CC8
CC9
CP
C01
MB1
MB2
MC1
MC2
MC3
MS1
MT1
MT2
M01
SS1
SS2
S01
WS1
WF1
WD1
WT1
WM1
W01
TOTAL

Figure 9.19 *Job Control Form—Part 2*

material estimated to be in place as determined on the job progress form in Figure 9.18 (thus the job progress form provides input for the job control sheet). The number placed in the column represents a cumulative total to date for the estimated material put in place.

The next column, labeled "ACT./EST.," is used for determining the *ratio* between the actual material used to date versus the material that it was estimated would be put in place to date. This division of estimated material into actual material provides us a means of determining a wastage factor, or the fact that some material has been put in inventory and not put in place.

The next column is critical in that it indicates a *percentage* of material put in place versus the total estimate of material to be put in place. The column is labeled "EST. MAT'L. TO DATE/TOTL. EST." Actually this column is merely the division of the second column in the second part of the form by the material estimated for the job and shown in the first column of Part One of the job control sheet. All the information for determining the percentage of completion is on the job control sheet. This column will also be used as the basis for evaluating the efficiency of labor performance to date.

Let us now turn our attention to the columns for the 7 different labor types. There is a column for each labor category and, under each of these, 2 additional columns, one labeled "HRS" and one labeled "%." In the HRS column the accumulated labor hours for that craft for a specific work item are shown. The source of the information is the weekly hours form shown in Figure 9.16. Every month, a new job control sheet is filled out to reflect the progress concerning material and cumulated labor hours for each of the applicable labor categories.

The entries for the percentage column under each labor type are determined by dividing hours shown in the HRS column by the total estimated labor hours for each of the crafts as shown in Part One of the job control sheet. The total estimated hours should reflect any additional hours due to change orders that are documented in Part One of the form. The result is a calculation of percentage of labor hours expended for each of the labor types for each work item versus the total hours estimated. Based on the assumption that labor hours expended should correlate with work put in place, these percentages should correspond closely to Column 4, which represents the percentage of material put in place versus the total estimate of materials to be put in place for the entire project. A difference in the labor percentage expended versus the estimated percentage of material put in place should be investigated. There may be a good reason for the variance; for example, it may be that the comparing of *estimated* material put in place to the *total* estimated material for the project is a misleading indicator in selected instances. Sometimes much of the labor that is required to put in a given amount of material is at the beginning of the work. However, this is not the rule in regard to construction work. It is more likely that the labor-expended percentage is higher than the work-performed percentage because that labor is inefficient or there is a productivity problem. By comparing the percentage of labor hours to work performed, the construction firm has a mechanism for determining inefficient operations and therefore taking corrective action. This can be done on each work item that is singled out as part of the project management system.

A column, labeled "TOTAL," is also shown for the accumulation of all the labor hours expended to date. The entry in the HRS part of the TOTAL column is merely the sum of the hours shown in each of the 7 different labor types that were used for a given work item. The entries in the % column are merely the hours indicated in the HRS column divided by the total labor hours for a given work item as established in part one of the job control sheet.

Finally, the last column, labeled "VR," can be used for several purposes. It is intended to be an indicator of a total projected variance in regard to the labor hours. It is beneficial to project the total labor hours based on the over- or underestimate of labor hours to date. For example, if the total estimated hours for a given work item were 100 and the work is currently 50% complete with an indicated 100% overrun, then the projected labor hours shown in the last column would be 200. The bottom row of the job control sheet can be used to calculate the total labor hours by craft for the entire project. This can serve as a check when compared to the hours documented in payroll for the individual crafts.

The job control sheet serves as an important link in the construction firm's ability to control and correct project inefficiencies. The project information that flags project progress as to material and labor is an outgrowth of the data collected in the field. Processing the information and filling out the job control sheet is a clerical function, and the mathematics is limited to addition and division. The monthly job control sheet can be filed in individual job binders which can serve as a single location for filing the project job estimate form, the weekly hours forms, the job progress form, and the job control sheet forms.

Work-Item File Form

The benefits obtained from the information placed on Part Two of the job control sheet are not limited to project control. In particular, the cumulated material and labor performance information can be used for estimating future projects. This is done by using the work-item file form shown in Figure 9.20.

The work-item file form is designed so that there is a single form for each work item. If the firm has defined 44 different work items for its project management systems, then there would be 44 different work-item forms. The first column on this form is used to identify the job for which the information in the following columns is being documented. The completion date of the project is placed in the second column of the work-item file form. The material put in place for the given work time, as determined from the final job control sheet, is placed in the third column. The next three columns, labeled "C = ," are used for documenting the total labor hours expended for 3 different crafts used on a specific work item. The 3 columns are in recognition of the fact that probably there will be no more than 3 labor types per work item, given the definition of work item we have employed. For example, when placing concrete we might utilize laborers, concrete finishers, and carpenters.

The productivity for each labor type is determined by dividing the total hours for that type of labor by the estimated quantity of material shown in Column 3. The resulting productivity figures would be entered in the columns labeled "PROD." Naturally, if fewer than 3 labor types are used on a specific work item, then less than 3 C columns would be used.

The next column shown on the work item file, labeled "TOTAL," is merely for summing the total hours for all the labor types for the project. The total productivity for all the labor types is the sum of all the labor hours divided by the estimated material for the work item for the project being entered.

The next column, labeled "PROD/CPRD," is used as an indicator column for measuring and evaluating the project's total labor productivity versus the typical productivity for this type of work item as indicated by past projects. Obviously, if this were the first project being entered into the system, there would be no number for the denominator in this calculation. However, as time went on and more projects were entered, an average cumulative productivity would be determined and used.

The next column on the work-item file has a checkmark in the heading. This column is used merely to indicate whether the data for a given job entry are to be made a part of the cumulative averaging of data for the work item in question. Sometimes a project is performed that has unique factors that make the related data unrepresentative of future performances. In that case, the firm may not want to average those data into other productivity data in establishing a norm for the operation. In that case, no entry would be made in the checkmark column. On the other hand, if the data are considered representative, a checkmark would be made.

Assuming that data from a project are to be entered into a cumulative or average set of information, the information would then be added to a running total of previous projects and averaged into calculations for specific labor-type productivities and a total labor productivity. The hours and productivity are indicated as cumulative hours and cumulative productivity by a capital "C" in the cumulative calculation columns.

When the firm is faced with determining an estimate for future projects, it can refer to the cumulative data in the individual work-item file to aid it in determining the number of manhours for an upcoming project. It may be that while the cumulative data are of interest, an upcoming project may have characteristics of a previous project that was entered into the system at a prior date. In that case, the information for that specific past project may be extremely useful in determining the estimate for the future project. The information on the individual work-item file form stays current by continuous

WORK ITEM FILE

PAGE _____   UNITS _____   WORK ITEM NAME _____   CODE _____

JOB	COMP	EST. MAT'L.	C.___ HRS	PROD	C.___ HRS	PROD	C.___ HRS	PROD	TOTAL HRS	PROD	PROD CPRD	√ ENT	CUM. MAT'L.	C.___ CHRS	CPRD	C.___ CHRS	CPRD	C.___ CHRS	CPRD	TOTAL CHRS	CPRD

Figure 9.20 *Work Item History Form*

updating from new projects. In addition, because the information is productivity data rather than unit-cost data, it probably will be current for a relatively long period of time. Productivity information is much less sensitive to change across time than is unit-cost information.

It should be noted that individual work-item file forms are used to collect data from several projects. Therefore, they differ from the forms previously described that collect data for a specific project. This means that the work-item file forms should be stored and filed in a different manner. In particular, the individual work-item file forms can be stored and filed as a group in an individual binder. Specific data regarding a work item can then be traced by merely turning to the specific form. As with work-item codes, the firm probably will want to establish codes for specific jobs so it can cross-reference the entries in the work-item files to the specific job binders. The form that we have discussed provides space for indicating a code for a specific job. These codes can be designed in many ways, although one with significance aids communication. For example, the job code may indicate in some manner the location of the job and perhaps the year in which the job is either started or completed.

The entire project management system that has been described and illustrated in Figures 9.12 to 9.20 repeats itself as the firm undertakes new projects. Data for the new projects come from the performance of past projects, while data from ongoing projects are entered in with the past project data. The cycle is a continuing one. As the firm continues to utilize the system, the data that are part of it become more accurate and reliable.

The project management system described in Figure 9.13 is now complete. Various miscellaneous forms may be designed to supplement the system; for example, the firm may want to design a change order form or an estimate recap form that includes overhead and profits.

The entire system is keyed to productivity; that is, much of the estimating and controlling function is tied to documentation of labor hours rather than labor cost. The benefit of this procedure has been discussed; however, it may be useful for the firm to cross-check a payroll cost against the system of documentation of labor hours. This cross-check can be made easily by adding the labor hours and multiplying by the relevant rate. The result should be a dollar value equal to the dollar value of payroll cost of the individual labor type. This provides a procedure for checking to determine that the payroll function and the estimating and control functions are being performed without error. Differences in labor dollars as documented in the labor-hour system versus the payroll process should be investigated and corrected.

EXERCISE 9.1

A mason contractor who is planning an upcoming project has determined that 112,500 standard bricks will have to be put in place in an 8-inch wall. The contractor attempts to integrate the estimating, activity planning, material ordering, cash-flow analysis, project planning, and project control by using a system that uses common cost objects for all of these functions. The "placing of standard brick in a straight wall" is identified as one of these cost objects or work items.

In addition to having performed the quantity take-off and established that 112,500 bricks have to be placed, the contractor has determined the following information.

Each brick will cost $.65, and half of these (56,250) will have to be purchased before the project starts. This invoice must be paid by the contractor on the first day the contractor starts the mason work. The second half of the required bricks must be purchased midway in the scheduled work span, and that invoice will be payable on the last day of construction of the brick wall.

The contractor will use 5 masons, 5 laborers, and 1 foreman each day to perform the work. Each mason is expected to place on the average of 750 bricks per day. It is assumed that the duration of the work is dictated by the masons' productivity. (Laborers and foreman can be considered support resources and their output is assumed to be dictated by the masons.) The hourly rate for each of the 3 crafts is $15 per hour for each mason, $12 per hour for each laborer, and $16 per hour for the foreman. Each worker is to be paid weekly—at the end of each week for the hours incurred during that week.

The contractor will also have to rent scaffolding equipment to perform the work. Because of the different sections of the wall for which it has to place bricks, it will need two different types of scaffolding. For the initial one-third of the work duration it will need a certain type of scaffolding that rents for $120 per day. The cost for renting this equipment is due in total on the first day the scaffolding is rented. For the last two-thirds of the work duration, a more complex scaffolding system is required that rents for $160 per day. This cost is also due in total the first day this scaffolding system is required.

The masonry work is planned such that once the contractor starts the work, he will work 5-day weeks, 8 hours a day, continually until the work is completed.

Subsequent to planning and estimating the project, further assume that the contractor receives the contract to perform the work. After 10 days of work, the contractor determines that 30,000 bricks have been placed.

Based on the above information and consistent with the concept of integrating the project functions of estimating, activity planning, resource scheduling, cash budgeting, and control, determine the following:

   a. Activity duration in workdays and workweeks.
   b. Total number of mason, laborer, and foreman hours required to perform the work.
   c. Total direct cost of the work to include the direct labor, material, and equipment for performing the work.
   d. The contractor's outflow of cash as a function of time for the project. Specifically determine the contractor's cash outflow at the end of each week during the duration of the work.
   e. Assuming that the contractor's productivity for the first 10 days (i.e., 30,000 bricks placed in 10 days) represents the same productivity that will occur for the remainder of the project, calculate the total over- or underprojected direct labor cost for the project.

EXERCISE 9.2

A contractor has determined that he will use a carpenter crew of 6 men and 2 laborers to place 10,000 square feet of forms (contact area). Assuming that the productivity of a single carpenter is 6 manhours per 100 square feet of contact area (sfca); that the 2 laborers simply assist the carpenters—i.e., they do not put up forms; and that the wage rate for a carpenter is $16 per hour and for a laborer is $15 per hour, calculate the following:

    a. The direct labor cost for doing the work.
    b. The duration required to do the work in manhours.
    c. The duration required to do the work in crew hours.
    d. The duration required to do the work in workdays.

EXERCISE 9.3

A contractor has a cost of goods sold that consists primarily of its labor cost, equipment cost, and subcontractor costs. Material that is normally a part of a contractor's cost of doing work is purchased directly by the contractor's project owner; the contractor, in effect, contracts to put the project owner's material in place.

Because of the nature of the work the contractor performs, he primarily uses only one labor craft. He maintains approximately 8 major pieces of equipment. His average project duration is on the order of 25 working days.

The contractor wants to develop an information flow or system for each of the 3 areas of labor, equipment, and subcontractor costs. His objectives are to develop a set of forms that allows him to plan work, resources, time, and cost, to control the work, resources, time, and cost, and to integrate all project management functions and accounting functions via the forms and information flow. With these objectives as the criteria, address the following problems.

The following problems relate to the contractor's labor costs:

1. The contractor has designed the labor-cost work items so that there are 20 work items that any single craft, such as carpenters, can charge labor hours to on any one day. The contractor believes that on any one day, a worker will not work on more than 5 work items. He further assumes that a worker will not work on more than 2 projects in any one day. The contractor wants to implement a daily time-card in order to enhance the collection of accurate field data.

    Given the above circumstances, design a timecard for the contractor.

2. The contractor's projects are of a relatively short duration, with no single project taking in excess of 30 days (6 weeks). The contractor wants to design a single form for the estimating of labor hours and the subsequent monitoring of the labor hours throughout the project. There is to be a single form for each project which is to report on the possible 20 work items a contractor performs. The monitoring

and control of the single labor craft—e.g., carpenters—is to be achieved by comparing the percentage of labor hours expended to total labor hours budgeted for each work item, to the percentage of material put in place to date. This control comparison is to be performed weekly via the form. The labor hours reported on the form are to come from the timecard designed in Problem 1 above. The weekly material put in place is to be reported weekly by the project superintendent via the use of a progress report.

Given these considerations, design a project estimate/control form for the contractor.

3. The contractor wants to design a single form that will assist in planning work on a daily basis in regard to the work item to be performed, the number of carpenters required on any one day, and the payroll cost expense (cash outflow) on a daily basis for carpenters. This project plan form is to be designed so it is compatible with the fact that the project could have as many as 20 work items and take as long as 30 days to complete. Further, the form is to be designed in a time-scale CPM/bar-chart format. This form will be used by the contractor both to plan a project and subsequently to monitor the project. The above information is to be placed on a single form. This form will utilize the same work items as those used for the timecard in Problem 1 above and the project estimate/control form in Problem 2.

Given the above considerations, design a project planning and scheduling form.

The following problems relate to the contractor's equipment costs. In order to determine accurate equipment costs for a project and subsequently control equipment costs, the contractor has decided to implement an equipment costing system that charges a project for the use of a piece of equipment and charges individual equipment accounts for actual equipment expenditures (see description of an equipment costing system described in Chapter 8 and illustrated in Figures 8.8 through 8.13). This system is to be implemented for 8 different pieces of equipment.

4. The contractor has established hourly equipment rates for each of the 8 pieces of equipment in his system. He wants to estimate the equipment cost for a project by estimating the number of hours required per piece of equipment on each of the 20 different work items that the firm might perform. (*Note:* these would be the same work items used for the direct labor-hour estimating and control.) Design a single form that can be used by the contractor for estimating the equipment cost for a project as a function of the hourly rate for each piece of equipment and the estimated hours of use on the project as a function of individual work items.

5. In order to charge a project for the use of a specific piece of equipment, the contractor wants to design a daily reporting form, to be filled out by a superintendent or foreman, that records the productive hours of use of equipment on a single project as a function of the work item or work items on which it is used in any one day. The contractor has established that any one piece of equipment will be used on only one project per day and no more than 4 work items per day. On the same form the contractor wants to have the superintendent or foreman report on actual maintenance or repair (or any other operating expenses) expenditures for

a specific piece of equipment on the day in question. The equipment hour reporting on this form will be used to aid in the control of equipment hours on the project and the establishing or revision of hourly equipment rates. The reporting of actual expenditures on this form will be used to establish or revise hourly equipment rates and ownership costs.

Design a daily reporting form consistent with the above set of considerations.

6. In order to establish and revise an hourly rate for each piece of equipment, the contractor wants to design a single form for each piece of equipment that will be used to keep track of actual productive hours of equipment use and all expenditures for the specific piece of equipment, and periodic depreciation entries and any interest cost associated with owning the equipment.

Design a single form for a piece of equipment that can be used for performing the above requirements for the contractor.

Most contractors subcontract a portion of their work to another contractor who is referred to as a subcontractor. One of the more difficult aspects of contractor control of construction schedules is getting subcontractors to comply with the contractor's overall schedule. In effect, the contractor is often at the mercy of his subcontractors in regard to adherence to a schedule. The successful contractor must "manage" his subcontractors if he is going to deliver a project on schedule. The following problem relates to the previously described contractor's ability to monitor and control his subcontractors.

7. Assume that the contractor usually awards approximately 5 subcontracts per project. He has decided that, rather than relying on the subcontractors to prepare their own schedules, he, the contractor, will provide each subcontractor for a project with a form that shows the work that the subcontractor is going to do on a given day. The objective will be for the contractor to help each subcontractor fill out the form at the start of a project; subsequently, the contractor will monitor the progress of the subcontractor via the use of the form. The contractor will use his common list of work items (i.e., those used for estimating and control) for performing this function. He assumes that no subcontractor will perform more than 6 work items on any one project. He further assumes that no one subcontractor will be at a project for in excess of 10 days. The contractor would also like to use this subcontractor plan form to calculate the cost of the subcontractor's work on a daily basis in order to determine the accuracy and timeliness of subcontractor billings.

Design a form to accomplish the above considerations.

10

Sitework and Excavation

*Excavation Work for a Building Project   *   Types of Excavation Work: Definition of Work Items   *   Excavation Take-Off   *   Pricing Excavation/Sitework   *   The Queuing Model: An Aid in Selecting the Optimal Excavation Method*

EXCAVATION WORK FOR A BUILDING PROJECT

The terms "sitework" and "excavation" are often used interchangeably in the construction industry. Nevertheless, some prefer to use the term "sitework" to refer only to surface work that includes cut-and-fill dirtwork at a project site, grading for parking lots and/or sidewalks, and landscaping type of work, including removal and planting of trees and sodding of grass. These same builders reserve the term "excavation work" to designate the removal or backfilling of earth for the substructure of a proposed project. For example, this may entail the bulk excavating of a basement for a proposed building or the digging of a trench for a foundation wall. In this chapter, we will not differentiate between the terms sitework and excavation for earthwork. The type of work involved, regardless of whether it is grading work or the digging of a deep foundation, is classified as Division 2 work in the 16-division classification system of the Construction Specification Institutes (CSI).

The amount of excavation work that must be performed by a contractor for a building construction project varies from almost none to an extremely expensive and complex excavation of rock-type material or the fabrication of caissons in conjunction with the construction of deep foundation piers. More often the excavation work for a building construction project is somewhat standard to include cut-and-fill work, grading for concrete parking lots and sidewalks, and the removal of earth for the building foundation.

The amount of work and the cost of the work for excavation as a percentage of the total project cost tends to be significantly different for building construction projects and heavy and highway projects. Excavation work for a heavy and highway project is usually a significant part (in some cases as high as 50%) of the total project cost. Many cubic yards of earthwork may need to be cut or filled to make way for a proposed roadway, bridge, or dam.

Large and expensive construction equipment fleets including scrapers, bull-dozers, and blasting equipment may be needed to perform the work.

With some exceptions, the cost of excavation work for a building construction project is seldom much greater than 10 or 15% of the total project cost. If the proposed building project involves a very large or tall construction requiring a large and/or complex foundation, this excavation cost as a percentage of the total project cost increases significantly.

While the total cost of the excavation work for a building project may be relatively small relative to other costs, such as for concrete, masonry, or carpentry work, the need to accurately take off and price this work remains important. An estimate for any project is only as accurate as the accuracy of each of the component parts, which include the earthwork estimate.

A somewhat unique aspect of excavation work and its pricing is the fact that, unlike other kinds of work, there often is no material cost component for it. Frequently, excavation work merely entails moving dirt from one location to another, or the excavation and removal of dirt from the building site. The contractor incurs equipment costs and labor costs in performing the work. Some of the labor costs relate to the manhours required to operate equipment, whereas other labor hours relate to the use of manual labor in digging or backfilling dirt.

Some building projects, however, involve purchase material costs for excavation work. For example, a contractor may be required to purchase a high-quality granular type of material for use as backfill around a building foundation. But in any case, the purchase material cost is less significant for excavation work than it is for most other construction-related types of work. The focus of attention in regard to an excavation estimate centers around the *labor cost* (sometimes referred to as the installation cost) and the equipment cost.

Excavation work may or may not be performed by the general contractor for a project. Some general contractors own equipment that enables them to perform what might be considered "standard" type of excavation work. This would include cut-and-fill work, grading, and the digging of a small or medium-sized foundation. Equipment needed to perform these types of work would likely include a back-hoe and perhaps a dozer and blade.

Given the relatively high cost of construction equipment and the relatively short duration in which excavation work takes place for a building project, many general contractors have judged it more feasible to subcontract excavation work. A subcontractor specializing in excavation often can get more utilization from its excavation equipment than can the general contractor, and therefore has a lower cost per hour of use of the equipment. Nevertheless, given the fact that some general contracting firms do in fact perform the excavation work, we cover the take-off and pricing of this work in this chapter.

TYPES OF EXCAVATION WORK: DEFINITION OF WORK ITEMS

As was discussed earlier in the book, fundamental to the preparation of an accurate estimate is the definition of specific types of work—i.e., work items to be taken off the drawings. The definition of work items has to be made consistent with both the quantity take-off function and the subsequent pricing of the work. The cost of excavation work consists in great part of equipment costs. Because of this, and given the fact that varying types of construction equipment perform only certain types of excavation work (e.g., a blade is somewhat limited to cut-and-fill work), the type of equipment needed to perform work should strongly influence the definition of specific excavation work items.

Cut-and-Fill Work

Owing to the need to preserve the drainage of surrounding property, building construction sites are seldom drastically changed in regard to the sitework around the building. However, it is common for the contractor to have to perform some dirtwork around the building for proper drainage or improvement of the appearance of the building lot. This work usually entails moving a relatively thin layer from one location on the site to another. This "leveling" or changing of the contour of the land is commonly referred to as *cut-and-fill work*.

Cut-and-fill work is not performed underneath a proposed building. The work is limited to the area around the building. A dozer with a blade that can scrape and push a relatively thin layer of the soil is used for the work. Cut-and-fill work is commonly defined as a work item.

Cut-and-fill work is taken off in the measure of cubic yards. The amount of cut—the dirt that has to be removed from a location—should not be netted or subtracted with the amount of required fill—dirt that has to be put in place at a location—to determine the total cubic yards of cut-and-fill work. Instead, the total cubic yards of cut work required should be added to the total cubic yards of fill required in order to determine the total take-off quantity. This method of calculating the cut-and-fill quantities is in recognition of the fact that two operations are required to cut at one location and subsequently fill at another location. Obviously, the fact that a contractor has to cut 50 cubic yards of dirt from one location and fill another location at the same site with the 50 cubic yards of cut material does not mean that *no* work is required; 100 cubic yards of cut-and-fill work would be required (50 cubic yards of cut plus 50 cubic yards of fill).

Grading

Somewhat similar to cut and fill work is the work related to ready a site for the pouring of concrete or placement of a bituminous material for parking lots or walkways outside the proposed building. The contractor might also utilize a blade to perform this construction work. This type of work is referred to as *grading*.

Grading work usually has to be performed more precisely than cut-and-fill work. While it may be permissible for a contractor to be only relatively close in matching up the cut-and-fill work to the exact elevations shown on working drawings, the grading work for concrete or bituminous slabs must be quite exact. Concrete or bituminous parking lots, driveways, or sidewalks typically tie to floor elevation (e.g., the first-floor elevation, parking-lot drainage structures, etc.); consequently, the grading work needs to be performed carefully and precisely by the equipment operator. Because of this need for care and preciseness, the cost of grading work is typically more expensive than cut-and-fill work (assuming that both types of work are taken off using the same units of measure, which may not be the case). The result is that the estimator usually defines grading as a work item in preparing an excavation estimate.

Excavation grading work is frequently taken off in units of square feet of area instead of cubic yards. This is in recognition of the fact that the cost of doing the work is usually most dependent on the area to be graded. Grading is usually performed after cut-and-fill work has been completed on the area that is to be graded. The result is that only a thin layer of dirt needs to be "leveled" in the grading operation. It takes an equipment operator approximately the same time to grade or level 2 inches of dirt as it does to grade or level 1 inch. So we can see that the grading work effort is for the most part independent of the volume of material. The area to be graded is more indicative of the work effort and costs that will be required.

Landscaping

Besides the cut-and-fill work and the grading work, a contractor may have to perform other non-foundation type of excavation work. Included in this work may be the removal of trees, the planting of trees, and the laying of sod. However, this type of sitework improvement is commonly subcontracted to a landscaping specialty contractor. Thus, plus the fact that the take-off and estimating of these items present few problems, we will not discuss this area further.

Trenches

Another type of sitework that may be required for a project is the digging of *trenches* on the building lot for the placement of utilities, including water and disposal pipes. The digging of these trenches and the placement of the utility piping may be the responsibility of the contractor or the utility company. Often the contractor has the responsibility of placement from the building to the property line. The utility company in turn may do any additional work necessary to bring the lines or pipes to the connecting lines.

Digging an excavation for the placement of pipe is commonly done with a trenching machine. A trenching machine, in one sweep, digs a relatively narrow trench to the desired depth. The width of a trench can be varied by using trenching machines of varing capacity or by changing the attachment blades on a single trenching machine.

The cost of digging a trench is somewhat independent of the width of the trench. For example, the equipment and labor time required to dig a 6-inch-wide trench is not significantly different from the cost to dig an 8-inch-wide trench. Therefore an estimator might choose to take off and price trenching in units of linear feet.

This measure is only valid if the trenches to be dug are of somewhat comparable widths and depths. While there may be little cost difference in a 6- or 8-inch-wide trench, a 2-foot-wide trench would result in a greater cost per linear foot because a larger-capacity trenching machine would be needed. The same problem in regard to the use of a linear foot measure would occur if the depth of the trenches varied significantly. If he must deal with widely varying trench depths and/or widths, the estimator is better served by taking off and pricing the trenching work in cubic yards.

In addition to the need to dig trenches for utility type of work, the contractor may also have to dig trenches in connection with the foundation for the building itself. In particular, if the substructure is designed as a continuous wall foundation without a basement slab, or if it involves what is commonly referred to as a **grade beam**, the foundation can be most economically excavated by means of trenching.

A continuous wall foundation and footing without a basement is illustrated in Figure 10.1. If this type of design is required, it is not feasible for a contractor to dig out the entire volume of material under the ground-floor slab, then form and pour concrete in the foundation or walls and footings, and then backfill the entire volume of dirt. Instead, he would more likely dig a trench along the line of wall. The width of the trench would be dictated by the footing width. In order to make possible the placement of concrete forms and the pouring of the concrete (or alternately, the placement of concrete blocks), the trench width would likely be a foot or more wider than the footing width.

Grade-beam construction is illustrated in Figure 10.2. A grade beam is similar to a continuous foundation wall except that it is essentially supported by individual footings, piers, or perhaps piles, rather than having a continuous footing. Grade-beam construction is commonly used when the contractor must deal with widely varying soil conditions. For example, on one side of a proposed building the soil may be very weak and thus piles are

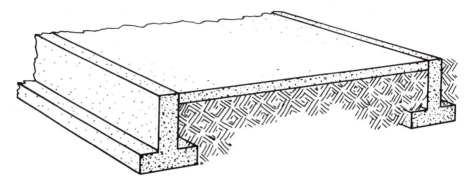

Figure 10.1   *Example Slab-on-Grade*

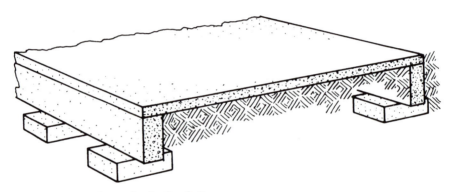

Figure 10.2   *Example of a Grade Beam*

needed to support the grade beam. If there is strong soil on another side of the same building, only a small footing may be needed to support the grade beam.

Independent of the reasons for the design of the foundation systems shown in Figures 10.1 and 10.2, both are commonly trenched. The equipment that can be used for digging the types of trenches needed for these foundation systems, as well as those for utilities, is not limited to trenching machines. A back hoe would provide the contractor another alternative. However, because the productivity and the capacity of a back hoe varies from those qualities for a trenching machine, it is necessary for the estimator to correctly conceptualize the equipment that will indeed be used in order to accurately price the work.

Bulk Excavation

The one excavation work item that is common to practically every building construction project involves the process of excavating earth for the basement of the proposed building. This excavation work is commonly referred to as *bulk excavation* or mass excavation. The bulk excavation work for a project usually entails digging out the entire area of earth underneath a building to a depth equal to the bottom of the foundation or basement floor slab.

The contractor commonly utilizes a back hoe to do the bulk excavation work. The work is taken off in cubic yards. The amount of cubic yards of bulk excavation work that must be dug for a typical building foundation is considerable. For example, a 10-foot-deep foundation for a building that has a cross-sectional area of 40 feet by 60 feet (*Note:* this can be considered a small building) would require 24,000 cubic feet or 889 cubic yards of bulk excavation.

Because of the relatively large amount of work that is typical of bulk excavation work and because the work is repetitious in regard to the work process, the cost of bulk excavation work is relatively low compared to other types of excavating work items. However, if there is difficult soil (e.g., rocky), or if water enters the foundation, the cost of bulk excavation work will increase significantly.

Bulk excavation work for a building is actually usually dug beyond the perimeter of the building itself. One reason for this is that the contractor has to provide space to allow the workers to form walls, place reinforcement, pour concrete, or place masonry block. Secondly, a relatively deep foundation cannot be cut vertically. Owing to the properties of most soils, the earth cannot structurally stand vertically. It tends to slope to what is defined as its "angle of repose." Various soils have different angles of repose, as illustrated in Figure 10.3.

The point is that because an excavated soil cannot usually be cut vertically, the properties of the soil must be identified, and then the soil sloped as

REPRESENTATIVE VALUES OF THE
ANGLE OF REPOSE FOR DRY SAND

	Round Grains, Uniform	Angular Grains, Well Graded
Loose	28 degrees	34 degrees
Dense	35 degrees	46 degrees

Figure 10.3 *Example of Angles of Repose*

part of the bulk excavation work—and this incurs further expense. This sloping of the bulk excavation is shown in Figure 10.4.

If a bulk excavation is not sloped, an alternative is to **shore** the cut sides of the excavation. A shore is in effect a vertical form that "holds back" the earth pressures. This technique is commonly used when a contractor finds it

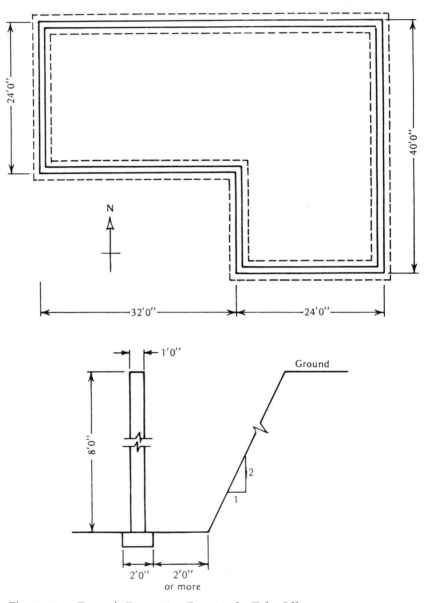

Figure 10.4 *Example Excavation Drawing for Take-Off*

impossible to slope the excavation because adjacent buildings or roadways prevent digging the foundation much beyond the building perimeter. Shoring of the excavation walls might also be done if there is a need to prevent water seepage into the excavation during construction. This seepage might take place if the excavation is of a depth considerably deeper than the water table.

Consideration of the water table, adjoining buildings and roadways, and the type of soil to be excavated would all in part dictate whether the contractor sloped the bulk excavation, cut or shored the sides, and it would also dictate the distance of the excavation beyond the building perimeter. In the absence of particular considerations, we will assume a vertical bulk excavation cut 2 feet beyond the outside perimeter of a proposed building when the bulk excavation work item is being taken off. This extra 2 feet will account for the possible need for the cut to slope so that there is more than a 2-foot distance on the top and less than this amount on the bottom of the excavation.

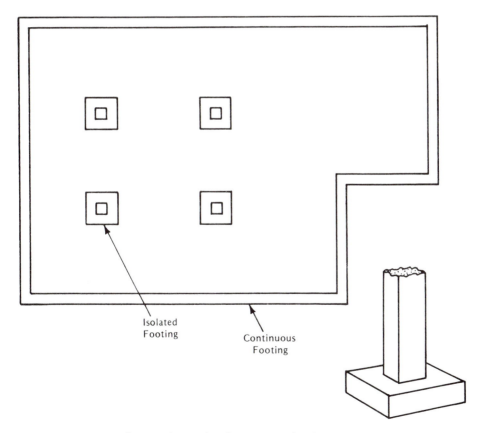

Figure 10.5 *Example Use of an Isolated Footing and Column*

Isolated Footings

Another type of building foundation element that needs to be excavated are *isolated footings*. A footing is essentially a means of distributing a concentrated load over a larger soil area. Footings often in turn support columns, as is illustrated in Figure 10.5.

An isolated footing is also referred to as a *pad footing*, or sometimes a *column footing*. Footings vary in size. For relatively small building projects, the footing depth is shallow—e.g., 1 or 2 feet. The cross-sectional area of the footing might be as small as a few feet square or as large or larger than 10 feet square.

Isolated footings are either excavated with special equipment or by manual labor. If the footings are large or there are many footings, the excavation is typically dug with equipment. On the other hand, for a building requiring only a few small footings, the contractor may choose to manually dig the footings.

Isolated footing excavation is usually measured in cubic yards of excavation. If all the footings are of equal depth, the estimator might alternatively choose to take off the excavation in units of square feet. When a cubic yard measure is used, the cost per cubic yard for digging the footing decreases as the footing size increases. This reflects the fact that the smaller footing requires as much "set-up" time to dig as does the larger footing.

Isolated footings are often designed to be part of a slab foundation system. This is illustrated in Figure 10.5. The individual footings are in effect part of a built-up slab system. The slab and built-up slab (i.e., the isolated footings) are poured in a single concrete pour. The built-up slab or isolated

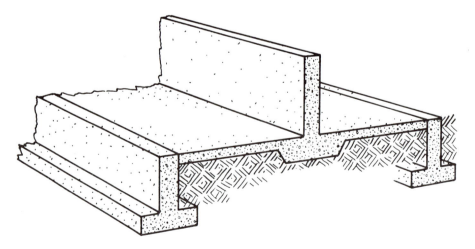

Figure 10.6   *Example of a Built-Up Slab*

footings shown in Figure 10.6 typically support columns that sit on the slab at the location of the footings.

The built-up slab construction shown in Figure 10.6 demonstrates how important it is for the estimator to be able to conceptualize the way the contractor will excavate the required work. In digging the foundation system shown in Figure 10.6, the contractor has at least two choices. He could bulk-excavate under the entire building to a depth defined by the bottom of the isolated footings. Having formed the footings, the contractor could backfill the soil inside the footings to the bottom of the slab. This would require considerable work—which would be somewhat redundant because some of the bulk-excavated earth would then need to be returned (backfilled). On the other hand, this approach to digging the foundation would eliminate the need to dig out the individual footings. This is an advantage in that it requires significant manhours or equipment hours to dig out individual footings to the exact size.

A second alternative to digging out the design shown in Figure 10.6 is for the contractor to bulk-excavate the area under the building to the bottom of the slab itself. He could then in turn dig out the individual built-up sections (the isolated footings). More likely than not, he would utilize this approach. Seldom would a contractor choose to excavate material only to have to subsequently have to put much of it back. However, the point here is that the estimator must be able to conceptualize the actual construction process to be used to be able to properly take off the quantity of the work and price the work.

Backfilling

The term "backfilling" has been used several times in the previous paragraphs to describe various types of excavating work. Backfilling is essentially the placement or replacement of a material around a building foundation member. As has been discussed, the contractor typically has to cut back an excavation further than the actual structural foundation member will go in order to provide craftsmen room to work. This space has to be filled up again after the structural member is placed (i.e., after the concrete is poured or the masonry blocks are placed). Often the material that is put back, or backfilled. is the material that was originally excavated. However, sometimes the need for a higher-quality or more or less pervious material may

Cut and Fill Work	Trenching
Fine Grading	Hand Excavation
Bulk Excavation	Backfilling

Figure 10.7 *Example List of Work Items for Excavation/Sitework*

result in the contractor having to purchase backfill material. Independent of whether material has to be purchased or not, the backfilling operation has to be recognized as a cost in the estimate. As shown in Figure 10.7, it is defined as a work item.

Hauling

Usually there will be excess earthwork left after a building foundation has been excavated and backfilled. This extra material has to be hauled from the job site by the contractor to a location where it will be dumped. The actual hauling cost is determined by the distance the material has to be hauled as well as the quantity of material that must be hauled. Given excess earth from the excavation process, the estimator has to recognize hauling as a work item in the excavation estimate.

Miscellaneous Work Items

In addition to the common excavation work items listed in Figure 10.7, the contractor may have to perform less frequent types of excavation work, including removing water from the building site when it is necessary to make possible the construction of the foundation system; driving piles under footings; or excavating for large piers. In addition, the contractor often has to "dampproof" the fabricated foundation with a bituminous material to protect it from water. All these items are also considered part of the excavation estimate. However, given the fact that these types of work result in few estimating problems and may be somewhat infrequent in occurrence, we will not discuss their take-off or pricing in this chapter.

EXCAVATION TAKE-OFF

When he is taking off any type of quantities from the drawings, to include excavation work, the estimator has to decide upon the degree of accuracy needed in calculating the quantities. Normally, he does not have to be as precise in his take-off of excavation work items as he does for other types of work. This does not mean that the estimator should be careless in his take-off calculations for excavation items. However, given a trade-off between time and accuracy, and given the uncertainties that characterize excavation work, the estimator should know that it isn't necessary to spend quite as much time striving for accuracy in the take-off of excavation work.

There are essentially two reasons for this decreased need for accuracy regarding sitework and excavation. As was discussed earlier, the exact boundaries of any excavation work item are not always set out on the drawings. For example, a building basement is dug out beyond the outside perimeter of the proposed building in order to leave space for the craftsmen to work. The actual distance the excavation is "dug back" will be determined only when the excavation is actually made. The point is that the estimator has to use "educated guesswork" to a certain extent in his calculation of quantity of work.

A second reason is that the cost of performing a unit of excavation work relative to the cost of other non-excavation work items is less, as we will see in a following section on the pricing of excavation work. In fact, the cost of performing excavation work is usually only a few dollars per cubic yard of material. This compares to well in excess of $100 per unit of work for some other work items. Thus the impact of a slight quantity take-off error will not usually significantly affect the total project estimate. Neither will it adversely affect the purchasing function, because ordinarily there is no need to purchase excavation material.

In spite of our discussion about not being overly concerned about the accuracy of the excavation take-off, the estimator *should* seek to utilize efficient take-off procedures that promote all possible accuracy within the time allotted. Given this objective, we will discuss specific efficient practices relevant to individual excavation work items.

Efficient Take-Off Practices

Procedures for Cut-and-Fill Work: Perhaps the one excavation work item that requires the most take-off time and effort is the cut-and-fill work. As we saw earlier in the chapter, the amount of cut-and-fill work can vary from virtually none to a significant amount. Unlike building construction projects, heavy and highway projects almost always involve considerable cut-and-fill work. Determination of the amount of cubic yards of cut-and-fill-work required for a heavy and highway project is usually accomplished by what is referred to as the **end-area method**. Using a topographic and profile plan such as the one illustrated in Figure 10.8, end areas at regular intervals (referred to as sections) are drawn. By means of an averaging process between sections, a volume of cut or fill can then be determined. Because this method is used almost exclusively for heavy and highway work, it will not be emphasized here.

Similar to the use of the end-area method for determining the cut-and-fill work for heavy and highway construction is the use of the **four-point method** for building construction. To illustrate the use of this method, let us consider the plot plan for a proposed building and site as shown in Figure 10.9. The existing contour lines in this figure differ from the proposed contour lines—that is, some of the existing land will have to be cut and/or filled to the proposed elevations.

The first step in determining the cubic yards of cut and fill work is for the estimator to draw a grid on the drawing shown in Figure 10.9. This procedure is illustrated in Figure 10.10. In effect, the grid breaks down the entire site plan requiring various cuts and fills into smaller grids or pieces of the overall drawing, each requiring an amount of cut-and-fill work.

The grids shown are equal on all four sides—that is, they are squares. The size of the grids is left to the discretion of the estimator. However, the

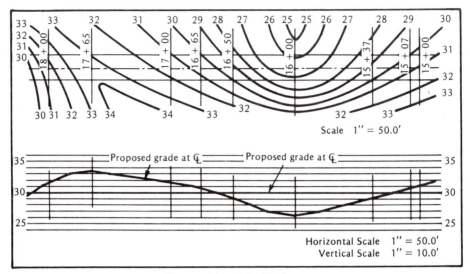

Scale 1″ = 50.0′

Horizontal Scale   1″ = 50.0′
Vertical Scale   1″ = 10.0′

Topographic and Profile Plans.

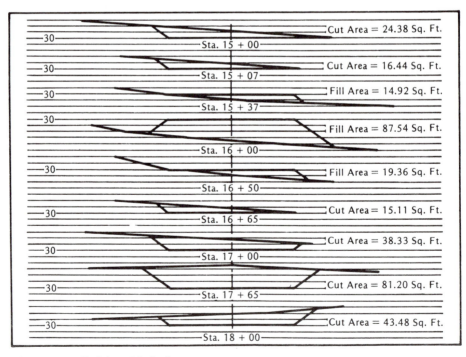

Figure 10.8   *End Area Method*

Property
Line

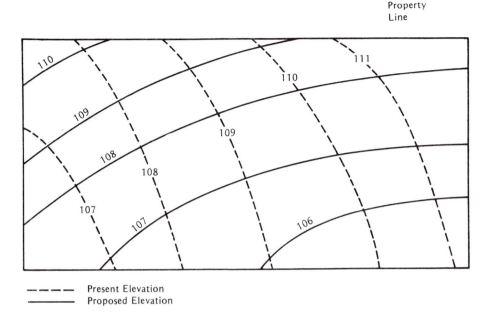

— — — — Present Elevation
————— Proposed Elevation

Figure 10.9 *Example Plot Plan*

size should be drawn with the objective of having the corner of the grids somewhat intersect adjacent contour lines. If the grids are too large, the subsequent calculation process will be inaccurate because their corners may overlap too many contour lines. On the other hand, if the grids are too small, too many corners will fall between adjacent contour lines and the estimator will have to make too many approximations, the result being wasted take-off time and detail relative to the accuracy attained.

Given the scale of the construction drawing shown in Figure 10.10, the grid is selected as 100 feet long on each side. With the drawn grids and using the four-point method, we can calculate the quantity of work to be performed for the cut-and-fill work item. Let us consider one of the grids shown in Figure 10.10—in particular, grid #101. By observing the contour lines on the drawing shown in Figure 10.10, we can determine a required cut or fill at each of the corners of grid #101 by comparing the elevation of the existing contour at the corner of the grid in question to the proposed elevation. Usually the estimator interpolates to a cut or fill depth equal to one-half the distance between adjacent contour elevations. For example, if contour lines are drawn at intervals of a foot, then the cut or fill depths would be interpolated to the nearest one-half of a foot.

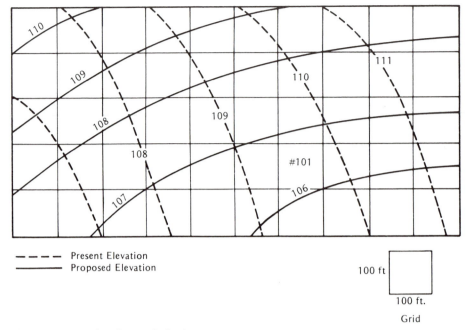

Figure 10.10 *Plot Plan with Grids*

In any one grid section it is possible that all the corners are cuts, all are fills, or some are cuts and some are fills. A drawing showing these possibilities is illustrated in Figure 10.11. To determine the cubic yards of cut and/or fill in any of the sections shown in Figure 10.11, formulas can be developed as follows:

$$Vc = \frac{L^2 \times (Hc)^2}{108 \times (Hc + Hf)}$$

$$Vf = \frac{L^2 \times (Hf)^2}{108 \times (Hc + Hf)}$$

Where:

Vc = Volume of cut in cubic yards
Vf = Volume of fill in cubic yards
Hc = Sum of cuts on four corners of grid squares
Hf = Sum of fills on four corners of grid squares
L = Length of side of grid squares (in feet)
108 cubic yards = Factor to convert to cubic yards

These formulas are what constitute the four-point method. The application of these two formulas to any of the grids shown in Figure 10.11 will yield the total cubic yards of cut and fill. Naturally, if all the corners of a grid are cuts, Hf will be equal to zero and therefore Vf will equal zero because the numerator of Vf term will equal zero. The same is true in regard to Vc – i.e., if all the corners are fills, Vc will equal zero.

Let us now apply the four-point formulas to grid number (2) shown in Figure 10.11. The calculated cubic yards of cut and fill are as follows:

$$Vc = \frac{100^2 \times 3^2}{108 \times (3 + 5)} = 104.2 \text{ cubic yards}$$

$$Vf = \frac{100^2 \times 5^2}{108 \times (3 + 5)} = 289.4 \text{ cubic yards}$$

The same process could now be applied to each and every grid shown in Figure 10.9. The total cubic yards of cut and the total cubic yards of fill can then be added for all the grids to yield the total cut-and-fill quantity of work to be performed.

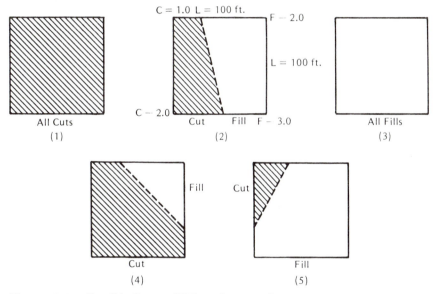

Figure 10.11   *Possible Cuts and Fills within a Grid*

Take-Off Procedures for Grading Work: While the same type of construction equipment might be utilized by a contractor to do cut-and-fill work and grading work, the quantity take-off for grading work is easier and less time-consuming than for cut-and-fill take-off. Grading work entails the leveling of surface material to the bottom of the elevation of a to be placed slab on grade. The slab is usually poured flat so that its bottom is at a level elevation. This differs from cut-and-fill work, which might be finished to varying contour elevations.

Given the fact that grading work is brought to a level elevation, the quantity take-off merely entails calculating the difference between the existing grade and the proposed level grade. Often this type of excavation work is of a somewhat common depth. Therefore, grading work is sometimes taken off in square feet rather than a volume measure. Assuming constant depth and a small depth of cut—e.g., 1 foot or less—a square-foot measure may prove more indicative of the cost of the work than a volume measure.

In taking off grading work, the estimator is often confronted with the need to calculate a quantity of sub-base material to be placed under the proposed slab, as well as the quantity of excavation material. The sub-base material usually consists of gravel that must be purchased and compacted by construction equipment. The calculation of this quantity of material presents few problems for the estimator because the depth and location of the sub-base material are easily identified on the drawings.

The largest excavation quantity of work to be taken off a set of building construction drawings is usually the bulk excavation work. Most buildings are designed to include a basement substructure. The cross-sectional area of the basement dictates the work to be excavated. Owing to the need for space in which the craftsmen can work, the perimeter of the excavation is usually extended 2 to 3 feet beyond the building foundation perimeter. The depth of the basement—the area that has to be bulk-excavated—is dependent on the building loads and the quality of the soil.

The calculation of the bulk excavation merely entails multiplying the cross-sectional area of the excavation by the depth of the excavation. These dimensions are usually stated on the drawings in feet. Therefore the estimator has to divide the calculated quantity by 27 to yield a cubic-yard quantity (the units in which bulk excavation work is costed).

The take-off of bulk excavation work does present one potential problem for the estimator. Because the depth of the bulk excavation may cause cave-ins of the excavated wall, the contractor will likely have to either *shore* the cut vertical walls or *slope* the excavation from the bottom of the excavation outwards. Shoring entails the vertical driving of sheet piling or the fabrication of plywood and/or sheathing against the cut walls. When this process is required, the estimator must be careful to include the work in his take-off. This point is especially important, given the fact that the shoring is not delineated on the drawings.

When the contractor must slope the cut walls—an approach to bulk excavation illustrated in Figure 10.4—he must determine the angle at which the embankment must be sloped. This depends on the type of soil. (The angle at which various types of soil will be stable were listed in Figure 10.3.)

In digging the bulk excavation as illustrated in Figure 10.4, the estimator can determine the volume of material in the sloped part of the excavation by first calculating the area of the triangular cut portion. The area of a triangle is given by the following formula:

$$A = \frac{b \times h}{2}$$

Where

    A = area in square feet
    b = base or width of the triangle in feet
    h = height of the triangle in feet

Once the area is determined, the result can be multiplied by the excavation perimeter to yield a volume quantity. Assuming a 1 to 2 slope and a 2-foot craftsmen working area, the quantity take-off for the bulk excavation of the building plan illustrated in Figure 10.4 is shown in Figure 10.12.

One other concern that might need to be addressed in the bulk excavation take-off is the possibility that the contractor might be confronted with groundwater in the bulk excavation work. If there is a high water table and a deep excavation, water may flow into the excavation, preventing the construction of the substructure. In this case the contractor would need to remove the water from the job site by means of constructing a series of well sites, and/or drive shoring around the perimeter and below the depth of the excavation to prevent water seepage. Removal of the water by either of these means is relatively expensive. If the project conditions are such that this work is necessary, the estimator must include these costs in his estimate. Inspection of boring drawings included in the construction working drawings would aid the estimator in his determination of potential water problems.

Take-Off Procedures for Trenching: The quantity take-off of trenching excavation presents few problems. Given the temporary nature of trenches and the shallow depth of a trench relative to the depth of bulk excavation work, there is seldom a need to slope the excavation. Trenches are usually dug with a trenching machine. Trench excavation is taken off in cubic yards or in linear feet. The linear-foot measure is sometimes used because when trenching machines are used their productivity is somewhat independent of the depth or width of the trench. This assumption is not valid if the trench is unusually deep or wide.

Take-Off Procedures for Footing: The take-off of footing excavation entails the calculation of the cubic yards of volume to be excavated. The one exception to this is that sometimes continuous wall footings might be taken off in a linear foot measure. This might be done if the wall footing is of a continuous standardized size—e.g., 1 foot deep and 2 feet wide. Given the construction process of digging a continuous footing, the productivity/cost is sometimes more dependent on the length of the footing than on the volume of the earth to be excavated.

~~PRACTICAL~~
FORM 314

QUANTITY SHEET

PROJECT	EXCAVATION PROJECT	ESTIMATOR	JKA	ESTIMATE NO	101
LOCATION	ANYWHERE	EXTENSIONS	DJA	SHEET NO	1
ARCHITECT ENGINEER	JJA	CHECKED	SLL	DATE	9-12-X0
CLASSIFICATION	SITEWORK				

DESCRIPTION	NO.	Length	Width	Depth				cu. ft.	$27\frac{ft^3}{yd^3}$	ESTIMATED QUANTITY	UNIT
02201 BULK EXCAVATION											
NORTHWEST SECTION (TO OUTSIDE FTG.)	1	24.0'	32.0'	8.0'				6144			
NORTHEAS SECTION (TO OUTSIDE FTG.)	1	40.0'	24.0'	8.0'				7680			
WORKING SPACE (VERTICAL)	1	192.0'	2.0'	8.0'				3072 (SUPER)			
WORKING SPACE (SLOPED)	1	192.0'	1.0'	8.0'				1536 (SUPER)			
									18432 ÷ 27		683	cu yd
02202 FOOTINGS-CONT.	1	192.0'	2.0'	1.0'				384 ÷ 27		14	cu yd
02203 BACKFILL											
WORKING SPACE (VERTICAL)	1	FROM ABOVE						3072			
WORKING SPACE (SLOPED)	1	FROM ABOVE						1536			
ABOVE FOOTINGS	1	192.0'	0.5'	8.0'				768			
									5376 ÷ 27		199	cu yd

MFG. IN U.S.A.   FRANK S. WALKER CO., PUBLISHERS, CHICAGO

Figure 10.12

The fact remains that usually footing excavations are taken off in cubic yards. For isolated footings (also referred to as pad or column footings), the take-off process entails the multiplying of the cross-sectional area of the footing by the depth of the footing.

The taking-off of footing excavation does provide the estimator an opportunity to use some estimating take-off shortcuts that were discussed in Chapter 8. In particular, footings are often of a common thickness—for example, 1½ feet deep. Given several footings of the same thickness, the estimator can gain efficiency and enhance the take-off accuracy if he will use the thickness as a common dimension in the calculation process. The area of each footing can be determined, the areas added, and then multiplied by the common dimension.

It is also possible that several footings have the same cross-sectional area. When this is the case the estimator should only determine the area once, and then multiply by the number of footings. These two procedures for enhancing the efficiency and accuracy of the footing take-off are illustrated in Figure 10.13.

In rare instances a footing is dug out to the exact size of the to-be-placed concrete footing. This might be done when the soil is very stable and the contractor chooses not to form the concrete footing. However, usually a footing excavation is cut approximately one-half or one foot beyond the footing perimeter to allow for the forming of the footing. This increased cross-sectional area should be included in the calculated footing excavation.

PRICING EXCAVATION/SITEWORK

Estimating Cost of Excavation and Equipment

The costing or pricing of excavation/sitework presents the estimator two somewhat unique situations relative to costing other work on a building project. For one, as was noted in an earlier section of this chapter, the determination of the cost of excavation work is usually absent of a material cost component. Other than when the drawings indicate that backfilling or grading material must be brought to the proposed building site, excavation work usually only entails the removing or relocating (cut-and-fill work) of earth.

The common absence of an excavation material cost component does not cause the estimator difficulty. In fact, this characteristic eases his task. However, the matter of pricing isn't as simple for equipment.

Virtually every type of excavation work entails the contractor's use of relatively expensive construction equipment. This equipment might vary from a small back hoe to a very large scraper. More often than not, the typical building construction project requires the use of an average-sized back hoe, and perhaps a dozer or a blade.

PRACTICAL
~~PRACTICAL~~
FORM 116

QUANTITY SHEET

PROJECT	EXAMPLE OF USING COMMON DIMENSIONS			ESTIMATOR		ESTIMATE NO	EXAMPLE
LOCATION	ANYWHERE			EXTENSIONS		SHEET NO	
ARCHITECT ENGINEER	JSA			CHECKED		DATE	
CLASSIFICATION	EXAMPLE						

DESCRIPTION	NO.	DIMENSIONS				Sq. ft.	× Depth'	ft³	ESTIMATED QUANTITY	UNIT
		Length	Width	Depth						
02201										
FOOTING EXCAV.										
Ftg No. 1	1	5.0'	4.0'	1.5'		20.0				
" " 2	1	5.0'	3.0'	1.5'		15.0				
" " 3	1	4.0'	4.0'	1.5'		16.0				
" " 4	1	4.0'	6.0'	1.5'		24.0				
" " 5	1	4.0'	7.0'	1.5'		28.0				
" " 6	1	4.0'	8.0'	1.5'		32.0				
" " 7	1	3.0'	6.0'	1.5'		18.0				
						152.0'	1.5'	229.5		
" " 8	3	5.5'	4.5'	1.2'				74.3		
								303.8		
							÷ 27		11.3	yd³

Figure 10.13 Example of Using a Common Dimension

310

The equipment cost—whether it is for rental or for ownership—for excavation work may exceed the labor cost for performing the work. The result is that the estimator has to be as attentive to accurately determining the equipment costs for excavation work as he does for determining the labor costs. The estimating of equipment costs via the use of an equipment costing system was discussed in Chapter 8. The implementation of such a system entails the determination of an hourly rate for individual pieces of equipment and the determination of the number of hours each piece of equipment will be needed on the project being estimated. The reader is referred back to Chapter 8 for a review of the estimating of equipment costs for a project.

As an alternative to implementing an equipment costing system to determine the equipment cost for a project, the estimator may choose to utilize widely circulated cost books that delineate equipment productivity and/or equipment hourly rates. For example, average productivity rates for various items of construction equipment are listed in Figure 10.14. With these productivity rates plus the hourly rate for renting or owning specific pieces of equipment, along with excavation quantity as determined from the take-off process, the equipment excavation cost can be determined. Equipment rental rates can be obtained from an equipment dealer. Naturally, if the contractor owns his own equipment, the hourly rate should be determined based on the ownership costs to include depreciation, maintenance and repair costs, and finance costs. (An example calculation of hourly ownership cost of a piece of equipment was illustrated in Chapter 8.)

Type of Equipment	
3/4 cubic yard power shovel	150 cubic yards/hr.
7 cubic yard scraper	80.0 cubic yards/hr.
90 horsepower bulldozer	1200 square ft./hr. of clearing
90 horsepower bulldozer	40.0 cubic yards of backfill
Trenching machine	120 linear feet/hr.

Figure 10.14 *Example Equipment Excavation Productivity*

Variables Affecting Cost: The variables or uncertainties that affect the productivity and/or cost of performing excavation work are for the most part the same independent of the specific type of work involved. Naturally, the cost of performing excavation work is dependent on *weather and seasonal considerations.* Given the fact that rain significantly affects the soil conditions, it follows that the cost of excavation work is more dependent on the weather than any other type of building work.

The productivity and/or cost of excavation work is also very dependent on the *type of soil.* In other words, the equipment and labor effort required to excavate a specific amount of one type of soil—for example, 1,000 cubic yards of a clay-type soil—will be different than the effort required to excavate 1,000 cubic yards of, for example, a gravel-silt type of material. Figure 10.15 shows a table illustrating the dependence of equipment productivity on the type of soil being excavated. The type of soil is of particular concern in regard to bulk excavation work, which usually entails the digging of large quantities of earth to relatively deep excavations. Given these two characteristics, the cost of the work can be significantly affected by the soil conditions. As was discussed in an earlier section, the cost of deep excavation work can also necessitate procedures for removal of water and/or shoring. These too can significantly increase excavation work.

Average Output for 3/4 Cubic Yard Power Shovel	
Type of Soil	*Cubic Yards/Hour*
Moist Clay	170
Sand and Gravel	150
Common Earth	130
Hard Clay	100
Rock	60

Figure 10.15 *Equipment Productivity as a Function of Soil Conditions*

Another factor that can affect the cost of excavation work, in particular the equipment cost, is the *physical layout and adjacent structures*. The productivity of a piece of construction equipment and its operator is affected by the presence of structures that may curtail the maneuvering of the equipment or the operation of the equipment at optimal level. For example, if an adjacent building prevents a crane or power shovel from operating with its boom at an optimal angle, the hourly output potential of the equipment will be curtailed. (Figure 10.16 illustrates the effect the angle of swing has on the productivity of a power shovel.) The fact that the output of this and other

Angle (%)	*Output as a Function of Ideal*
30	1.0
45	0.95
60	0.90
75	0.85

Figure 10.16 *Output of Equipment as a Function of Angle of Swing*

types of excavation equipment is dependent on the working area and adjacent structures emphasizes the estimator's need to visit the job site before the estimate is prepared. At the very least, the estimator should investigate these conditions through reference to the drawings or communication with individuals acquainted with the site conditions.

While weather, soil type, and working conditions all affect the cost of performing excavation work, the fact remains that its cost remains most a function of the type of equipment utilized to perform the work. As has been discussed earlier, numerous types of equipment, each available in varying work capacities (e.g., a 6-cubic-yard dump truck versus a 10-cubic-yard dump truck), are available to contractors. To some degree, a contractor has an almost infinite number of options available to him in performing a specific type of work. Each of these options results in a different unit cost for performing the work. Fortunately, the estimator's difficulty in estimating costs is lessened by the fact that the contractor usually tries to make use of his own equipment fleet in performing the proposed work. Even though the equipment may not be optimal in regard to productivity for performing the proposed work, its unit cost because of ownership may prove less than rental of "ideal," top-performing pieces of equipment. However, it is well to remember that one of the more difficult aspects of preparing the excavation estimate relates to the estimator's identification of the method that will be used to perform the work.

THE QUEUING MODEL: AN AID IN SELECTING
THE OPTIMAL EXCAVATION METHOD

As is true for other types of construction work, the contractor has several available alternative methods to choose from when he is considering excavation work needs for a project. Because choice of the method significantly affects the cost of doing the work, it is important that the estimator be able to (1) select the most economical method available; (2) communicate this method to the project production personnel; and (3) determine the cost for the method. The selection of the optimal construction method for performing a given type of construction work might be referred to as **method or productivity analysis**. As was discussed in Chapter 2, this analysis is a fundamental ingredient in estimating.

Several scientific or mathematical models for analyzing a construction method were discussed in Chapter 2; some of these will be illustrated by means of examples in subsequent chapters.

Another scientific or mathematical model that is available to the estimator for evaluating alternative methods of construction is the *queuing model*. The queuing model has been especially utilized in analyzing the production interface between construction equipment. With the objective of providing the estimator a means of choosing an optimal excavation method for an amount of work determined in the excavation take-off, we will now illustrate the use of the queuing model.

Almost everyone has experienced the queuing problem in the form of a waiting line at a grocery checkout, a line to purchase tickets, or a line to purchase gasoline at a service station. This same feature is also characteristic of the production of construction activities, especially excavation work. This type of problem may be classified as queuing problems.

The components of such a class of problems include one or more arrival units being serviced by one or more service units. The queuing model attempts to analyze the interface between arrival unit and service unit. The arrival units "arrive" at the service units and either wait to be served or are served. If there is no arrival unit in the queue, there is no production. On the other hand, too long a waiting line may prove uneconomical.

The mathematical queuing model may take one of several different forms, depending on the characteristics of the problem. The number of arrival units may be finite or infinite. The arrival units may arrive at a pre-determined rate or according to some probability distribution. In addition, the arrival patterns may either vary with time or be in a continuous flow. The service patterns may also be varied. For example, the service rate may be constant, it may be probabilistic and subject to a distribution (i.e., an exponential distribution) or it may be affected by the queue length. There may also be more than one servicer. Moreover, the queue discipline may vary from one problem to the next. For example, the first arrival unit may be the first to be served or the last arrived may be served first. Some arrival units may be given priority in the queue. The model may also have to consider balking; this situation occurs when an arrival unit may not join the queue line if the line is too long.

The described types of problems and their variations have been covered extensively in queuing theory literature. In each case, a different mathematical model has been derived to represent the variables of the problem and the interrelationships between them. The mathematics of the model becomes more complex as number and interrelationships of the variables increase. The fact that the queuing mathematical model may be derived for increasingly complex problems is evidence of the capability of mathematics to represent reality.

Several of the derived queuing models have been applied to the productivity study of excavation activities. Owing to the different types of mathematical models required by different construction productivity queuing problems, however, most of the work done has been aimed at specific problems rather than an overall general model. For example, queuing models have been derived to study the productivity of scrapers and a service unit hauling earth on a construction site, the productivity of transient concrete trucks with one or more concrete pavers, and the productivity of dump trucks being loaded with a back hoe. The queuing models can aid in seeking the optimal method of performing such activities.

Consider the following construction activity. A construction contractor is studying the productivity of a bulk excavation of a building. To do the work he is using a single back hoe which digs the earth and loads it into the dump trucks. These trucks carry and dump the earth at a nearby location. The receiving location has a large capacity in comparison to the number of trucks it receives, so that no truck ever has to wait to be unloaded. However, this is not the case at the interface of the back hoe and its dump trucks. There may be a waiting line, or the back hoe may be sitting idle. The contractor is trying to determine the optimal number of inputs (trucks) to use to satisfy his objective. Let us assume that he has the choice of using either 4 or 5 trucks. His objective is to maximize his daily profits from the activity. He has the following information:

Cost of renting and operating a dump truck = $90 per day
Cost of renting and operating a back hoe = $250 per day
Income from work = $2 per cubic yard of earthwork
Average service rate of back hoe (exponential distribution) = 10 services per hour
Average arrival rate of a dump truck (Poisson distribution) = 3 arrivals per hour
Capacity of a truck = 6 cubic yards of earthwork
Workday = 8 hours

After studying the given information, the problem is recognized as a queuing problem with a single server (the back hoe) and arrival units (the trucks). It may be modeled as shown in Figure 10.17.

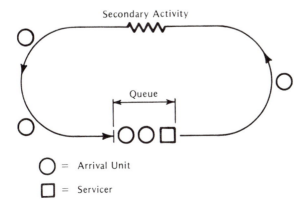

Figure 10.17   *The Queuing Model*

To solve the problem, a mathematical queuing model may be developed for a single server, with an exponential service rate, a finite number of arrival units with the Poisson arrival rates, and a queue discipline of "first-come first-served." Actually, the problem is one of determining how much work may be performed using 4 trucks versus using 5 trucks. This may be determined by finding the percentage of the time the service unit is actually working (availability of an arrival unit) for each of the alternatives. If the amount of time it was not working (in an 8-hour day) could be determined, then, by subtracting this figure from 8, the amount of time it was working could be found. Realizing that an arrival unit must be in the queue to produce work, we can calculate the work or production as follows:

W = (8 hr.) (6 cu. yd. service) (10 services/hr.) (1-probability of no truck in queue)

Therefore, the problem is to determine the percentage of time the service unit is not working. Let P(0) stand for the probability of having no dump truck in the queue.

Using probability theory, a mathematical equation can be derived for the solution of the probability of no arrival unit being in the queue—i.e., P(0)—as a function of the arrival rate of an arriving unit, the service rate of the service unit, and the number of arrival units. We will not derive the equation. However, the derivation would yield the following result:

$$P(0) = \left[1 + \frac{n!}{(n-1)!} \left(\frac{\lambda}{\mu} \right)^1 + \frac{n!}{(n-2)!} \left(\frac{\lambda}{\mu} \right)^2 + \frac{n!}{(n-3)!} \left(\frac{\lambda}{\mu} \right)^3 + \ldots + \frac{n!}{1!} \left(\frac{\lambda}{\mu} \right)^{n-1} + \frac{n!}{0!} \left(\frac{\lambda}{\mu} \right)^n \right]^{-1}$$

The original problem we presented may now be solved. First, P(0) is calculated for the 4 trucks versus the 5 trucks method:

4 trucks:
P(0) = 1/4.122 = 0.242

5 trucks:
P(0) = 1/7.184 = 0.139

The production for 8 hours for each of the systems may then be determined as follows:

4 trucks:

$$W = (8 \text{ hours}) \frac{(6 \text{ cubic yards})}{(\text{service})} \frac{(10 \text{ services})}{(\text{hour})} (.758) = 364 \text{ cubic yards}$$

5 trucks:

$$W = (8 \text{ hours}) \frac{(6 \text{ cubic yards})}{(\text{service})} \frac{(10 \text{ services})}{(\text{hour})} (.861) = 413 \text{ cubic yards}$$

Finally, the optimal solution (given the objective of maximizing daily profit) may be found by determining daily revenues and costs for the 4 truck versus the 5 truck system:

4 trucks:
Revenue	=	2(364)	= 728
Costs	=	4(90) + 250	= 610
			118

5 trucks:
Revenues	=	2(413)	= 826
Costs	=	5(90) + 250	= 700
			126 (optimal)

In the described manner, the formulation of the problem as a queuing model, and its solution, have yielded an optimal answer to the problem.

The queuing model approach to studying productivity is feasible in that solutions are exact, and the interface between the various inputs may be properly taken into consideration. The difficulty that arises with such an approach is that the mathematics becomes extremely complex as the model becomes larger. When more than 2 variables or inputs are considered, the mathematics becomes too complex even for advanced mathematicians. Computers help alleviate this problem. Thus, it may be that the feasibility of extending the queuing model depends on the ability to handle the mathematics or access to a computer.

The actual solution process is accomplished by a computer program. Having such a program available, the contractor merely submits the input information according to a defined format, and receives solutions from the computer printout. Sample computer results are shown in Figure 10.18.

Another type of "black box" queuing model that can be used by the contractor, is in the form of charts which show the production accomplished by varying combinations of inputs. These charts, often determined by the computer model, are easy to use and save the contractor the time of performing cumbersome mathematical queuing calculations. An example of one of these queuing production charts is shown in Figure 10.19.

Naturally, the results of the 4 or 5 dump truck analysis might yield different results with a different set of variables. However, each situation could be solved more objectively through the use of the queuing model. In effect, the queuing model becomes another estimating ingredient; an ingredient that while not as obvious as the ability to interpret construction drawings, may in fact prove equally important to an accurate and efficient building construction estimate.

Program for 1 C/Y Production Unit
05.0012.50007.00001.0008800.000.00000.00001.0
Truck Fleet = 7; Production = 123.75
State Probabilities

0 = 0.0099
1 = 0.0279
2 = 0.0670
3 = 0.1341
4 = 0.2146
5 = 0.2576
6 = 0.2060
7 = 0.0824

No of Tks	Production Rate	Duration	Cost
1	35.71	246	0
2	66.03	133	0
3	89.72	98	0
4	106.26	82	0
5	116.28	75	0
6	121.47	72	0
7	123.75	71	0

Program for 0.5 C/Y Production Unit
05.007.50007.010.00001.0008800.000.00000.00001.0
Truck Fleet = 7; Production = 74.94
State Probabilities

0 = 0.0007
1 = 0.0035
2 = 0.0141
3 = 0.0470
4 = 0.1255
5 = 0.2510
6 = 0.3347
7 = 0.2231

No of Tks	Production Rate	Duration	Cost
1	30.00	293	0
2	51.72	170	0
3	64.92	135	0
4	71.40	123	0
5	73.93	119	0
6	74.73	117	0
7	74.94	117	0

Figure 10.18   *Computer Output from the Queuing Model*

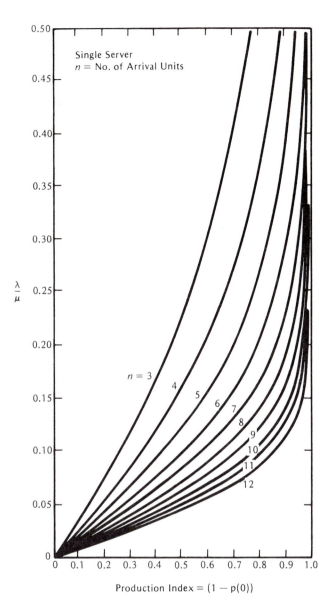

Figure 10.19 *Charts for Use with the Queuing Model*

EXERCISE 10.1

Given the site plan shown in the illustration, and given the grid pattern shown, use the four-point method described in the chapter to calculate the total cubic yards of cut-and-fill work required.

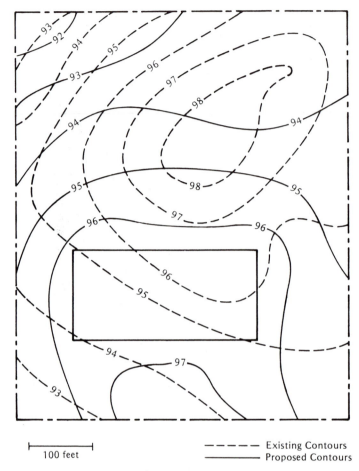

|———————| 100 feet

– – – – – Existing Contours
————— Proposed Contours

Suggested Grid Lines @ 100' cts

Exercise 10.1

EXERCISE 10.2

Assume a contractor has to dig a massive excavation for a building foundation. He uses 4 dump trucks and a large back hoe. The back hoe excavates material and in turn loads a dump truck with 5 cubic yards of material. When a truck is loaded it leaves the excavation site, hauls the material to a dump site, and then returns empty to be loaded again. Assume that once a truck leaves the excavation site, it takes 20 minutes to return. Further assume that the contractor has determined that the back hoe can load a dump truck in 6 minutes, assuming a truck is available to be loaded.

Using the queuing model that was described in this chapter, and assuming that the back hoe rents for $80 per hour and that each truck rents for $40 per hour, determine the estimated cost for excavating 20,000 cubic yards of material.

EXERCISE 10.3

A contractor who has to perform a considerable amount of excavation work is well advised to analyze the options of renting versus buying equipment to perform the work. By renting equipment, the contractor does not require the large investment that is associated with buying equipment, and the firm does not commit itself to future interest expense associated with financing purchased equipment. On the other hand, rental equipment may result in a higher hourly cost versus the hourly cost of owning the equipment.

Assume that a contractor estimates that over the next year, he will have to perform 450,000 cubic yards of excavation work. Assume that the firm determines that a back hoe that will perform the work can be rented for a monthly cost of $8,500. Another option would be to purchase the equipment at a cost of $120,000. The purchase of equipment would be by means of a $20,000 equity investment, and the securing of a 5-year loan for $100,000 at an interest rate of 15% compounded annually. Assume that the equipment can be depreciated over a 5-year life at an accelerated rate of 175%.

When the equipment is rented, the contractor has no obligation for the maintenance cost of the equipment. This ownership cost is estimated at $15,000 per year.

Assume the contractor is in the 40% income tax bracket. Given the fact that rental expenses, finance costs, depreciation, and maintenance costs are tax-deductible, on the basis of a 1-year analysis, determine the unit cost per cubic yard of performing the work both with the rental equipment and by means of purchased equipment.

11

Concrete: Forming, Reinforcing, and Placement

The Use of Concrete in Buildings * *Types of Concrete: Definition of Work Items* * *Taking-Off Concrete Work* * *Pricing Concrete Work* * *Productivity Analysis and Estimating*

THE USE OF CONCRETE IN BUILDINGS

Perhaps the most common construction material, one that is part of almost every project, is concrete. Concrete is an especially prominent material in heavy and highway projects, including roadways, bridges, and engineering types of projects. Concrete can also be found on most building construction projects. Substructures of modern buildings usually consist of either concrete or masonry units (concrete blocks). Concrete is also used extensively as a building material for part of a project's landscaping. Sidewalks, driveways, and loading docks are examples of concrete construction. To a lesser degree, concrete can be found in the superstructure (i.e., the part of a building above ground) of buildings. The amount of concrete in a building superstructure varies from none for most residential units to a significant quantity and dollar amount in high-rise commercial or industrial projects.

A designer utilizes concrete in a structure for one of two reasons. First, concrete is an excellent load-bearing material. Owing to its relatively high compression strength and the ability to cast concrete in varying shapes, designers utilize concrete members to carry the live and dead load of a building. **Live loads** are in the form of people, fixtures, equipment, and snow and wind loads. **Dead loads** are the weights of the structure itself, including other beams, columns, and so forth. Concrete members designed to carry live and dead loads include beams, columns, slabs, and walls.

Concrete is also utilized as an architectural material. For example, a designer may utilize a non-bearing concrete panel, referred to as a curtain wall, on the outside wall of a steel skeleton building. The use of concrete as an architectural material recognizes the fact that concrete can be shaped and finished to yield an attractive appearance. Besides being used in architectural design, concrete is a good fireproofing material and thus may be designed to cover and "insulate" less fire-resistant materials.

In regard to construction estimating, perhaps the most unique aspect of concrete is the vast number of different types of concrete work that may be part of a proposed project. Because concrete is often designed to be "molded" to a unique design for each project, almost every use and placement of concrete is different for every project. The estimating of concrete work is made more difficult by the fact that concrete work for a project usually requires the performance of 3 distinct operations; the forming for the concrete, the placement of reinforcement, and the actual placement of the concrete.

Owing to the relatively high cost of concrete work compared to other types of work, and the difficulty in determining its cost, the contractor must be especially careful in estimating such work. Errors in either the quantity take-off or the pricing of concrete work can result in a significant error in the overall estimate for a project. Both the taking off of concrete-related work and the pricing of the work are discussed in this chapter.

TYPES OF CONCRETE: DEFINITION OF WORK ITEMS

The task of preparing a concrete estimate varies significantly depending on whether the designated concrete work for a project is to be cast-in-place concrete or precast concrete. When concrete is cast in place, the contractor must usually perform the 3 aforementioned operations (forming, placement of reinforcement, and placement of concrete). On the other hand, the use of precast concrete usually means the contractor has to perform a single operation, which consists of the placement of the precast concrete members.

While the majority of concrete work for a structure continues to involve cast-in-place concrete, there has been a recent increase in the use of the precast method. This is due in part to the reduced on-site labor installation cost. The use of precast concrete also reflects the fact that relative to the cast-in-place method, precast construction work can proceed more quickly and hence shorten construction schedules, and also has the potential to result in an improved quality.

Simply put, precast concrete is concrete that is designed and molded at a location other than its final location of placement. Often precast concrete is purchased by a contractor from an external manufacturer. A contractor might also pre-cast his own concrete at the job site. For example, both *tilted concrete walls* (i.e., concrete walls poured on the ground and tilted up in the hardened stage to form the walls) and *lift slab construction* (i.e., slabs poured on the ground and lifted in a hardened stage to a suspended elevation) can be considered types of precasting. However, the term "precast concrete" is usually considered to refer to the use of concrete members purchased in their final form from an external manufacturer.

In precast concrete, as in cast-in-place concrete, steel is usually embedded in the material to carry the tension stresses that a member must sustain. The steel might consist of reinforcing bars or steel tendons. In the case of precast concrete, **tendons** (several steel wires bound together) are commonly used because precast concrete is often prestressed concrete. Prestressed concrete is made by superimposing loads on a member and transferring the loads into wire tendons placed in the materials. When the wire steel tendons are placed in the members before the concrete is cast, the process is referred to as ***pretensioned precast concrete***. The placement of the tendons in holes formed in the cast-in-place concrete after the concrete has been cast is referred to as ***post-tensioned*** process.

Pre-cast concrete is not *by definition* prestressed concrete. However, the process of manufacturing precast concrete often goes hand in hand with prestressed concrete and hence precast concrete is often prestressed concrete.

Estimating the cost of precast concrete work creates little difficulty. The take-off function is limited to determining the number and size of precast members. The cost of purchasing a precast member is usually determined by obtaining a quote from a supplier/manufacturer. The significant estimating variable becomes the cost of placing the precast members. Because it is a different type of work, precast concrete should be designated as a specific work item in preparing an estimate.

Estimating cast-in-place concrete is, however, difficult. The previously listed three operations entailed in the process, especially the placement of forms and the placement of the concrete, are subject to significant variability in regard to alternative work processes and the productivity of performing the work. The technology of each of the three steps will now be discussed. The objective of this discussion is to enable the reader to understand the rationale for describing specific work items.

Forming

The forming cost component of concrete work can be significant. For relatively complicated concrete work, labor and material forming costs may be equal to as much as 70% of the total cost. Concrete forming is characterized by the numerous alternative forming systems available to a contractor to perform a specific type of work. There may be literally over 100 alternative forming systems available to a contractor to form a defined concrete member. Perhaps there is no other area of construction work that offers the contractor more flexibility in regard to a work method.

Because of this, one of the more difficult estimating problems the estimator faces is selection of the most effective forming system for the work in question. A related problem is the need for the estimator to be correct in his assumption that the forms he estimates will be the forms the superintendent will in fact use. If this is not true, there is a good possibility that the cost of forms put in place will vary significantly from the cost of those estimated.

Concrete forming differs from other types of construction work because the put-in-place forms are temporary. Unlike the other types of construction work that is placed permanently in the structure, concrete forms usually are erected and subsequently taken apart (this is referred to as the *stripping of the forms*). The temporary characteristic of forming creates two somewhat unique estimating problems. First, the estimator must be careful to include in his labor cost the erection, stripping, and any related cleaning and/or storing costs. The labor time required to strip, clean, and store forms may approach the labor time required to erect the forms.

Second, a more difficult aspect of estimating formwork—a problem that relates to the temporary characteristic of the in-place forms—is the fact that forms are often re-used several times for a single project. It is common for the constructor to re-use a single form 4 or more times on a single project. This means that the estimator must be able to estimate the number of actual re-uses of concrete forms on a project as well as the material cost of the forms and the labor cost of erecting them. A form that is re-used twice on a project in effect has a material cost twice that if the form were re-used 4 times. That is, the material cost per use is dependent on the cost of the form and the number of times the form is used.

Formwork is most commonly measured or costed in units of square feet of contact area. A concrete wall that is 8 feet high and 10 feet long that needs to be formed on both sides will require 160 square feet of contact area (sfca). Sometimes forming is taken off in 100 square feet of contact area units.

Concrete forming is characterized by many alternative systems that are available to the builder. While many forming systems are available for different types of members (for example, walls, beams, columns, slabs), the various sytems are often characterized as being either **job-built** or **pre-built**. A job-built forming system is essentially custom-designed at its placement of location for the concrete. Plywood or sheathing, boards, and nails are fabricated at the job site in a job-built forming system. An example job-built forming system for a wall is illustrated in Figure 11.1.

Job-built forming systems are commonly constructed with sheathing and plywood. Given the flexibility to cut the materials, job-built systems can be custom-designed to almost any shape or design. These systems require a significant amount of labor time and cost to fabricate. The labor cost for building the forms is typically greater than the material cost for the

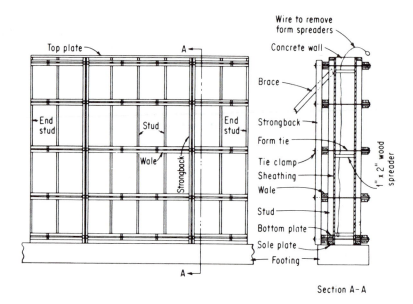

Figure 11.1 *Example of a Job-Built Form*

materials. Job-built forms made from sheathing or plywood are commonly used 4 to 8 times on a single project. Job-built forms are used to form walls, beams, columns, slabs, or any other member.

Contrasted to job-built forms are pre-built or pre-fabricated forms. Unlike many job-built forms, a contractor typically re-uses a pre-built form many times, much as a truck or a crane. In other words, a pre-built form might be considered as equipment by the contractor rather than material. Pre-built forms are often made of a combination of steel or a related metal and a wood product, including plywood. An example pre-built form is illustrated in Figure 11.2. Numerous manufacturers have patented a pre-built system similar to the one illustrated; many of the systems compete with one another.

Pre-built forms made from a combination of steel and wood are sometimes referred to as a *steel-ply system*. While steel-ply pre-built forms are predominant for wall systems, fiberglass and paper products are often used for pre-built beam and column systems. Metal pan systems are extensively used for pre-built slab forms.

Unlike job-built forms, the relatively high cost of a pre-built system often results in a higher material cost for pre-built forms than the labor cost for erecting the form. The re-use factor and perhaps a lower erection cost are the reasons why a pre-built form may cost less than a job-built form for a specific application. The economics of job-built versus pre-built forming systems is

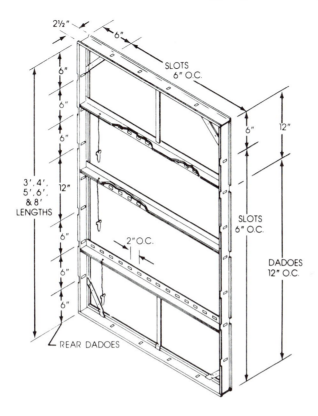

Figure 11.2 *Example of a Steel-Ply Form (Symons Co.)*

dependent on many factors, including the amount and type of forming to be done.

By now it should be apparent that there is the potential for an estimator to define numerous separate work items when taking off quantities and pricing them. However, given estimating time constraints, too long a list is not feasible. In defining concrete forming work items, the estimator should consider the following factors in delineating separate work items:

1. Type of member or structure being formed.
2. Specific type of form that is to be used (i.e., pre-built versus job-built, etc.).
3. Characteristics or location of the forming that causes it to be unique.

It is true that some other considerations also likely affect the price of forming and thus; might be considered in delineating specific work items. For example, the quality of the concrete finish that is to be obtained for a wall surface

Code	Description
03101	Job Built Forming-Isolated Footings
03102	Job Built Forms-Continuous Footings
03103	Job Built Forms-Walls
03104	Job Built Forms-Columns
03105	Job Built Forms-Beams
03106	Job Built Forms-Slabs
03107	Job Built Forms-Special Forms
	Measured in Linear Feet
03111	Pre Built Forms-Walls
03112	Pre Built Forms-Columns
03113	Pre Built Forms-Beams
03114	Pre Built Forms-Slabs
03115	Pre Built Forms-Steel Pan Systems

Figure 11.3   *Work Items for Forming*

might be considered in estimating formwork. However, this and other factors can be recognized in the pricing of specific work items rather than as a basis for work-item definition.

An example list of work items for a building construction type of estimate is provided in Figure 11.3. As we noted in earlier chapters, some estimators would be more detailed in their list of work items—and unfortunately, many would be significantly less detailed.

Reinforcement

The majority of concrete used in building construction is designed to be a structural bearing material. Given the fact that concrete is relatively weak in regard to resisting tension loads, most concrete applications include a form of structural reinforcing, such as reinforcing bars, wire mesh, or steel wire tendons. The use of tendons is usually limited to pre-stressed concrete. The estimating of tendons usually causes the estimator few problems because pre-stressed concrete is often purchased as a completed unit.

Most construction projects call for reinforced concrete. Reinforcing, like concrete forming, is done before the concrete is placed. However, unlike concrete forming, the steel reinforcement is left in the concrete.

Reinforcing Bars: Steel reinforcement embedded in concrete is usually designed to carry the tension or bending loads imposed on the finished concrete member. Reinforcement sizes have been standardized and are usually quoted in the diameter of the bar in eighths of an inch. For example, a #5 bar is ⅝ of an inch of a diameter in thickness.

Owing to the uniqueness of designed concrete members, there are no standardized amounts or sizes of reinforcement to be placed in concrete members. The result is that the estimator cannot merely take off the quantity of concrete to be placed and use a factor to determine the reinforcement to be purchased and fabricated. Instead, more often than not, the estimator has to take off the reinforcement separately from his determination of the form-work or concrete quantity calculation.

The reinforcement work to be performed is designated on the construction drawings by the designer; it specifies the length of reinforcement to be placed and the size of the reinforcement (e.g., #5 bar), along with its location. While designated via a listing of its diameter, reinforcement is purchased and priced via a *weight* calculation; the contractor usually purchases reinforcement at a price per ton. This price per ton may also reflect the amount of factory bends that are to be made, the distance the reinforcement has to be shipped, and the amount of variation in the number of sizes ordered. However, the total weight of the reinforcement is the predominant factor in determining the purchase price.

There is a need to convert a reinforcement take-off from the length factor into a weight calculation. In this regard, various reinforcement sizes are combined to yield a total weight of reinforcing required. In order to convert reinforcing sizes and lengths into a weight calculation, it is necessary for the estimator to use a conversion factor. A table for making this calculation is illustrated in Figure 11.4. Given the length of a specific size of reinforcement,

Standard Sizes and Weights of Concrete Reinforcing Bars

Nominal Dimensions,—Round Sections

Unit Weight Pounds Per Foot	Inches Diameter	Cross Sectional Area Square Inches	Perimeter Inches
0.167	.250	0.05	0.786
0.376	.375	0.11	1.178
0.668	.500	0.20	1.571
1.043	.625	0.31	1.963
1.502	.750	0.44	2.356
2.044	.875	0.60	2.749
2.670	1.000	0.79	3.142
3.400	1.128	1.00	3.544
4.303	1.270	1.27	3.990
5.313	1.410	1.56	4.430
7.650	1.693	2.25	5.320
13.600	2.257	4.00	7.090

Figure 11.4 *Areas of Reinforcing Bars*

the estimator merely multiplies the length of the reinforcement required by the weight per foot shown in the table. For example, if 150 linear feet of #8 reinforcement bars are required for a project, the table enables the estimator to determine that 400.5 pounds of reinforcing will be required (150 feet × 2.670 pounds/foot).

While it does not affect the preparation of an estimate in that each and every reinforcement bar must be taken off by the estimator, small-sized reinforcement bars such as #3, #4, or #5 are commonly referred to as *stirrups* or *tie bars*. A stirrup or tie bar is usually utilized in part to support larger reinforcement for a beam, as illustrated in Figure 11.5. A stirrup or tie, owing to their small diameter, are light and thus easily handled by the ironworker. This means a smaller labor cost for placing a specific length of a stirrup relevant to a larger reinforcement bar.

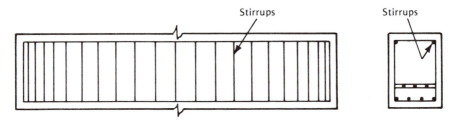

Figure 11.5   *Example Beam with Reinforcing*

The taking off and pricing of reinforcement presents a "unit" problem for the estimator. As was discussed in the preceding paragraphs, the material cost for reinforcement is most dependent on its weight. However, the labor cost relevant to placing the reinforcement is dependent on several factors. While the weight of the reinforcement affects the productivity and in turn the labor cost for fabricating the reinforcement, the labor cost is usually more dependent on the *number* of reinforcing bars that have to be placed.

Let us consider the placement of two reinforcement bars by a worker. Assume that both bars are 10 feet long, but one is a #4 bar and the other is a #8 bar. For the defined length, the #8 bar is twice the weight of the #4 bar. However, it is doubtful that twice as much labor time or cost will be required to place it. While the placement of the #8 bar may require a bit more labor effort owing to its weight than the #4, the increase will be marginal. In other words, calculation of the labor cost for placing reinforcement that is dependent on weight alone is likely to be erroneous.

The labor cost for placing reinforcement is usually most dependent on the length of the reinforcing to be placed or the number of reinforcement bars that are to be placed. In fact, the number of reinforcement bars to be placed is perhaps the best measure of the cost. An ironworker fabricating reinforcement for a concrete member can place a 20-foot length of reinforcement in approximately the same time it takes to place a 10-foot-long bar. The most time expended in placing a reinforcing bar is devoted to make the end connections. A reinforcing bar has two ends no matter what its length is.

While the number of bars may be a better measure of the labor cost of placing reinforcing than the total length of reinforcing required, the fact that the length of reinforcing bars is somewhat standardized (because of requirements for easy transport), results in a strong correlation between the number of reinforcing bars and their total length. Therefore, taking off reinforcing bars in length required represents a good basis for determining the labor cost.

It can be concluded that there are two separate bases for determining the material and labor cost of reinforcement for concrete. If the estimator accepts this valid consideration, his estimating take-off process should accommodate the calculation of both the weight and the length of reinforcement required. Another approach would be to identify reinforcing material and labor as 2 distinct work items and take off the first in units of weight and the second in units of length. This is the approach taken in the example list of reinforcement work items listed in Figure 11.6. Efficient means of taking off reinforcement to accommodate a calculation of both weight and length is discussed in a following section of this chapter.

Code	Description
03201	Reinforcing Bars-Material
03202	Reinforcing Bars-Labor
03203	Special Reinforcing Bars-Material
03204	Special Reinforcing Bars-Labor
03211	Wire Mesh

Figure 11.6 *Work Items for Reinforcing*

Steel Mesh Reinforcement: The placement of reinforcement for concrete is a highly labor-intensive operation. Given the fabrication of relatively complex reinforcement layouts, and considering that the work is often performed in a variable and uncertain weather environment, the contractor performing the reinforcement work is subject to considerable labor cost risk. In an attempt to reduce this, the contractor may utilize the use of reinforcement steel mesh as an alternative to individual reinforcing bars.

Given the fact that reinforcement mesh may be more effective in distributing forces than individual fabricated reinforcement bars, a project's architect/engineer designer may designate the use of steel mesh instead of individual reinforcing bars. Mesh is especially prevalent in use within concrete slabs. Usually, mesh is more effective in minimizing temperature stress cracks in a cast-in-place concrete slab than are individual reinforcing bars.

Steel mesh, often referred to as wire mesh, is essentially a pre-fabricated steel reinforcement. Unlike individual reinforcing bars that must be wired together at the job site to form continuous load-bearing members, wire mesh is welded at designated spaces at a manufacturing plant and shipped to a job site as an entire mat. Mesh is typically designated by the spacing between the steel and the gauge (which is a measure of the size) of the steel making up the mesh. For example, a mesh designated as 4 × 6 × ¼ refers to a wire or steel mesh that has wires of a gauge size of ¼ inch that are spaced every 4 inches in one direction and every 6 inches in the other direction.

Wire mesh is taken off in units of area of the mat—for example, in square feet. The area measure provides a good basis for both the material and labor cost because mesh is priced as a function of the area purchased and the labor cost is usually most dependent on the gauge of mesh and the spacing of members. In other words, each size of mesh required might be designated as a separate work item. This possibility for numerous work items owing to numerous possible sizes and gauges of mesh is minimized by the fact that where required, it is common for the same size and gauge of mesh to be required throughout a project.

Placement of Concrete

When we think of concrete construction work, we often focus on its actual placement, whether this entails pumping the concrete, placing it with the use of a crane bucket, or using an alternative production process. However, as has been pointed out in this chapter, the placement of concrete is only one of three processes that are typical of concrete construction. In fact, in regard to the labor cost component of the work, the placement of the concrete is typically less time-consuming and expensive than the forming or placement of reinforcement. Nonetheless, concrete placement remains an important work process in regard to the need to accurately estimate its cost. If it weren't for the need to place concrete there would be no forming or placement of reinforcement work processes.

As noted earlier, concrete is fabricated for a building either in the pre-cast or the cast-in-place state. Pre-cast concrete is essentially a totally pre-fabricated unit that includes the embedded reinforcing. The estimating of pre-cast concrete causes few difficulties. The quantity of work is usually easily determined via the project drawings, and the purchase price is usually well defined because the pre-cast members are commonly purchased from an outside vendor. Therefore, perhaps the only variable at issue is the labor time and cost required to erect the pre-cast members. Given the specialty nature of the work, pre-cast concrete work should be delineated as a separate work item from cast-in-place concrete work. Because of its somewhat infre-

Code	Description
03301	Concrete-Isolated Footings
03302	Concrete-Continuous Footings
03303	Concrete-Walls
03304	Concrete-Columns
03305	Concrete-Beams
03306	Concrete-On Grade Slab
03307	Concrete-Suspended Slab
03401	Precast Members

Figure 11.7 *Work Items for Concrete Placement*

quent usage, it may be sufficient to identify a single work item for pre-cast concrete. This is what has been done in the listing of example work items for the placement of concrete illustrated in Figure 11.7.

Owing to its more common usage and the relatively more complicated estimating process, the estimator likely needs to be more attentive to the estimating of cast-in-place concrete than to that of pre-cast concrete. Cast-in-place concrete is essentially concrete that is poured in a liquid state at the job site and finished and cured into a hardened state.

Concerning the estimating of cast-in-place concrete, two types of estimates may be required. For one, it might be the contractor's responsibility or decision to produce the concrete himself; that is, purchasing the concrete ingredients of sand, water, and cement, and mixing them to yield the concrete to be poured. The proportioning of and mixing of the ingredients of concrete to yield finished material is commonly referred to as **mix design**. Often the mix-design process is outside the responsibility of the building contractor. Instead, as the second alternative (which also affects the estimate), the contractor purchases the concrete already mixed from a manufacturer of the concrete. This is referred to as **ready-mix concrete**.

A construction project has to be of significant size to justify the contractor's setting up a concrete batching plant that is capable of proportioning and mixing concrete at the job site. While this practice may be somewhat common on heavy and highway projects, it is rare for a building project. When it is done, the estimator must in fact understand mix design in order to accurately estimate both the material and labor cost of producing the concrete. Because this process is rare in the building process, following paragraphs of this chapter will assume that the contractor will purchase ready-mix concrete for the cast-in-place process.

Even using ready-mix concrete, the estimator is still faced with the preparation of an estimate for cast-in-place concrete. But because the concrete is purchased from an outside vendor, estimating the concrete material cost presents few problems. Often a contractor enters into a contract with a ready-mix producer for a fixed price per cubic yard of concrete before he starts work on a project, or even before the submission of a bid for the project in question. Thus, perhaps the only difficulty the estimator has in determining the material cost is to estimate the quantity from the construction drawings and then to determine the material wastage factor.

Both the material cost and the labor cost for placing concrete are related to the volume of concrete to be placed. It is for this reason that cast-in-place concrete is usually taken off in cubic yards of material required.

The labor cost required to place the concrete is more difficult to estimate. It is dependent upon the factors that affect any type of construction work, including labor morale, degree of supervision, weather conditions, and so on. In addition to being subject to these conditions, the determination of the labor cost for placing concrete is made more complex by the fact that there are numerous different *methods* of placing concrete. For example, a crane, pump, or perhaps a conveyor might all be used to place a specific concrete pour.

The contractor is almost always given the freedom to use whatever method, equipment, and labor crew he wants in regard to making a concrete pour. Obviously, his choice is somewhat constrained by the availability of equipment or labor resources. However, given the availability of rental equipment, the contractor usually has several alternative feasible methods. Even if he limits his choice to a specific type of construction method—say, a crane—numerous types of cranes with varying production capacities are likely to be available. Each of these and accompanying required labor crews involve different costs.

The use of alternative means of placing concrete presents the estimator with several problems. For one, the estimator has to be able to know the production rates for the feasible alternative methods.

A second problem relates to the possible lack of communication that sometimes exists between a contractor's estimating personnel and project production personnel. There is a possibility that the estimator might assume the use of one construction method in preparing his estimate, only to find later that the project superintendent has chosen a different method. For example, the estimator might assume the use of a crane in preparing his estimate, whereas the superintendent in fact plans to use a pump to place the concrete. Given different costs for the two methods, the result is that the estimate is incorrect in regard to the cost of performing the work.

It should be obvious that there is a strong need for communication in regard to the estimating and production functions. Naturally, this is true in regard to all the construction work that is to be estimated and built. However, with the numerous possible alternatives for a specific type of concrete placement, the need for the communication is emphasized for the type of work.

TAKING OFF CONCRETE WORK

As noted in earlier paragraphs, the taking off of concrete work actually entails the taking off of three somewhat independent types of work; concrete forming, placement of reinforcement, and the placement of the concrete. While some estimators might determine two of these take-offs as a function of the other (e.g., multiply the determined concrete quantity by a factor to determine the weight of steel required), this process usually proves inaccurate because there is not a precise relationship between the quantities of forming, reinforcement, and concrete for every building design.

As in any take-off, the concrete work should be estimated with the objectives of accuracy and efficiency. The take-off also has to be performed to yield units of work that are consistent with the subsequent pricing functions. Primarily with the efficiency concern in mind, it usually proves advantageous to take off the concrete placement quantities before determining the forming or reinforcement quantities. The concrete placement take-off usually results in the determination of mathematical calculations and quantities that can be subsequently used for the calculation of formwork and reinforcement quantities. In other words, the estimator usually "learns" the most from the concrete placement take-off in regard to data that can be then transposed to the take-off of the other two categories. We will now proceed to discuss the take-off of concrete work in the order suggested for an efficient determination of quantities.

Concrete Placement

As discussed in the earlier paragraphs, concrete is a flexible material that is custom-designed to designated design shapes. For purchasing and placing concrete, the material cost is for the most part independent of the type of member (other than when a different quality or strength of concrete is designated for separate members). On the other hand, when placement cost includes labor and equipment, there are significant variations depending on the types of concrete members to be used (e.g., an on-grade concrete slab versus a concrete pour for an elevated wall section).

Concrete is usually taken off in units of cubic yards of volume. An exception to this is concrete slabs, which are usually taken off in square *feet* of surface area. On occasion, an estimator might also take off beams in linear feet. However, given the need to purchase concrete in a volume quantity, the take-off calculations always have to be structured to make possible a calculation of the cubic yards of concrete required.

Compared to many building materials, the sum of the material and placement cost for concrete is very high. Therefore the estimator needs to be relatively precise in his determination of the concrete quantities. As a general rule, he should take off concrete quantities to the nearest half of a cubic yard. For example, calculations for the cubic yards of beam concrete may yield a quantity of 14.4 cubic yards. The estimator would round off this quantity to 14.5.

Because the placement of concrete is usually performed in several separate pours, and because it is difficult for the field personnel to order the precise quantity of concrete needed for a specific pour, a relatively high wastage factor can be expected. The actual wastage amount as a percentage of the total concrete estimated is dependent on the type of member being placed and even more dependent on the size or quantity of the pour being made. For example, an estimator can expect a higher wastage factor as a percentage of the total concrete pour if a 10-cubic-yard slab is being placed versus a 50-cubic-yard slab. While some of the material wastage results from the concrete overflowing or seeping through the forms, most stems from the excess material left over after the last pour of the day.

It is common for an estimator to take off the quantity of concrete from the dimensions on the drawings independent of the fact that some of the volume of the concrete will be replaced by reinforcing. The volume of the reinforcement is often approximately 2% of the volume of a concrete member. Knowing this, and knowing that a 2% concrete wastage is common, the estimator often assumes that the wastage and the overcalculation of concrete owing to the presence of reinforcement compensate each other. The result is that a zero wastage factor is used in estimating concrete; the wastage and overcalculation of concrete quantity net to zero. This "zero" wastage factor should be adjusted for a specific project, depending on the unique factors of a project, including the use of non-reinforced concrete. A zero wastage factor will be assumed in the following take-off examples.

Example Take-Off: An example take-off of concrete footings, slab, wall, and columns will now be illustrated using the construction drawing shown in Figure 11.8. It is assumed that work items are defined in regard to the concrete placement take-off, including the continuous footing, the walls, the columns, and the slab.

Let us first address the calculation of the continuous wall footing and the foundation wall quantities. To find the cubic yards of concrete in the continuous footing or wall, it is necessary to multiply the length by the width by the height in feet to find cubic feet. The cubic feet are then divided by 27 to yield cubic yards. There are several different ways to calculate the total length of either the footing or the wall. In particular two alternative methods, **the unit method** and the **center-line method** yield the same result.

To illustrate the unit method, lines indicating the walls shown in Figure 11.8 are broken into separate areas in Figure 11.9. Note that if we add the perimeters shown in Figure 11.9—i.e., 26.0 ft. + 12.0 ft. + 16.0 ft. + 36.0 ft. + 42.0 ft. + 24.0 ft. = 156.0 ft.—this represents the outside perimeter, but it does not represent the true stretch-out area of the foundation wall. Instead,

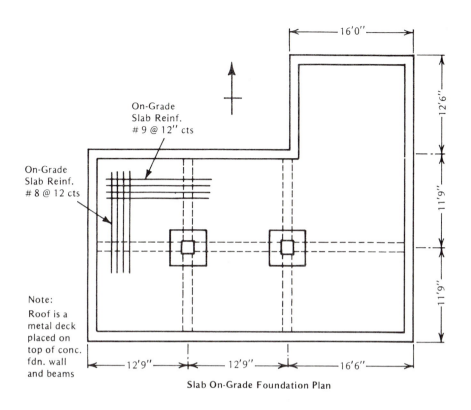

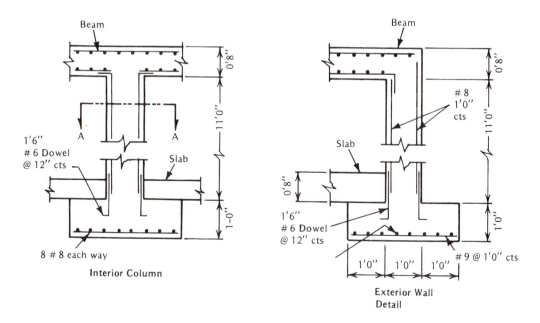

Figure 11.8 *Example Drawing for Concrete Take-Off*

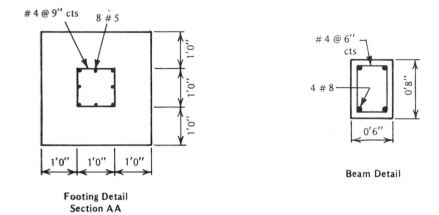

Footing Detail
Section A A

Beam Detail

Figure 11.8 Continued

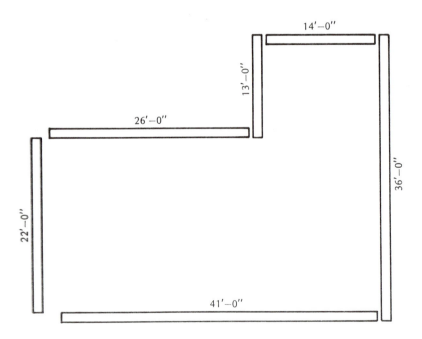

Figure 11.9 Wall Broken in Sections

let us look at the stretched-out areas in Figure 11.9. Here we add 26.0 ft. + 13.0 ft. + 14.0 ft. + 36.0 ft. + 41.0 ft. + 22.0 ft. = 152.0 ft., the actual and correct stretch-out of the walls. In effect what we have done is to subtract the width of the wall every time we have an external corner (e.g., where the 24.0-ft. length and 26.0-ft. length come together in Figure 11.9), and subtract a thickness of the wall at an interior corner (where the 26.0-ft. length intersects the 12.0-ft. length in Figure 11.8).

Another quick and easy method of finding the stretch-out of footings and walls is to assume a center line all around the centers of the walls or footings. For example, Figure 11.10 shows the center line of the walls illustrated in Figure 11.8. Adding the center-line dimensions we obtain the following = 26.0 ft. + 12.0 ft. + 15.0 ft. + 35.0 ft. + 41.0 ft. + 23.0 ft. = 152.0 ft. In effect, this method has eliminated the need to count the corner areas twice.

As can be seen from the calculations in Figures 11.11, the same correct length is determined using either method. Given equal wall thicknesses, the use of the unit method is usually quicker in regard to the time required to do the calculations. This is because of the common practice of the designer indicating outside dimensions (and not center line) on the drawings.

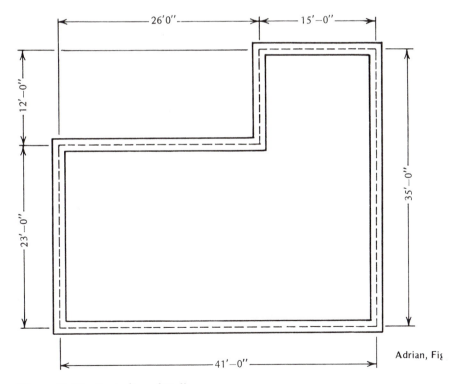

Adrian, Fig

Figure 11.10 *Centerline of Walls*

PRACTICAL	QUANTITY SHEET													cu.ft.	÷27 $\frac{ft^3}{yd^3}$	ESTIMATED QUANTITY	UNIT

QUANTITY SHEET

FORM NO				
PROJECT	CONCRETE PROJECT	ESTIMATOR JKA	ESTIMATE NO 101	
LOCATION	ANYWHERE	EXTENSIONS PJA	SHEET NO	
ARCHITECT ENGINEER	JJA	CHECKED SLL	DATE 9-12-X0	
CLASSIFICATION	CONCRETE			

DESCRIPTION	NO.	DIMENSIONS						cu.ft.	÷27 $\frac{ft^3}{yd^3}$	ESTIMATED QUANTITY	UNIT
PRELIMINARY CALCULATIONS:											
LENGTH OF FOOTINGS											
		Method 1					*Method 2*				
		26.0					26.0				
		12.0					12.0				
		16.0	156.0				16.0				
		36.0	-4.0 (corners)				36.0				
		42.0	152.0'				38.0				
		24.0					41.0				
		156.0'					23.0				
							152.0 (Super)				
03301											
FOOTINGS - CONT.	1	152.0' (Super)	2.0'	1.0'			304 ÷27		11.3	cu.yd
03302											
FOOTINGS - COLUMN	2	3.0	3.0'	1.0'			18 ÷27		0.7	cu.yd
03303											
SLAB ON-GRADE											
SOUTHWEST SECTION	1	22.0'	26.0'	0.67'			381.3			
EAST SECTION	1	14.0'	34.0'	0.67'			317.3			
							698.6 ÷27		25.9	cu.yd
03304											
FOUNDATION WALL	1	152.0'	0.667'	11.667'			1182.3 ÷27		43.8	cu.yd
03305											
COLUMNS	2	1.0'	1.0'	11.0'			22.0 ÷27		0.8	cu.yd
03306											
BEAMS											
EAST/WEST BEAMS	1	40.0'	0.5'	0.667'			13.3			
EAST/WEST BEAMS	1	14.0'	0.5'	0.667'			4.7			
NORTH/SOUTH BEAMS	2	22.0'	0.5'	0.667'			14.7			
							32.7 ÷27		1.2	cu.yd

MFD. IN U.S.A.   FRANK R. WALKER CO. PUBLISHERS, CHICAGO

Figure 11.11 Example Concrete Take-Off

Once the length of the footing or wall is determined, the cubic yards of concrete required can be determined by merely multiplying the determined length times the width and height of the footing or wall. These calculations for both the continuous footing and foundation wall shown in Figure 11.8 are illustrated on the general estimate sheet in Figure 11.11. Note that once the length of the foundation wall is determined, there is no need to repeat the calculation for the foundation wall because the result is the same. Therefore, as can be seen in Figure 11.11, the result of the length of the foundation wall is labeled a "super" item on the general estimate page with the intent that the number for that foundation wall calculation will be used again.

Given the fact that the wall area will likely be used again (e.g., in determining the surface area of wall forming required), it would be advantageous for the estimator to set out the calculation of the wall area as another common dimension or super item. Thus, he should perform the mathematics for calculating the cubic yards of concrete in the wall as shown in Figure 11.11.

Let us calculate the concrete in the slab in Figure 11.8 as an example. To begin with, the overall surface area is divided into rectangular areas, as indicated by diagonal lines shown in Figure 11.12. This create two areas, Area 1 and 2. The thickness of the slab is 8 inches. To get the slab dimensions, it is necessary to subtract the thickness of the wall, which is 2 feet from each dimension at both ends. The 26.0-ft. dimension is not changed because at the right end of the dimension 1 ft. is subtracted while at the left end of the

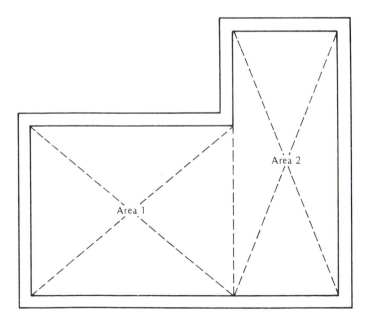

Figure 11.12 *Slab Broken into Areas*

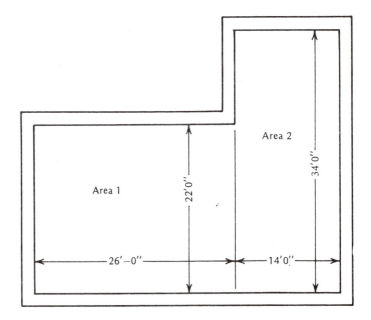

Figure 11.13   *Calculated Lengths of Segmented Areas*

dimension 1 ft. is added. Once these dimension calculations are made, the square feet of areas for Areas 1 and 2 and the cubic yards of concrete for the slab are shown in Figure 11.13. Note that the slab thickness is treated as a common dimension to gain estimating efficiency.

The quantity take-off of beams and columns usually presents the estimator with few problems. The process is usually only a matter of multiplying the specified length, width, and height to yield a cubic-foot and cubic-yard quantity. However, on occasion the geometric shape of a column or beam is something other than a rectangle. Consider the beam shown in Figure 11.14. The sectional area is not rectangular. Instead, it is what is referred to as a trapezoid—the two bases are of different length (14 inches and 8 inches). This example illustrates the need for the estimator to have knowledge or access to formulas for figuring the area or volume of irregular shapes. In the case of a trapezoid, the cross-sectional area is determined via the following formula:

$$A = \frac{H(b + B)}{2}$$

Where A is the area
      H is the height of the trapezoid
      b is the length of one of the bases
      B is the length of the other base

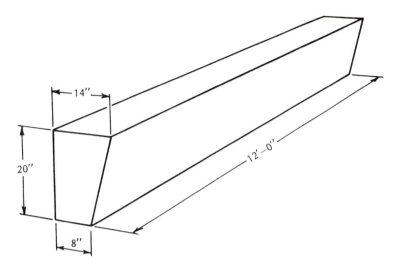

Figure 11.14 *Example Beam*

Applying this formula to the cross-sectional area of the beam shown in Figure 11.14 yields the following.

$$A = \frac{20 \text{ in. } (14 + 8 \text{ in.})}{2} = 220 \text{ sq. in. } = 1.53 \text{ sq. ft.}$$

Assuming there are 30 beams as shown in Figure 11.14 on a specific project, the total cubic yards of concrete for beams is calculated in Figure 11.15.

Concrete Forming

Given the fact that several concrete calculations yield area and volume quantities that also pertain to the concrete forming take-off, it is advantageous to perform the forming take-off upon the completion of the concrete take-off. Unlike most other work items that are taken off the drawings, the construction drawings do not actually specify the location or quantity of forms to be erected. Perhaps there is no type of work to be taken off by the estimator that requires more knowledge of the means and methods of construction. The estimator has to be able to determine what concrete has to be formed, along with the need for extra braces, scaffolds, or similar forming members not indicated on the drawings.

Concrete forming is usually taken off in square feet of contact area. The estimator calculates the actual contact area independent of whether or not the forms extend beyond the contact area. For example, an 8-foot-high piece of plywood might be used to form a 6-foot-high concrete wall. The estimator would use the contact height—i.e., 6 feet—when determining the square feet of contact area for the forming.

PRACTICAL FORM N6	QUANTITY SHEET														
PROJECT CONCRETE BEAMS				ESTIMATOR JKA		ESTIMATE NO 101									
LOCATION ANYWHERE				EXTENSIONS DJA		SHEET NO 3									
ARCHITECT ENGINEER JJA				CHECKED SLL		DATE 9-12-80									
CLASSIFICATION CONCRETE															

DESCRIPTION	NO.	Length	DIMENSIONS WIDTH	Depth				cu. ft	$27 \frac{ft^3}{yd^3}$	ESTIMATED QUANTITY	UNIT
PRELIMINARY CALCULATION: BEAM WIDTH											
			$(1.167' + 0.667')/2 = .917'$								
BEAMS	30	12.5'	0.92'	1.67'				550.0 ÷ 27		20.4	cu yd

Figure 11.15 Calculation of Beam Concrete

The cost of forming is in part dependent on the type of concrete member that is to be formed. Note that a different type of form is not used for each and every form. For example, a 2-foot by 8-foot steel-ply form might be used for forming a wall, column, or beam. However, varying degrees of difficulty are required to form different members. For example, the labor cost for placing a square foot of contact area of wall forms is usually considerably less than the cost of placing a similar quantity of beam or column forms. The result is that it is common to take off wall, beam, slab, and column forming as separate work items.

The estimator should calculate the total forms to be placed independent of how many times a single form will be utilized. The re-use factor is brought into consideration in pricing the formwork, not in taking it off.

There are 2 possible approaches an estimator might take when taking off formwork. One approach is to take off each and every board and piece of auxiliary hardware that makes up the formwork. This would entail listing plywood, various support members (commonly referred to as walers and/or strongbacks), and boards that serve to support the forms such as braces. This process would be very time-consuming.

Fortunately, the numbers and amount of hardware and support members (walers, strongbacks, and braces) is somewhat constant for a specified amount of forming. Given this fact, most estimators would follow the alternate approach: They would simply take off the contact area of forming. The required hardware and boards would be recognized in the unit cost per square foot of forming area that would be used by the estimator.

To illustrate the calculation of concrete forming, let us consider the foundation walls illustrated in Figure 11.8. In order to place the concrete, both sides of the wall have to be formed. The reader will remember that we considered the wall area as part of the concrete quantity calculation. This is illustrated as part of the "place concrete foundation wall" work item in Figure 11.11. It is easy to convert the calculated wall area to the total square feet of contact area (sfca) of forming by multiplying the area by 2, as shown in Figure 11.16.

You might observe from Figure 11.16 that the calculation of the sfca is not precise. The interior corners of the foundation walls intersect so that there is less sfca on the interior of the walls than on the exterior side. However, this slight error in the calculation is usually ignored. There is not the need to be as precise in the determination of forming area as there is in the determination of the concrete quantity. This is because forming material is not as expensive as concrete; but even more importantly, given all the uncertainties of forming, including the need to cut and shape various boards, and the numerous sytems that might be used, it would be a waste of time to try to be too precise in the take-off. In other words, given all the uncertainties that characterize the pricing of formwork, one should not worry about a slight over- or underestimate of the contact area. As a guideline, concrete forming should be rounded to the nearest 5 feet of contact area.

PRACTICAL
FORM 316

PROJECT	CONCRETE PROJECT	ESTIMATOR	JKA	ESTIMATE NO	101
LOCATION	ANYWHERE	EXTENSIONS	DJA	SHEET NO	104
ARCHITECT ENGINEER	JJA	CHECKED	SLL	DATE	9-12-X0

CLASSIFICATION CONCRETE FORMWORK

DESCRIPTION	NO.	DIMENSIONS						sfca			ESTIMATED QUANTITY	UNIT
03101												
FORM CONT. FOOTINGS	2	152.0'	1.0'								304.0	sfca
03102												
FORM COLUMN FTG	2	12.0'	1.0'								24.0	sfca
03104												
FORM FOUNDATION WALL	2	152.0'	11.67'								3,548.0	sfca
03105												
FORM COLUMNS	2	4.0'	11.0'								88.0	sfca
03106												
FORM BEAMS												
EAST/WEST BEAMS	1	1.83'	40.0'					73.2				
EAST/WEST BEAMS	1	1.83'	14.0'					25.6				
NORTH/SOUTH BEAMS	2	1.83'	22.0'					80.5				
											179.0	sfca

MFD. IN U.S.A.   FRANK R. WALKER CO., PUBLISHERS, CHICAGO

Figure 11.16 *Take-Off of Concrete Formwork*

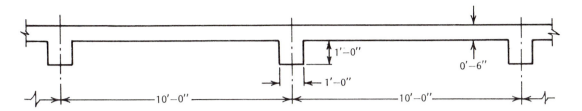

Figure 11.17   *Slab and T-Beam Construction*

The calculation of the square feet of contact area for slabs presents few problems. Perhaps the 2 areas of greatest concern are the calculation of the required area with a T-beam slab construction and the question of how to estimate required scaffolding for the slab forms. A T-beam slab construction section is illustrated in Figure 11.17. This design shape is common in slab design owing to the strength and relatively lightweight features of the design. The quickest and most accurate means of calculating the concrete in the cross-sectional area shown is for the estimator to assume a rectangle with a base equal to the distance from center line to center line of the adjacent beams and a height equal to the distance from the top of the slab to the bottom of the beam. After calculating this cross-sectional area, an area equal to the rectangular void area bordered by the 2 beams can be subtracted to yield the cross-sectional area relevant to the volume calculation. In other words, the total area is collected and then a "deduct" area is subtracted. This calculation is shown in Figure 11.18.

Occasionally a prefabricated steel metal pan forming system is utilized for forming slabs similar to the one shown in Figure 11.17. When this is the case, the manufacturer of the pan system often indicates in its company literature the void area shown in Figure 11.17. This void area is sometimes referred to as the **displacement**.

Scaffolding (sometimes referred to as **falsework**) consists of temporary structural members utilized to hold forms in place. For example, the placement of vertical boards to hold a horizontal slab form in place is common in almost every building project. Scaffolding is custom-designed for each application.

Scaffolding can be estimated by 2 different methods. The estimator, in cooperation with the individuals who set out the scaffolding design, can in effect take off every scaffold member. Subsequently, as part of the pricing function the material cost for the materials and the labor cost for erecting them would be estimated.

Many estimators, however, include the cost of needed scaffolding (labor and material) in the price used to estimate the slab forms that are supported

Concrete: Forming, Reinforcing, and Placement

PRACTICAL FORM H6		QUANTITY SHEET								

PROJECT EXAMPLE OF CALCULAT. T-BEAM AREA **ESTIMATOR**   **ESTIMATE NO** EXAMPLE

LOCATION ANYWHERE   **EXTENSIONS**   **SHEET NO**

ARCHITECT ENGINEER JJA   **CHECKED**   **DATE**

CLASSIFICATION EXAMPLE

DESCRIPTION	NO.	DIMENSIONS				ft²				ESTIMATED QUANTITY	UNIT
		Length	Width	Depth							
CONCRETE BEAM											
TOTAL AREA	?	10.0'	1.5'			15.0					
– DEDUCT AREA	?	9.0'	1.0'			9.0					
						6.0					

Figure 11.18 *T-Beam Area Calculation*

by the scaffolding. In effect, the cost of the slab forms is multiplied by a factor to yield the cost of the slab forming and the scaffolding. This second method of estimating scaffolding is based on the assumption that there is a defined relationship between the cost of the scaffolding and the forms the scaffolding support. For relatively standardized slabs, those of a common height and carrying what might be considered to be standard loads, this assumption is for the most part valid. However, for unusual form designs it may be in error. Obviously the estimating of scaffolding as a function of the forms supported is far easier than taking off individual members. However, the estimator should also be aware that this process does not have the potential for accuracy that an individual member take-off would yield. As a general rule, the labor and material costs for scaffolding supporting slabs are approximately 25% of the labor and material costs for the slab forms themself.

As was noted earlier, concrete forming is usually taken off in square feet of contact area. There are a few exceptions to this. Occasionally the length of the forming to be placed rather than the square feet of area is more indicative of the cost of the forming work to be performed. Consider the two concrete members shown in Figure 11.19. The first section illustrates what is referred to as a **keyway** and the second section shows a protruded concrete section which might serve as a ledge for bricks that are to be placed on the exterior surface of the concrete.

A keyway is a connection between two concrete members. The keyway shown in Figure 11.19 keeps one member from sliding off (**shearing**) the other member. By forming an indentation in one member and subsequently

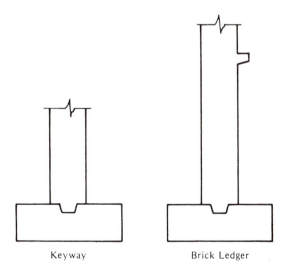

Keyway   Brick Ledger

Figure 11.19   *Examples of Measuring Forming in Feet*

pouring concrete in the connecting member into the indentation, the two members structurally act as one.

To form the type of keyway shown in Figure 11.19, the builder typically fabricates 3 boards, 2 along the sides and 1 on the bottom of the keyway shown. These boards might be 4 inches or 6 inches wide. The cost of either of these sizes or the labor cost for placing them is somewhat the same. In other words, the cost of placing the keyway forming is independent of the contact area of the forming. The cost of both the boards and placing them is most dependent on the length of the keyway. The result is that this work item (if it is delineated as a work item) should be taken off in linear feet.

The brick ledge concrete section shown in Figure 11.19 is formed by placement of horizontal and vertical board . As is the case of the keyway, the cost of forming is most dependent on the length of the forms—that is, it is somewhat the same independent of the width of the horizontal or vertical boards.

Reinforcing

Perhaps the most tedious quantity take-off process the estimator has to perform relates to steel reinforcement. Steel reinforcement (also referred to as *reinforcing bars*) consists of many individual bars of varying sizes and lengths. The quantity and lengths of reinforcement placed in concrete vary from one member to another.

Because reinforcement bars are somewhat unique for each building, and both their material purchase costs and the labor costs for erecting them are large relative to other work items, an accurate estimate requires the estimator to take off every reinforcing bar from the drawings.

As discussed earlier, reinforcing bars are taken off in the units shown on the drawings—i.e., a measure of length—and converted into a weight quantity. Because of the number of reinforcing bars that must be taken off, there is a strong need for the estimator to implement efficient take-off procedures.

The estimator has two options available to him in taking off reinforcing bars. Reinforcing bars are identified on the construction drawings by their length and bar size (e.g., #5, #6, etc.). One method the estimator might follow is to search for all bars of specific sizes identified in any concrete member on the drawings. These would be listed on a take-off sheet, and their lengths would be added and converted to a weight quantity. This process has the advantage of requiring the least number of mathematical conversions. It also identifies the total length of a specific bar size as part of the calculation process.

A second procedure is to take the reinforcing bars off in an order consistent with the specific members identified on the drawings. For example, the estimator might initially calculate the quantity of all the reinforcing bars in the footings, then proceed to calculate all the reinforcing bars in the columns, then the beams, etc. However, using this method, the estimator is likely to have to take off a certain size of bar several times in the estimating process; #8 bars, for instance, may be embedded in beams, the columns, slabs, etc.

Even so, this latter method is usually both more efficient and more accurate. Here, the estimator has only to review the drawings once as each member is reviewed for reinforcing requirements. Taking off by bar size, however, requires several redundant inspections or reviews of the drawings (one for each bar size). A "member approach" is also the means of taking-off the concrete or required forming. The fact that an inspection of the members for the reinforcing follows this same format usually lessens the possibility for overlooking any reinforcing bars.

The author prefers to combine the two different methods in order that both the efficiency and accuracy objectives are satisfied. The modified or combined method is illustrated in Figure 11.20. In the quantity take-off sheet shown, the estimator takes off the reinforcing in an order dictated by the type of member. For example, all the reinforcing bars in footings are listed, then the reinforcing bars in the columns, and so forth. The specific order might follow the order that the estimator utilized in taking off the concrete and/or formwork.

Inspection of Figure 11.20 also indicates that the reinforcing bars are "sorted out" as they are taken off of the drawings; this is done via a listing of the reinforcing bar sizes on the top of the quantity take-off sheet. This process accommodates the grouping of the bars by size so that the total length of a specific-sized bar can be determined and the reinforming bar length quantity can be easily converted to a weight quantity.

The reinforcing take-off illustrated in Figure 11.20 represents the quantity take-off of the construction drawing shown in Figure 11.8. The take-off process simultaneously takes off the reinforcing based on an order of member type and bar size.

While it is possible to manufacture reinforcing bars to the exact length specified on working drawings, this is seldom practical, especially if exceptionally long bars are required. A reinforcing bar longer than 15 or 20 feet is difficult to transport as well as handle at the job site; thus there is often a need to fabricate 2 reinforcing bars at the job site to have them behave structurally as a single bar. There are also other reasons for the need to "join" 2 reinforcing bars together. For example, two adjoining bars in two walls meeting at a corner may need to be connected to form a continuous wall structure. The joining of 2 reinforcing bars together to form a single bar is accomplished by "overlapping" 2 bars and in turn tying them together with a thin wire. The required length of the overlap is dependent on the size of the reinforcing bars to be joined.

QUANTITY SHEET

PRACTICAL FORM H6

| PROJECT | CONCRETE PROJECT | | ESTIMATOR | JKA | ESTIMATE NO | 101 |

| LOCATION | ANYWHERE | | EXTENSIONS | DJA | SHEET NO | |

| ARCHITECT ENGINEER | JJA | | CHECKED | SLL | DATE | 9-12-X0 |

CLASSIFICATION: Concrete (REINFORCEMENT)

DESCRIPTION	NO.	FT #4	FT #5	FT #6	FT #7	FT #8	FT #9	ESTIMATED QUANTITY	UNIT
03801 REINFORCING									
REINF-Cont. Ftg									
(12") Perimeter	8					1216'			
(1.5') Overlaps	10					15'			
(2.5') Horizontal Bars	152					380'			
Dowels (1.5')	300			450'					
REINF-Isol. Ftg									
(2.5')	32					80'			
REINF-SLAB									
East/West (40.1)	22					880'			
East/West (16.0')	12					192'			
Overlaps (1.5')	31					51'			
North/south (22.5')	25				563				
North/south (36.0')	16				576				
Overlaps (1.5')	41				61				
REINF-Fdn. Wall									
Vertical (11.33')	304				3444				
REINF-Columns									
Vertical Bars (11.0)	16		176'						
Ties (4.0')	29	116'							
REINF. - Beams									
Horiz. Bars (98.0')	4				392				
Stirrups (2.5')	196			490'					
Overlaps (1.5)	28				42				
TOTAL LENGTHS		116'	176'	940'	0	5158	2734		
WEIGHT PER FT.	lbs	.668	1.043	1.502	2.044	2.670	2.712		
TOTAL WEIGH (lb)		77.5	183.6	1411.9	0	13,776.9	7414.6		
							lbs =	22,857.5	
							2000 lb/Ton	11.4	Tons

MFD. IN U.S.A.   FRANK R. WALKER CO., PUBLISHERS, CHICAGO

Figure 11.20 Reinforcing Take-Off

The need to overlap reinforcing bars affects the estimator's quantity take-off. Each overlap requires ironworker time. In addition, the length of reinforcing required depends on the number of overlaps required and the length of the overlaps.

Construction working drawings seldom indicate the number or location of reinforcing overlaps. Instead, they simply indicate the total length of the bar required. For example, the drawings might indicate that the reinforcing in a continuous wall is to be 60 feet long. The length of reinforcing bars ordered to make up this 60 feet is left to the discretion of the contractor. If he decides on purchasing 6 10-foot bars, 5 overlaps will be required; 4 15-foot bars will need 3 overlaps; and so forth. The point is that the estimator has to interpret the reinforcing requirements, including a determination of the number of bars the firm will purchase, and thus the required number of overlaps. The number of overlaps will in turn affect the total length of reinforcing required.

Lacking the identification of the number or location of reinforcing overlaps or splices on the drawings, the estimator must develop some "rules of thumb" to determine the added length of reinforcing bars required. As a general rule, an average of 18 inches of overlap is required when 2 reinforcing bars are spliced together. Without specific identification of the length of bars to be purchased or the location of the splices, an estimator might assume that bars more than 15 feet long will need splicing. Additionally, each time 2 bars interconnect at a corner location, a splice will be required. Given these three assumptions, the number of splices and the added length of reinforcing required can be calculated as follows:

Number of end connections   =

+ Bars of length greater than 15 ft.   =   + _____

= Number of splices   =

× 1.5 ft. per splice   =   × _____

= Length of additional reinforcement   =

In addition to the numerous reinforcement bars that are placed in beams, columns, etc. for resisting loads placed on the concrete members, a building design usually also includes several reinforcing bars that serve to transfer a load from one member to another or to connect two members. These reinforcing bars are commonly referred to as **dowel bars**. A common location for them would be between 2 concrete pours (e.g., between a continuous footing and a foundation wall). The dowel bars are embedded into the first concrete pour—say, the continuous footing—and extended beyond so that the second pour—say, the foundation wall—can be poured over the extended bars.

The quantity take-off of dowel bars presents few or no problems. The length and size of the bars are usually well defined on the construction drawings. Perhaps the only estimating problem relates to the delineation of each of these bars from the drawings.

In an earlier section of the chapter, wire reinforcing mesh was identified as an alternative to the use of individual reinforcing bars. Wire mesh is used often in slabs. Its take-off presents little or no problem. The estimator merely takes off the area where the mesh is to be placed. This area is usually calculated the same as the slab, even though the mesh is usually placed only within a few inches of the slab area.

PRICING CONCRETE WORK

As is true of estimating most, if not all, construction work, *pricing* the take-off work is usually more difficult and the results more subject to error than the take-off process itself. The total estimate is only as good as both the quantity take-off and the pricing of the work; if one or the other is inaccurate, the total estimate will be wrong.

The pricing of concrete work, as we have noted, entails the pricing of three somewhat independent types of work; placement, forming, and reinforcement. Each of these three has at least two cost components—that of labor and material. The placement of the concrete often has a third cost component—equipment costs.

The pricing of the reinforcement, including the determination of the material cost and the labor fabrication cost, usually presents the estimator less difficulty than estimating the concrete or the forming costs. For the most part reinforcing is of a standard size. Unlike concrete that can be purchased with numerous varying qualities and strengths, more often than not a standardized type of reinforcing steel is specified. Therefore, there is a well-established purchase price.

The labor cost, while highly sensitive to labor productivity, is usually dependent on a standardized labor fabrication process. Such standardization of material type and work methods for placing the material result in reduced estimating risk.

Concrete Placement

The estimator seldom has to be concerned with the determination of the cost of "proportioning" or making the concrete. As we have already mentioned, the proportioning of the ingredients of concrete, including cement, water, and fine and coarse aggregate, is commonly referred to as *mix design*. Only on very large projects would a contractor proportion his own concrete. Instead, he is likely to purchase ready-mixed concrete. The "mixing" cost and the costs of the ingredients are included in the cost charged per cubic yard of concrete.

The cost per cubic yard of concrete for a project is usually agreed upon by the contractor and the ready-mix supplier before the start of a project. The price or cost is well publicized and often the ready-mix supplier agrees to a guaranteed price for an entire project duration. The result is that the cost determination of the concrete material presents few problems.

Determination of the cost of placing concrete presents the estimator with two somewhat unrelated problems. First, there are numerous alternative means available for concrete placement. In addition to varying labor crew sizes, various alternative equipment (cranes, concrete pumps, conveyors, and so forth) is required to place a specified concrete pour. Because each individual method results in a different cost, the estimator has to know which method will be used in each construction project. Productivity studies performed by the author indicate the "average" calculated productivities for the various concrete placement methods shown in Figure 11.21. It should be noted that: (1) these are average numbers and need to be modified given knowledge of the specific project conditions; and (2) it is implied that the productivities shown represent expected output or productivity when the method is in fact the appropriate method to be used. In other words, if an improper or ill-fitted technology is chosen for a specific concrete placement, the estimator should not expect to obtain the productivities listed. If a crane, for example, is utilized when in fact a concrete pump is more appropriate for the concrete pour in question, the productivity for the crane method shown in Figure 11.21 will not be realized.

Hours Required for Placing 1.0 Cubic Yards

Method	Laborer	Operator
1. Place concrete directly from ready mix truck	0.4	—
2. Place concrete from ready mix truck into wheelbarrows and placing	1.3	—
3. Place concrete from ready mix truck into concrete buggies	1.0	.25
4. Hoisting concrete in buggies vertically with a hoist	1.0	.25
5. Hoisting concrete in a crane bucket and placing	0.5	.50
6. Hoisting concrete in a crane bucket and placing in a hopper	1.7	.25
7. Placing concrete with a concrete pump	0.3	.25

Figure 11.21 *Example Productivities for Placing Concrete*

The actual cost of performing concrete placement work is dependent on both the productivity of the resources utilized and the cost of the utilized resources as a function of time (e.g., the hourly rate of a concrete finisher). In order to convert the productivities illustrated in Figure 11.21 into a cost figure, the estimator must be aware of various factors that can affect both the productivity and the cost of the resources. Some of these "adjustment" factors are listed in Figure 11.22.

Exactly how the estimator would recognize the existence or non-existence of each of the listed factors and adjust his calculation accordingly is also an individual matter. For example, in order to adjust the concrete productivity data shown in Figure 11.21 for the fact that proposed work to be performed will be performed using an experienced versus a relatively unexperienced superintendent is not quantified via the use of a formula. Obviously, the more historic data the estimator has collected from past projects subject to varying conditions, the more accurately he can predict productivity and cost as a function of varying conditions.

Quantity of Concrete to Pour	Batch Processing or Ready Mix Concrete
Elevation of Concrete Pour	Available Water for Curing
Physical Surroundings of Concrete Pour	Air Entrainment Required
Strength of Concrete Required	Expected Temperature
Quality of Surface Area Required	Equipment to Be Used to Place Concrete
Availability of Concrete to the Site	Expected Wind and Humidity
Relative Dimensions of Width to Height and Length	Union Requirements Regarding Crew Composition

Figure 11.22 *Example Variables Affecting the Cost of Concrete and Its Placement*

Perhaps of all the factors that are listed in Figure 11.22, the ones that are most unique and perhaps most critical to pricing of the concrete are (1) the quality of the concrete to be used, (2) the quality of the finished concrete as required by the specifications, (3) the quantity of the concrete to be placed, and (4) the method to be used for placing the concrete.

In regard to Factor 1, the required quality of the concrete greatly affects the price of the purchased concrete. For example, a 6,000-psi (pounds per square inch) concrete requires more cement than a 4,000-psi concrete. The cost of the concrete is in turn mainly dictated by the quantity of cement required. The required finish of the concrete in part dictates the labor cost of finishing the concrete which in turn affects the cost of performing the work.

As is true of most construction operations, the contractor incurs considerable "start-up" costs in the placement of concrete. It is common to observe workers spending considerable time "anticipating" the concrete pour. This non-productive start-up time, as well as the costs of readying tools and equipment, can represent a large percentage of the actual concrete placement costs, particularly if the pour is small. As the concrete pour gets larger, this start-up cost decreases as a percentage of total cost. The size of the concrete pour also somewhat dictates the proper equipment to be utilized. What might appear to be a cost-effective piece of equipment in Figure 11.21 may in fact prove unfeasible if a proposed pour is too small. As noted in Figure 11.21, many types of technologies can be used to place concrete; this is likely the biggest variable in regard to determining what method to use for concrete placement and its costs.

Formwork

It is more difficult to price concrete formwork than to price its placement. In an earlier section, we mentioned that numerous forming systems have evolved for each specific member. In the broadest sense, two types of systems are available: job-built forms and pre-built forms. Several vendors manufacture and produce their own pre-built forms.

The estimator's and contractor's selection of the most cost-effective system for a specific usage entails the consideration of several factors. Let us consider wall forms as an example. There are 2 cost components pertinent to wall forming: material and labor or fabrication costs. The material cost is made more complex in terms of forming because forms are usually re-used. The material cost when a form is used 6 times on a project is less than when it is used 3 times. The result is that the number of uses of a form, the initial cost or the rental cost, and the labor cost to include the setting, bracing, aligning, stripping, and cleaning costs all affect the choice between two or more possible forming systems. A graph illustrating the cost of two different

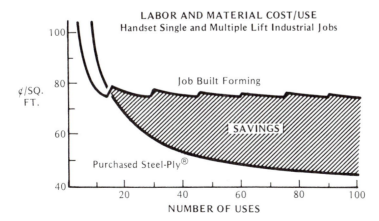

Figure 11.23 *Cost of Forms as a Function of Use*

forming systems as a function of the re-use factor is illustrated in Figure 11.23. A step-by-step formulation criteria for aiding in the choice of two methods is illustrated in Figure 11.24.

It may appear that the analysis being suggested in Figures 11.23 and 11.24 is removed from the estimating function. It may appear to be field-personnel–oriented because the superintendent may select the forming system. However, as we have noted, it is necessary for the estimator to be able to recognize the optimal method and/or the method that will be used in order to determine an estimate that best represents real costs. Figure 11.25 lists factors that relate to initial cost and the re-use factor, plus others. Representative productivities for forming various types of concrete members are listed in Figure 11.26.

Perhaps second to the re-use factor in regard to the impact on the cost of concrete forming is the matter of its *uniformity*. Irregularities or interruptions in the forming of a member are, as we said earlier, called *blockouts*.

Let us consider the forming of foundation walls for a building. If the walls are straight with few or no "holes" for windows, doors, pipes, or vents, the forming of the walls can proceed efficiently; labor productivity will be high. On the other hand, if there are blockouts, the on-site carpenters will have to form around and custom-design each of them. This process is time-consuming, and results in lowered productivity.

Forming irregularities or blockouts that interrupt the flow of production can also curtail the productivity associated with forming beams, columns, or slabs. Thus the estimator must recognize the uniformity or regularity of the forming process as well as the type of forming system that is to be used when he is calculating the cost of concrete formwork.

Reinforcement

As noted earlier, the pricing of material and labor for reinforcing usually causes the estimator fewer problems than for concrete placement or forming. This is in part due to the repetitiveness or streamlining of the fabrication process. Example labor productivity for the placing of reinforcement is illustrated in Figure 11.27.

Reinforcing is usually purchased via a weight quantity. The contractor will also have to pay a "premium" if the fabricator must perform work on the standard members such as making bends, fabricating beam seat plates on members, and so forth. A premium may also be added if the steel is to be transported for a considerable distance. But because fabricating or transportation distances are known before construction work proceeds, the estimator has no difficulty obtaining price quotes from the reinforcing supplier, so pricing presents few problems.

GENERAL JOB INFORMATION

1. _____ ⟦⟧ Total contact area to be formed.
2. _____ Pours planned.
3. _____ Pours scheduled/month.
4. _____ Months total form usage (#2 ÷ #3).
5. _____ ⟦⟧ Formwork required — min. (#1 ÷ #2).
6. _____ ⟦⟧ "Extra" formwork (form over construction joint, extra side for better cycling, etc.).
7. _____ ⟦⟧ Total formwork planned on (#5 + #6).
8. $ _____ /Hr. labor cost — avg. of one laborer/carpenter.
   NOTE: Include overhead, fringe benefits + profit.
9. Distance from formwork material sources:
   _____ Miles from Symons.
   _____ Miles from job built alternative.
   $ _____ /Hundredweight — freight rate.
10. _____ Average wall width.

STEEL-PLY® COST SUMMARY

FORMWORK
Purchase
11. $ _____ /⟦⟧ For STEEL-PLY®(per bill of material).
    × _____ ⟦⟧ Form required (#7).
12. $ _____ STEEL-PLY® purchase value.
13. –$ _____ Salvage value, based on (#14.) _____% recovery value.
    NOTE: Take into account resale value, future applications + renovation cost.
15. $ _____ Net form material cost — Symons (#12 − #13).
16. $ _____ Waling, aligning, scaffold + bulkhead lumber.
17. –$ _____ Salvage value lumber at (#18.) _____%.
19. $ _____ Net cost waling, aligning, scaffold + bulkhead lumber (#16 − #17).
20. $ _____ Total net STEEL-PLY® form cost (#15 + #19).

Rental
21. $ _____ Total STEEL-PLY® purchase value (#12).
22. $ _____ First month rent @ 11%.
23. $ _____ Second + subsequent months rent @ 9%/month, $ _____/month
    × _____ months.
24. $ _____ Total rental cost (#22 + #23).

Ties
(Same for purchase or rent).
25. TYPE TIE: Standard ☐ Heavy Duty ☐
26. _____ ⟦⟧ Contact area carried/tie.
27. _____ Total ties required (#1 ÷ #26).
28. $ _____ Avg. cost/tie.
29. $ _____ Total tie cost (#27 × #28).

Form Release Agent
(Same for purchase or rent).
    NOTE: Assume that on the average plastic coated plywood will require 50% less form release than standard, uncoated plywood.
30. _____ ⟦⟧ /Gallon coverage.
31. _____ Gallons required (#1 ÷ #30).
32. $ _____ /Gallon cost — MAGIC KOTE.
33. _____ Total form release cost (#31 × #32).

Form Layout
34. $ ____0____ With STEEL-PLY.®
    NOTE: Layout—Planning—is a very significant cost associated with job building forms. If ignored, the subsequent additional labor costs will far exceed the layout costs. STEEL-PLY® layout costs average 3-4% of rental income. Symons absorbs this cost.

    ASSUME: Job built layouts will cost an average of 50% more than a corresponding STEEL-PLY® layout due to: Non-standard tie spacing, special cuts, indeterminent lumber strength, layout inexperience.

Figure 11.24 Analysis of Formwork Needs (Courtesy of Symons Co.)

Concrete: Forming, Reinforcing, and Placement

Form Labor — STEEL-PLY®

35. _____ [/]/Man hour — set, align, strip, clean, move.
    NOTE: 38 [/]/man hour is the documented average for cut-up, multi-lift, handset industrial work.
36. $ _____ /Hour labor cost (#8).
37. $ _____ /[/] Contact area labor cost (#36 ÷ #35).
38. _____ [/] Total C.A. (#1).
39. $ _____ Total formwork labor (#37 × #38).

Freight

40. _____ Pounds.
41. $ _____ /Hundred weight delivered (from #9).
42. $ _____ Total est. freight (one way).
43. $ _____ Est. rental freight (double #42).

CLEANING — DAMAGES — SHORTAGES (Rental Only)

44. $ _____ Estimate — will vary with contractor and job conditions.

SUMMARY — STEEL-PLY® COSTS

		Purchase		Rental
Formwork	(#20)	$ _____	(#24)	$ _____
Ties	(#29)	$ _____	(#29)	$ _____
Form Release	(#33)	$ _____	(#33)	$ _____
Layout	(#34)	$ _____0_____	(#34)	$ _____0_____
Labor	(#39)	$ _____	(#39)	$ _____
Freight	(#42)	$ _____	(#43)	$ _____
Clean — Damage		$ _____0_____	(#44)	$ _____
Total Cost		$ _____		$ _____

For _____ Uses

JOB-BUILT FORMS — COST SUMMARY

Formwork

$ _____ /[/] include: Face material _____ /[/]
× _____ [/] Form required (#7). Studs _____ /[/]
45. $ _____ Form purchase value. Walers _____ /[/]
                      Top + bottom plates _____ /[/]
                      Strongbacks _____ /[/]
                      Aligning lumber _____ /[/]
                      Nails _____ /[/]
                      Std. brackets/wedges _____ /[/]
                      Corner brackets _____ /[/]
                      Misc. specialty hdwre. _____ /[/]

46. _____ % Allotment for waste/scrap/special cuts/one time use/etc.
47. $ _____ Material cost (#45 + #45 × #46).
48.–$ _____ Salvage value, based on (#48a.) _____% recovery value.
    NOTE: Compare this % with that for STEEL-PLY® in #14. to make sure the two are equitable.
49. $ _____ Total net formwork material cost.

Ties

50. _____ Pounds ultimate load ties.
51. _____ [/] Contact area carried/tie.
52. _____ Ties required (#1 ÷ #51).
53. $ _____ /Tie average cost.
54. $ _____ Total tie cost (#52 × #53).

Form Release

55. $ _____ = Double (#33). See assumption — STEEL-PLY® section on Form Release.

Figure 11.24 *Continued*

Form Layout

See assumptions (#34).
Job built layout = 1.5 × est. S/P layout cost.
= 1.5 × (3.5%) (#24).
56. = $_____ Layout cost.

Form Labor
57. _____ □ /Man hour.
NOTE: Estimating manuals report 5-7 □ /man hour. We assume a contractor experienced in building his own forms will *TRIPLE* the estimating books and will average 15-21 □ /man hour.
58. $ _____ /Hr. labor cost (#8).
59. $ _____ /□ C.A. labor cost (#58 ÷ #57).
60. _____ □ C.A. (#1).
61. $ _____ Total formwork labor (#59 × #60).

Freight
62. _____ Pounds.
63. $ _____ /Hundredweight (from #9).
64. $ _____ Total estimated freight.

Scrap Disposal
65. $ _____ /Pound disposal cost.
66. _____ Pounds scrap = (#1 − #48a.) × _____ lb. (#62).
67. $ _____ Total scrap disposal (#65 × #66).

SUMMARY — JOB BUILT FORMS

FORMWORK	(#49)	$ _____
TIES	(#54)	$ _____
FORM RELEASE	(#55)	$ _____
LAYOUT	(#56)	$ _____
LABOR	(#61)	$ _____
FREIGHT	(#64)	$ _____
SCRAP DISPOSAL	(#67)	$ _____
TOTAL COST		$ _____

For _____ Uses

SUMMARY — STEEL-PLY® vs. JOB BUILT

System	Total Cost	Divided by _____ □ Contact area (#1)
STEEL-PLY®—		
Purchase	$ _____	$ _____ /□ C.A.
Rent	$ _____	$ _____ /□ C.A.
JOB - BUILT —		
_____	$ _____	$ _____ /□ C.A.
_____	$ _____	$ _____ /□ C.A.

Figure 11.24 Continued

Height of Forming	Weight of Forms
Number of Corners	Types of Crafts Required to Place
Types of Corners	Ability to "Gang" or "Fly" Forms
Vertical or Curved Sections	Type of Ties Required
Number of Blockouts	Quality and Type of Support Hardware
(e.g., Windows, Pipes, etc.)	Required

Figure 11.25   *Factors Affecting Productivity of Placing Concrete Forms*

Type of Forming	Productivity
Job Built Plywood Wall Forms	4 Manhours/100 SFCA
Job Built Sheathing Wall Forms	7 Manhours/100 SFCA
Symons Wall Forming	2 Manhours/100 SFCA
Plywood Column Forming	16 Manhours/100 SFCA
Job Built Beam Forming	9 Manhours/100 SFCA
Flat Slab Forming	8 Manhours/100 SFCA

Figure 11.26   *Example Placement of Formwork Productivities*

A potential troublesome factor that can affect the pricing of reinforcement is the need for temporary storage of the bars. Reinforcing bars are usually transported to a project site in one or more deliveries. In other words, rather than bring the reinforcing to the job site as they are needed, the fabricating vendor often brings a large order. Because the materials are substantial in size, the contractor is often faced with the problem of where to

Method	Ironworker Hours Required Per Ton of Reinforcing
1. Placing Bars Where it is Not Necessary to Tie the Bars	22.0
2. Placing Bars Where it is Necessary to Tie the Bars with Tie Wire	25.0

Method	Ironworker Hours Required Per 100 Sq. Ft. of Wire Mesh
Placing Wire Mesh	1.1

Figure 11.27   *Example of Productivities for Placing Reinforcement*

temporarily store them. Given limited space at a job site, he may have to temporarily store some or all of the reinforcing at a considerable distance from the location in which they will be installed. This "doubling" of handling or transportation can almost double the total labor cost. It follows that the estimator must be able to foresee how and when the reinforcing will be delivered to the job site, and where it is to be stored, in order to accurately estimate the reinforcing labor cost. This is another example of the need for the estimator to be able to conceptualize the actual construction process when he is preparing an estimate.

PRODUCTIVITY ANALYSIS AND ESTIMATING

In order to illustrate the importance of the estimator's ability to recognize alternative methods of construction, their productivity and their costs, in preparing his estimate, we will now consider the use of two alternative construction methods to perform a specific amount of concrete work. It is intended that the productivity/cost analysis should be part of estimator's knowledge relative to selecting and pricing the work determined as part of the take-off process.

Let us assume an estimator is faced with estimating a project that includes the placement of 20,000 cubic yards of slab work. The contractor owns a crane which he would like to use to place the concrete. However, two alternative construction methods using the crane are judged appropriate for the work. In Method A, using ready-mix trucks, the crane, and several powered horizontal buggies, the sequence of the work process would consist of the ready-mix trucks bringing concrete to the job site and unloading it into a crane bucket, which in turn would lift the concrete to the elevated floors and unload it into the powered buggies. These would be moved horizontally and unloaded at the location of the concrete placement.

Using Method B, the contractor would only use the ready mix trucks and the crane. However, because the slab areas are large, the crane's boom would have to be at a considerable angle in order to reach the location of concrete placement. Given the relatively large weight of concrete, the efficiency of the crane decreases as the angle of the boom increases, because less concrete can be carried in the crane.

As noted in Chapter 2, a contractor or his estimator can obtain productivity data by collecting accounting records or by developing a scientific productivity standard through the use of various management techniques or models. Let us assume that the contractor in question has studied the 2 possible work processes via the use of the Method Productivity Delay Model (MPDM) (presented in Chapter 2). MPDM data collection, processing, and structuring for Method A are shown in Figures 11.28 to 11.30. Assuming that each unit of productivity is 1 cubic yard, the structured data in Figure 11.30

Concrete: Forming, Reinforcing, and Placement

Page 1 of 2	PRODUCTION CYCLE DELAY SAMPLING			unit: sec.				
Method: Crane Bucket Buggy Pour					Production unit: Concrete Drop			
Production cycle	Production cycle time (sec.)	Environment Delay	Equipment Delay	Labor Delay	Material Delay	Management Delay	Notes	(Mean non-delay time)
1	165			70			95	73
2	309			100	100		109 conc.	217
3	162			68			94	70
4	168			80			88	76
5	201			100			101	109
6	372			60	205		107 conc.	280
7	186			90			96	94
8	345					251	94	253
9	174			83			91	82
10	354			104		153	97	262
11	228			137			91	136
12	204			111			93	112
13	108						108 ✓	16
14	288				192		96 conc.	196
15	102						102 ✓	10
16	96						96 ✓	4
17	90						90 ✓	2
18	99						99 ✓	7
19	138					45	93	46
20	93						93 ✓	1
21	99						99 ✓	7
22	102						102 ✓	10
23	108						108 ✓	16
24	309		126	90			93	217
25	88						88 ✓	4
26	324			233			91	232
27	429			166		170	93	337
28	114			21			93	22
29	96						96 ✓	4
30	84						84 ✓	8
31	189					93	96	97
32	486					398	88	394
33	81						81 ✓	11
34	78						78 ✓	14
35	81						81 ✓	11
36	108						108 ✓	16
37	153		62				91	61
38	135		40				95	43
39	102						102 ✓	10
40	87						87 ✓	5

Figure 11.28 *Data Collected Using the MPDM*

| Page 2 of 2   PRODUCTION CYCLE DELAY SAMPLING   unit: sec. |
| Method: Crane Bucket Buggy Pour   Production unit: Concrete Drop |

Production cycle	Production cycle time (sec.)	Environment Delay	Equipment Delay	Labor Delay	Material Delay	Management Delay	Notes	(Mean non-delay time)
41	237					142	95	3
42	87						87 ✓	5
43	87						87 ✓	5
44	93						93 ✓	1
45	102						102 ✓	10
46	120		28				92	28
47	78						78 ✓	14
48	102						102 ✓	10
49	78						78 ✓	14
50	87						87 ✓	'5
51	222				130		92 ᶜᵒⁿᵗ	130
52	84						84 ✓	8
53	81						81 ✓	11
54	84						84 ✓	8
55								
56								
57								
58								
59								
60								
61								
62								
63								
64								
65								
66								
67								
68								
69								
70								
71								
72								
73								
74								
75								
76								
77								
78								
79								
80								

Figure 11.28 Continued

MPDM. PROCESSING				
Crane Bucket Method: Buggy Pour			Unit: Second Production Unit: Concrete Drop	

	Units	Total Production Time	Number of Cycles	Mean Cycle Time	$\Sigma(\mid$ Cycle Time- Non-Delay Cycle Time $\mid)/n$
A)	Non-Delayed Production Cycles	2655	29	91.9	8.5
B)	Overall Production Cycles	8677	54	160.7	73.3

DELAY INFORMATION

		Environ-ment	Equip-ment	Labor	Mater-ial	Manage-ment
C)	Occurrences	0	4	15	4	7
D)	Total Added Time	0	256	1513	627	1252
E)	Probability of Occurrence*	0	.074	.277	.074	.130
F)	Relative Severity**	0	.40	.63	.98	1.11
G)	Expected % Time Per Prod. Cycle***	0	2.9	17.4	7.2	14.5

*Delay Cycles/Total Number of Cycles
**Mean Added Cycle Time/Mean Overall Cycle Time
***Row E Times Row F Times 100

Figure 11.29 *Processed Data for Method A*

MPDM STRUCTURE

Crane Bucket Buggy Pour

Production Unit: Concrete Drop

I Productivity Equation

Overall Productivity $= ($Ideal Productivity$)(1-Een-Eeq-Ela-Emt-Emn)$

22.4 Units/Hr. $= (38.6 \text{ Units/Hr.})(1-0-.029-.174-.072-.145)$

II Method Indicators

A. Variability of Method Productivity

Ideal Cycle Variability $= 8.5/91.9 = .09$

Overall Cycle Variability $= 73.3/160.7 = .46$

B. Delay Information

	Environ-ment	Equip-ment	Labor	Material	Manage-ment
Probability of Occurrence	0	.074	.277	.074	.130
Relative Severity	0	.40	.63	.98	1.11
Expected % Delay Time Per Production Cycle	0	2.9	17.4	7.2	14.5

Figure 11.30 *Structured Data for Method A*

indicates that the *average* productivity for this method is 22.4 cubic yards per hour and the *ideal* or possible productivity is 38.6 cubic yards per hour.

The results of Method B are those that were illustrated in the description of the MPDM in Chapter 2. As indicated in Figure 2.15, the *average* productivity is 18.5 cubic yards per hour and the *ideal* or possible productivity is 38.9 cubic yards per hour.

Comparison of the two ideal or possible productivities indicate that both methods yield potentially about the same productivity—approximately 38 cubic yards per hour. However, unless the delays that are part of the 2 methods can be eliminated, these ideal productivities will not be obtained.

Perhaps a more meaningful comparison is to evaluate the average or overall productivities of the two methods. The comparison indicates that Method A yields a higher productivity—approximately 22.4 cubic yards per hour—than does Method B—18.5 cubic yards. One might conclude that Method A is the preferred method. However, this conclusion may be erroneous.

The selection of the better (most economical) of the 2 construction methods should be based on the *unit cost* of the two methods. It is true that productivity is a component of the unit cost. However, the cost of the resources used for the 2 methods also has to be considered. Assume that summing the cost of the labor crews and the equipment indicates that the hourly crew cost for Method A is $300 and for Method B is $210.

Having determined the crew cost and productivity for the two methods, we can now calculate the unit cost of the two methods; this is shown in Figure 11.31. The cost of the resources per hour divided by the productivity yields the unit cost. We can see that the lower unit-cost method is Method B—i.e., $11.35 per cubic yard. In effect, the lower cost of the crew for this method offsets the fact that the productivity of the crane is curtailed by the angle of the boom.

Based on the above analysis, the estimator should select Method B and calculate the cost of the work accordingly. The placement cost would be calculated as 20,000 cubic yards times $11.35 per cubic yard, or $227,000. This number would become part of the contractor's overall estimate.

It should be noted that other conditions may exist that affect the choice of Method A and B. For example, the contractor may have to complete the work within a limited time schedule. If this is the case, Method B may be unfeasible because, owing to its lower productivity, more time may be required for it as against Method A. If this is the case, this fact should be recognized in the decision as to the method to be used.

Method A

Productivity = 22.4 cubic yards/hour
Crew Cost = $300.00/hour

$$\text{Unit Cost} = \frac{\text{Crew Cost}}{\text{Productivity}} = \frac{\$300.00}{22.4} = \$13.39/\text{cubic yard}$$

Method B

Productivity = 18.5 cubic yards/hour
Crew Cost = $210.00/hour

$$\text{Unit Cost} = \frac{\text{Crew Cost}}{\text{Productivity}} = \frac{\$210.00}{18.5} = \$11.35/\text{cubic yard}$$

Figure 11.31 *Unit Cost Comparison for Two Methods*

EXERCISE 11.1

For the construction drawing illustrated on the next page, take off the quantities for the following work items:

Footing concrete (cubic yards)
Wall concrete (cubic yards)
Slab concrete (cubic yards)
Wall forms (square feet of contact area)
Reinforcing (tons)

EXERCISE 11.2

The cost of concrete forming often comprises a significant percentage of the total cost of concrete work. There are many different types of concrete forms available to the contractor. For example, for forming a concrete wall, a contractor might use either job-built plywood forms or pre-fabricated steel-ply forms. Steel-ply pre-fabricated forms are often relatively more expensive; however, they may be re-used more times and may involve a lower placement labor cost.

Given the high cost of forming, it is advantageous for a contractor/estimator to analyze forming alternatives. Assume that a contractor determines that during the next year, his firm will need to place approximately 900,000 square feet of contact area of wall forms. Assume he is considering 2 alternative wall-forming systems to be used for the work. Information regarding these alternatives are as follows:

Job-Built Plywood Forms

Purchase cost for material = $1.50/sq. ft.
Labor cost for installation = $1.40/sq. ft.
Expected number of uses per form = 7

Pre-fabricated Steel-Ply Forms

Rental cost per month = $4.00/sq. ft.
Labor cost per installation = $ .35/sq. ft.
Expected number of uses = No limit

Assume that the contractor will form approximately 1,600 square feet of contact area at a time, then strip the forms and move and re-use them. Using these data, determine whether the job-built plywood forms or the pre-fabricated steel-ply forms yield a lower unit cost.

EXERCISE 11.3

A contractor has several options available when he is placing concrete. Let us assume that he is evaluating the use of either a concrete pump or a crane and bucket

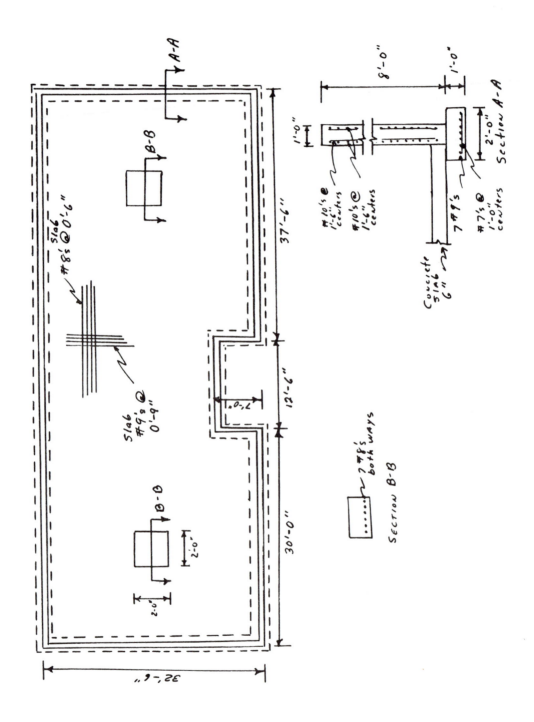

A-A

B-B

#8's Slab @ 0'-6"

Slab #9's @ 0'-9"

B-B

2'-0"

2'-0"

37'-6"

12'-6"

2'-0"

30'-0"

32'-6"

8'-0"

1'-0"

1'-0"

2'-0"

Section A-A

#10's @ 1'-6" centers

#10's @ 1'-6" centers

#7's @ 1'-0" centers

#9's

Concrete Slab 6"

7 #8's both ways

Section B-B

to place 1,000 cubic yards of slab concrete. He can rent a concrete pump that has an hourly output capacity of 100 cubic yards per hour. However, because of the productivity limitations of the finishing crew, the contractor estimates utilizing 40% of the pump's production capacity. The pump can be rented for $150 per hour plus $3 per cubic yard of concrete placed. The pump can be rented on an hourly basis—i.e., it does not have to be rented for consecutive hours. The cost of the accompanying labor crew needed to support the pump is estimated at $180 per hour.

As an alternative to using a pump, the contractor can rent a crane at a cost of $450 per day plus a $400 delivery cost. Using the crane, the contractor estimates a daily production of 150 cubic yards of concrete placement and an accompanying hourly labor crew cost of $250. Determine the least costly method of placing the concrete.

12

Masonry Work: Brick and Block

*Distinguishing Characteristics of Masonry Work *
*Definition of Masonry Work Items * Masonry Work
Take-Off * Pricing Masonry Work * Productivity
Analysis of Masons*

DISTINGUISHING CHARACTERISTICS OF MASONRY WORK

The amount of masonry work that is part of a building construction project varies from virtually none to a great amount for a building that uses masonry units for its entire foundation and for all of its exterior and interior walls. Thus the importance of the topics discussed in this chapter differ depending on the specific building design.

Masonry includes several types of work and/or materials. The most common and basic building units are masonry blocks and bricks. Less common materials include various types of stone and tile. Given the more common use of masonry blocks and bricks, this chapter will focus on these two masonry units.

Masonry units can be designed to be a structural material (i.e., load-bearing or non–load-bearing). Blocks are most commonly used; they can be designed as part of the load-bearing foundation or as an exterior or interior load-bearing wall. Masonry blocks are cut from a cement material. However, unlike concrete that is designed to be shaped in numerous sizes, masonry blocks are only available in a finite number of sizes. Not all masonry blocks are designed to be structural members. Non–load-bearing masonry blocks, often referred to as **cinder blocks**, are often used in non-bearing interior walls within a building design.

Many masonry materials are included in a building design because of the aesthetic appearance they add to the finish of the building. Masonry units also are durable and long-lasting in regard to their resistance to weathering. Given these appearance and durability characteristics, masonry units, including bricks and to a lesser degree various types of stone, are often used, especially on the exterior face of a building.

In addition to the fact that many types and quantities of masonry units might be included in a specific building design, the estimating of masonry work is perhaps characterized by 2 additional distinguishing features. For one, the cost of masonry work is especially dependent on labor productivity. There are few kinds of construction work on a building project that are more labor-intensive; the labor cost for placing masonry units can exceed the material costs of the units. This strong dependence on labor productivity draws attention to the estimator's need to focus on the collection and use of accurate mason productivity records and labor-cost data.

A second distinguishing aspect of estimating masonry costs is that seldom if ever is the actual number of masonry units delineated on the drawings. Instead, the designer merely specifies the location of the units on the construction drawings; it is left to the estimator in his take-off process to determine the actual number of units that will be needed in a specified area. This can be accomplished by calculating of the area of a single unit and dividing the quantity into the taken-off area. Naturally, the designer does not delineate every masonry unit on the drawings, because this would not only be tedious but, given the small size of individual units, perhaps impossible.

Sometimes a general contractor chooses to subcontract masonry work. When this is the case, the masonry subcontractor has the responsibility of performing the masonry take-off and subsequent costing. However, in this chapter we assume that the general contractor will perform the masonry work.

DEFINITION OF MASONRY WORK ITEMS

As we have said, the vast majority of masonry work entails the use of either concrete block (or cinder block) work or brickwork. To a lesser degree, various types of stonework may be required on a project.

A building is commonly designed to accommodate the use of a standardized concrete block size or a standardized brick. A standardized concrete block is illustrated in Figure 12.1 and a standardized brick is illustrated in Figure 12.2. While the block and brick shown in these figures usually make up the majority of the blocks and bricks included in a specific building design, the fact remains that concrete blocks and bricks are available in numerous sizes and qualities. Thus, any single building project may consist of several types and qualities of blocks and/or bricks. Some of these various concrete block sizes and shapes available are illustrated in Figure 12.3.

The labor effort and cost for placing a concrete block is for the most part independent of its size or shape. A mason has to lift, set, and grout each of the blocks shown in Figure 12.3.

While the labor cost may be independent of the size or shape of a concrete block, the material cost for the blocks varies considerably. In the

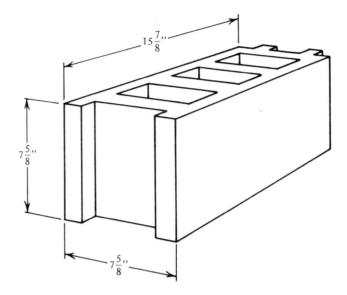

Figure 12.1 *Standard Concrete Block*

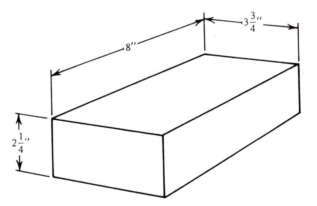

Figure 12.2 *Standard Brick*

manufacturing process, it costs less to make standardized blocks (Figure 12.1) than "non-standard" blocks (Figure 12.3). In effect, the manufacturer has to "disrupt" its production process in order to manufacture a "non-standardized" block. The result is that the purchase cost of the blocks varies. Given this variable material cost for each block, a strong argument can be made for identifying each block size or type of block—i.e., a concrete block

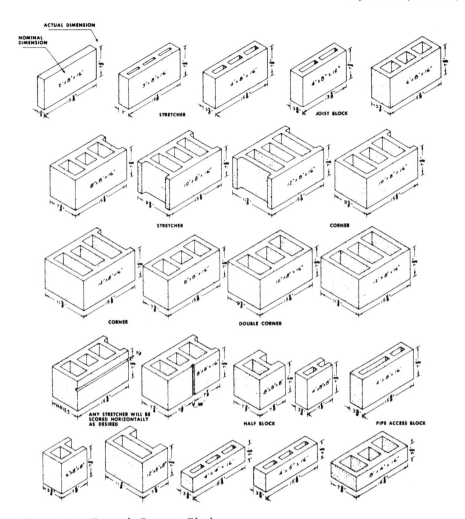

Figure 12.3 *Example Concrete Block*

versus a cinder block—as a separate work item for the quantity take-off function.

Defining each block size separately may seem cumbersome and time-consuming; however, an architect doesn't often design a building that requires a great many sizes or shapes of blocks. Therefore defining work items as individual block sizes or shapes is more feasible than would at first seem to be the case.

While bricks are also available in varying sizes, those for a specific building project typically vary less in size or shape than blocks. Often, a building project consists exclusively of the standardized brick size shown in Figure 12.2.

However, the *quality* of brick used often varies. Brick has several attributes, including its fire resistant capability and its durability; but one of the primary reasons for its frequent use is its aesthetic attribute. Depending on the aggregates used and on the production process of "surfacing" the outer face, the quality and appearance of bricks vary. The most common type of brick is referred to as **face brick**. Some other types are shown in Figure 12.4. Because the material cost for each of these types varies, the estimator is justified in defining each *type* of brick as a work item. But this doesn't usually make too great an estimating problem, because seldom are more than two or three different types used on a single project.

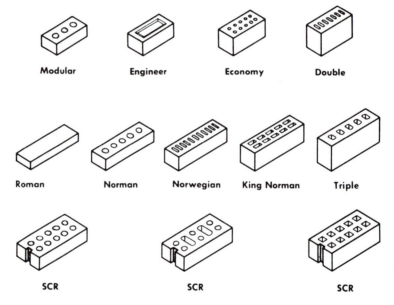

Figure 12.4 *Example of Various Types of Bricks*

It is the *size of block* and the *type of brick* which for the most part dictate masonry work items for the quantity take-off. Two other factors the estimator might consider in defining work items is the *location* and the "continuity" of the masonry work to be performed. The productivity of a mason is greatly affected by these factors. For example, masonry work performed at a height above ground level requires the construction of a working platform (scaffolding) and the lifting of blocks or bricks to the work location—the result is lower mason productivity and higher unit labor cost.

The labor effort and cost for placing masonry units also is dependent upon the *continuity* of the work. Masons can place block or brick for a straight wall with no blockouts at a near optimal productivity. However, when there are irregularities, the mason has to place the block or brick around the opening and perhaps construct a lintel. This "breaking" of the continuity of the wall area, and the need to align the masonry work to the opening significantly lowers the mason's productivity, perhaps representing as much as 50% of the straight-wall productivity.

Because labor productivity and cost vary as functions of the location and continuity of the work, it can be argued that these factors should be considered in defining work items. For example, perhaps straight-wall masonry work should be defined as one work item and masonry work around a window or door as a separate work item. The difficulty with this approach is that no 2 walls, whether they are straight walls or walls with windows or doors, are exactly alike. Even when confronted with a blockout, productivity

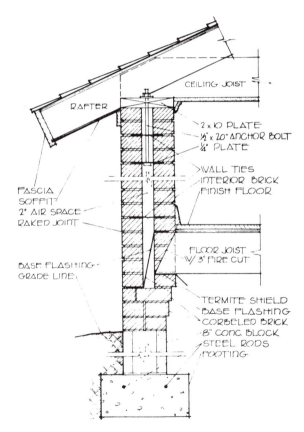

Figure 12.5 *Reinforced Masonry with Ties*

depends on the actual size and location. Similarly, productivity for placing blocks or bricks on a third-floor elevation would be lower than on the second floor, for fourth floor lower than third floor, and so on. In other words, almost every situation is somewhat different. The result is that it would be very difficult for the estimator to define an individual masonry work item for each and every situation.

A more practical approach to the recognition of the impact of location or continuity on productivity is to include these considerations in the *pricing* of the masonry work. For example, the unit price involved in determining the pricing cost of brickwork can be adjusted upward if the work entails considerable placement of bricks around windows or doors. (The adjustment of masonry unit prices as a function of the location or type of work performed will be discussed in a subsequent section in this chapter.)

One other factor relevant to the definition of masonry work items is whether or not the blockwork or brickwork is to be structurally reinforced. Owing to the need for the masonry to resist vertical and horizontal loads, a building may be designed to have two masonry units joined together by means of reinforcing rods. This type of construction is shown in Figure 12.5. Structurally reinforced masonry is especially common in cavity-wall types of construction—i.e., two masonry units separated by an air-space cavity.

The labor effort required to construct a reinforced masonry wall is considerably greater than for a non-reinforced wall. Thus it is advantageous for the estimator to differentiate between reinforced and non-reinforced masonry units when he is defining quantity take-off work items. An example list of masonry work items is illustrated in Figure 12.6.

```
Standard Block   8 X 8 X 16 — Concrete
Standard Brick   2 X 4 X 8 Face Brick (state bond type)
Standard Block   8 X 8 X 16 — Cinder
Other Block Sizes   (State Size)
Other Brick Sizes   (State Brick Size)
Other Brick Types   (State Type)
Stone
```

Figure 12.6   *Example Work Items for Masonry*

MASONRY WORK TAKE-OFF

The quantity take-off of masonry work is made complex by the fact that the project drawings do not delineate each and every masonry unit. The estimator has to convert areas designated on the drawings as masonry work into a number of blocks or bricks. In effect, take-off of masonry work is a 2-step process. First the estimator must identify the *locations* of the individual types of masonry units and convert these locations into an area of block- or brickwork. Second, he must consider the *size* of the individual block or bricks in order to convert the calculated areas into the required number of individual sizes or types of block or brick.

Fundamental to the conversion of wall areas into a number of blocks is the recognition of the surface area of an individual unit. The surface or exposed area of a standard concrete block is illustrated in Figure 12.7. For an individual standard block the surface area is calculated as 8 inches times 16 inches, or 128 square inches. Converting this to square feet (because the wall areas are usually determined in square feet), a standard block is calculated as 128 square inches divided by 144 or 0.8889 square feet.

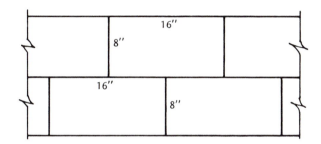

Figure 12.7   *Face Area of Blocks*

Given the calculated area of a single block, the number of blocks can be determined by dividing the area of a single block into the taken-off wall area. For example, for an 8-foot-high block wall that is 20 feet long, the surface area would be 160 square feet. Dividing 0.889 (the surface area of a single block) into 160 square feet yields a determination of 180 blocks.

It could be argued that the above calculation is in error because we have failed to recognize the joint between two adjacent blocks. This joint is filled grout. While this grouted joint does take space, the space is marginal relative to the surface area of the blocks. When he adds to this the fact that the groove at the end of each block is designed to take the grout so that two adjacent blocks virtually join together with marginal area between them, the estimator can ignore the joint space in the determination of the number of blocks.

While the estimator is correct in disregarding the joint space in calculating the number of concrete blocks, this is not an appropriate process when determining the number of bricks in a wall area. Owing to the smaller size of a brick, and given the technology of the joint between adjacent bricks, the joint space makes up a considerable area of the overall brick wall. The exact amount of area depends on the specific size of the joint constructed.

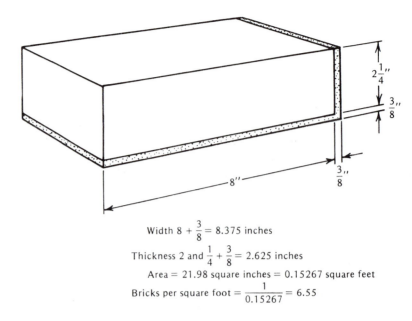

Width $8 + \frac{3}{8} = 8.375$ inches

Thickness 2 and $\frac{1}{4} + \frac{3}{8} = 2.625$ inches

Area $= 21.98$ square inches $= 0.15267$ square feet

Bricks per square foot $= \dfrac{1}{0.15267} = 6.55$

Figure 12.8 *Calculating the Number of Bricks Per Area*

The joint size constructed between two bricks varies from one building design to the next. The working drawings for the project specify the actual joint size. Given the specified joint size and given the size of the type of the brick used (usually a brick is standard in size), the estimator can calculate the area of the single brick and the joint space as shown in Figure 12.8.

As can be observed in Figure 12.8, only two sides of the joint space are included in the area calculation. This is in recognition of the fact that the adjacent bricks would enclose half of the joint space.

It is not necessary that the estimator calculate the brick area each time the joint space varies from the somewhat standard joint size shown in Figure 12.8. Instead, he can use the table shown in Figure 12.9 to determine the brick area as a function of the joint spacing.

With the table, it becomes easy for the estimator to convert a calculated brick area into a number of bricks. For example, given an area of 160 square feet of wall, and a standard brick with a 3/8-inch joint spacing, the number of brick in the wall area can be determined by dividing 160 square feet by 6.67 bricks per square foot or 24 bricks.

The placement of masonry units results in the contractor having material wastage. This wastage is partially the result of the need to break some blocks to fit a specific area or to serve as end blocks, and partially because a few of

Number of Standard Brick (8" × 2¼" × 3¾") Required for
One Square Foot of Brick Wall of Any Thickness
Vertical or End Mortar Joints Figured as ¼" Wide

Thickness of Wall	Number of Brick Thick	Width of Horizontal or "Bed" Mortar Joints					
		1/8"	1/4"	3/8"	1/2"	5/8"	3/4"
4" or 4½"	1	7.33	7	6.55	6.33	6.08	5.8
8" or 9"	2	14.67	14	13.33	12.67	12.17	11.6
12" or 13"	3	22.00	21	20.00	19.00	18.25	17.4
16" or 17"	4	29.33	28	26.67	25.33	24.33	23.2
20" or 21"	5	36.67	35	33.33	31.67	30.42	29.0
24" or 25"	6	44.00	42	40.00	38.00	36.50	34.8

Figure 12.9 *Table for Determining Bricks in a Square Foot of Area*

the delivered masonry units will prove defective. The amount of expected wastage is strongly dependent on the size or type of masonry units. A suggested wastage percentage factor for various brick and block sizes and types is illustrated in Figure 12.10.

The mortar that is part of masonry work is seldom taken off as a separate work item by the estimator. Instead, the cost of the mortar and the labor effort required for mixing and placing the mortar are included in the unit price for placing the masonry units. Given the joint size and the type of joint, tables are available that enable the estimator to determine the amount of mortar required for a specified area. The table illustrated in Figure 12.11 can be used for this purpose. While the use of this table may not be necessary to the task of preparing an estimate in that the mortar may be included in the masonry unit price, its use is necessary to determine the amount of mortar to be purchased and/or mixed — this is called the *procurement function.*

Occasionally walls are designed to have some of the blocks or bricks placed in a pattern other than the standard layout discussed earlier. This is

Type of Work	Wastage Factor %
Face Brick–Straight Walls	4%
Face Brick–Several Windows, Doors, Corners	10%
Concrete Block–Straight Walls	2%
Concrete Block–Several Windows, Doors, Corners	6%

Figure 12.10 *Example Wastage Factors for Masonry*

QUANTITY OF MORTAR REQUIRED TO LAY 1000 BRICKS

Width of Joint	1 Brick or 4" Wall	2 Bricks or 8" Wall	2 Bricks or 8" Backup
3/8" Joint 3-5/8" × 3-5/8" × 7-5/8" Brick	10.05 cu. ft.	13.80 cu. ft.	17.70 cu. ft.
3/8" Joint 1-5/8" × 3-5/8" × 11-5/8" Brick	11.70 cu. ft.	14.65 cu. ft.	
5/12" Joint 2-1/4" × 3-5/8" × 7-5/8" Brick	9.60 cu. ft.	12.15 cu. ft.	14.70 cu. ft.

Figure 12.11 *Mortar Required for Various Joint and Brick Sizes*

especially true of brick. Various designs are often identified as to a specific **bond design**. Several bond designs are illustrated in Figure 12.12. The designs shown that have one or more of the bricks turned so that the surface area is not the 2-inch by 8-inch side, presents the estimator somewhat of a problem. In order to determine the number of bricks in a specific wall area, he has to calculate the surface area for a combination of different brick surface areas. To some degree, each design is unique.

Occasionally, a masonry wall is 2 or even 3 units thick. For example, a specific design might include a double brick wall. This presents the estimator no problems in regard to the quantity take-off function. Assuming the same pattern in both thicknesses, the estimator merely doubles the number of bricks (or blocks) in a single wall thickness.

The estimator taking off masonry units often must deal with walls that include several windows and doors and various other blockouts. As a general rule, these areas should be deducted when the number of masonry units is being calculated. This can be most efficiently performed by calculating the total wall areas and subsequently subtracting the blockout areas.

If the blockout areas are small, it can be argued that, given the increased wastage of masonry units that occurs in the case of blockouts and the lower labor productivity of a mason working around a blockout, the estimator could make no deduction for the blockouts; instead, calculating the number of units assuming that the masonry units continue right through the blockout area. In theory, the added or extra masonry units that are calculated offset the wasted units and the lower productivity resulting from the need to place units around the blockouts. The validity of this assumption is dependent

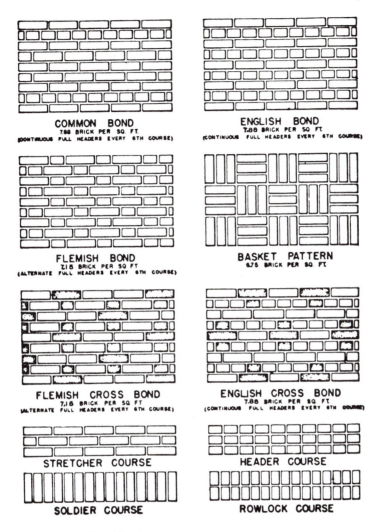

Figure 12.12 *Various Bond Designs*

upon the number and size of the blockouts. If, however, the area of the blockouts is considerable, this process would result in a calculation of too many masonry units. On the other hand, if there are only a few blockouts or if their area is small, the process may be appropriate. In general, many estimators ignore the blockouts and calculate the total area *if the area of the blockouts appear to be less than 15% of the total wall area.* They would not actually calculate the percentage. At best this 15-% rule is a rule of thumb—however, it provides a guideline. For example, confronted with a storefront wall with a great deal of window area, the estimator would calculate the area after the window areas were subtracted. However, in the case of a residential house with few windows (i.e., approximately 15% or less), the estimator would *not* deduct the window areas in calculating the number of breaks or blocks.

PRICING MASONRY WORK

Masonry work usually entails the placement of numerous units. Thus an error in pricing material or labor cost for an individual masonry unit can multiply to result in a significant estimating pricing, particularly when the area involved is large and requires many units.

As we discussed earlier, masonry units are usually taken off according to their number. This measure of work—i.e., number of blocks or bricks—is indicative of the cost of masonry work. Unlike some work items, both the material and the labor placement cost for masonry work are usually best related to or measured by the number of masonry units. Thus the estimator is not faced with having to take off masonry work in two different measures, one for pricing material and one for pricing labor.

The mortar that is part of masonry work is commonly taken off and priced as part of the estimating for the individual units. However, when unusual mortar joints or types of mortar are used, the estimator should calculate the costs of the ingredients and the mixing cost. Tables such as the one illustrated in Figure 12.13 are available to aid the estimator in such calculations.

Material	Quantity	
Lime	100	pounds
Cement	200	pounds
Sand	.67	cu. yds.

Figure 12.13 Mortar Ingredients for 1000 Bricks

While the estimator may occasionally have to price what might be considered irregular masonry work—e.g., mortar, or the building of a masonry lintel—the majority of masonry work entails the pricing of the individual blocks and bricks. Other than scaffolding, very little equipment is used in placing masonry units. Masonry work is essentially a labor-intensive operation; in other words, the material and labor costs comprise most if not all the cost of doing masonry work.

The material cost component of masonry work usually presents the estimator no difficulty. Often the contractor "buys out" the masonry work before construction begins. Established prices of blocks or bricks are available from suppliers, so few risks are included in making the material cost estimate. Perhaps the only variable or risk factor is the amount of material wastage that will occur. Some of the expected wastage factors based on the author's experience are illustrated in Figure 12.10.

The labor cost component of masonry work is the risk element. A mason's labor productivity is dependent on several factors, some of which are included in the pricing parameters listed in Figure 12.14.

Number of Ledgers	Type of Mortar
Type of Bond	Crew Combination Required
Height of Masonry Work	Location of Mortar
Type of Mortar Joint	Repetitiveness of Work

Figure 12.14 Factors Affecting Mason Productivity

Independent of the pricing and/or productivity variables, an estimator can use tables such as the one shown in Figure 12.14 to determine expected labor productivity for masonry work. These tables indicate average productivities as determined from the author's experience.

Perhaps the 2 factors or considerations that affect a mason's productivity, and hence the labor cost, more than any other factors are (1) the weather, and (2) the repetitious or continuous nature of the masonry work, including blockouts, lintels, and the like. Because most masonry work is performed outdoors, and not in an enclosed area, a mason is subject to varying temperatures, humidity, and precipitation. Several studies have been performed to measure the impact of some of these weather conditions on the productivity of a mason. Example data determined from studies performed by the author are illustrated in Figure 12.15. The estimator must acknowledge these types of considerations when preparing a masonry labor cost estimate.

The placement of masonry units, whether they are blocks or bricks, can be very repetitive. Given a straight wall with no blockouts for windows, doors, or similar irregularities, a mason can develop a "streamlined," repetitious production process. In addition to the development of a work pattern that aids productivity, the learning-curve principle of industrial engineering implies that a worker becomes more productive as he performs the same process over and over. The result is that the level of productivity that can be expected when units are placed in a continuous, non-interrupted wall will be significantly greater. The drop in productivity, and thus the increase in the masonry labor cost per unit, resulting from work on irregular areas is illustrated in Figure 12.16.

Month	Productivity Efficiency Factor
January	0.30
February	0.40
March	0.50
April	0.65
May	0.70
June	0.90
July	1.00
August	1.00
September	0.90
October	0.60
November	0.40
December	0.20

Figure 12.15 *Productivity as a Function of the Weather*

The vast difference between the productivities listed in Figure 12.16 has led some estimators to define two separate work items for a given masonry type—one for straight wall work and one for blockout work. The difficulty inherent in this approach is that no two walls may contain the same combination of straight non-interrupted wall and areas of blockouts; each wall is somewhat unique. The result of this approach is the need to define a separate work item for each and every one of these almost infinite number of wall combinations. This is not a feasible approach. Instead, the estimator is better served to utilize historical data such as that shown in Figure 12.16 and adjust

Type of Work	Bricks/Blocks Placed Per Mason Hour*
Straight Wall-Standard Brick	80–100
Straight Wall-Struck Mortar Joints Brick	60–80
Work Around Windows and Doors-Standard Brick	30–50
Standard Block- Straight Wall	13–18
Standard Block- Around Windows and Doors	8–11

*Assumes Assistance with 1 Laborer

Figure 12.16 *Example Mason Productivities*

the numbers shown based on his best judgment of the comparison of the wall being considered to those representative of the data illustrated in Figure 12.14. For example, for a wall with a few blockouts, an expected productivity somewhere between the 2 numbers shown in Figure 12.16 might be used to estimate the labor cost. The point here is that mason productivity, and hence expected masonry labor cost, is dependent on the repetitious nature of the wall being estimated.

PRODUCTIVITY ANALYSIS OF MASONS

The placement of masonry units is, as we have said, a very labor-intensive process. It follows that both its estimating and control entails careful analysis of the work of the individual masons and supporting laborers.

The work efforts of individual workers are often studied by observation at random points in time. What is defined as productivity or non-productivity is somewhat subjective and can vary depending on the judgment of the person collecting the data. A helpful definition might be as follows:

Productive: Whenever a worker is performing direct work such as placing a masonry unit, or doing support work, including lifting a block to a mason.

Non-Productive: Whenever a worker is idle, waiting, or performing unnecessary support work.

There are several different industrial engineering labor-intensive models that utilize the concept of productive and non-productive time. **Work sampling** is one of these models.

Let us assume that as part of a contractor's estimating of masonry work for a proposed project, he is studying the possibility of reducing his normal crew of 4 masons and 2 laborers to a crew of 4 masons and 1 laborer. While he recognizes that the labor cost will decrease, he is concerned that the productivity (blocks placed per mason) might decrease. Thus he is interested in determining the productivity/non-productivity of each of the 6 crew members to determine whether or not a laborer can be eliminated. (*Note:* it is assumed that there is no labor work rule that will prevent this change of crew composition.)

Work Sampling Model: Perhaps the simplest model to utilize for this analysis is the *work sampling model*. The model gives an indication as to how productive or non-productive a particular worker is on the crew. Such an indication can give insight as to the possibility of a crew-size reduction.

Worker #	1	2	3	4	5	6
	P N	P N	P N	P N	P N	P N
	P N	P N	P N	P N	P N	P N
	P N	P N	P N	P N	P N	P N
	P N	P N	P N	P N	P N	P N
	P N	P N	P N	P N	P N	P N
	P N	P N	P N	P N	P N	P N
	P N	P N	P N	P N	P N	P N
	P N	P N	P N	P N	P N	P N
	P N	P N	P N	P N	P N	P N
	P N	P N	P N	P N	P N	P N
	P N	P N	P N	P N	P N	P N
	P N	P N	P N	P N	P N	P N
	P N	P N	P N	P N	P N	P N
	P N	P N	P N	P N	P N	P N
	P N	P N	P N	P N	P N	P N
	P N	P N	P N	P N	P N	P N
	P N	P N	P N	P N	P N	N
	P N	P N	P N	P N	P N	N
	P N	P	P N	P N	P N	N
	P N	P	P N	P N	P N	N
	P	P	P N	P N	P N	N
	P	P	P N	P N	P N	N
	P	P	P N	P N	P N	N
	P	P	P N	P N	P N	N
	P	P	P N	P N	P N	N
	P	P	P N	P N	P N	N
	P	P	P	N	P N	N
	P	P	P	N	P N	N
	P	P	P	N	P	N
	P	P	P	N	P	N
	P	P	P	N	P	N
	P	P	P	N	P	N
	P	P	P	N	P	N
	P	P				N
	P					N
	P					

P = Productive
N = Non-Productive

Figure 12.17 *Data Collected at Job-Site*

$$\text{Worker \#1 — Rating} - \frac{\text{Productive}}{\text{Total}} - \frac{36}{57} - 63.2\%$$

$$\text{Worker \#2 — Rating} - \frac{\text{Productive}}{\text{Total}} - \frac{34}{53} - 64.2\%$$

$$\text{Worker \#3 — Rating} - \frac{\text{Productive}}{\text{Total}} - \frac{32}{58} - 55.2\%$$

$$\text{Worker \#4 — Rating} - \frac{\text{Productive}}{\text{Total}} - \frac{25}{57} - 43.9\%$$

$$\text{Worker \#5 — Rating} - \frac{\text{Productive}}{\text{Total}} - \frac{32}{60} - 53.3\%$$

$$\text{Worker \#6 — Rating} - \frac{\text{Productive}}{\text{Total}} - \frac{17}{51} - 33.3\%$$

Figure 12.17 *Continued*

The work sampling model is simple to implement in the field. The observer simply observes a particular worker randomly and records whether the worker is in a productive or non-productive state when he is first observed. The number of observations is up to the observer, but of course, the more the number, the better the accuracy. After all the observations are taken, the productive samples are divided by the total number of samples to produce a productivity rating.

Let us assume that the mason contractor collects data regarding productivity or non-productivity for the 6 crew members (as illustrated in Figure 12.17). Workers 1 through 4 are masons, and workers 5 and 6 are laborers.

Figure 12.17 shows that 1 of the 2 laborers is productive only around 33% of the time, and that the other is productive 53% of the time. The conclusion one might draw is that the contractor can eliminate one of the laborers, and have the other laborer pick up the slack. In the 6-member crew and the possible 5-member crew, the laborers/laborer support the masons. These activities include carrying and lifting blocks, moving scaffolding, and such tasks.

Analysis of the 6-member crew via the work sampling model might lead the contractor to select a 5-member crew. The estimate would be prepared accordingly. Given the fact that the crew size is changed, the contractor would find it advantageous to study the crew again, perhaps via the work sampling process, after the change has been made.

The user of the work sampling model can also utilize mathematical statistical tables to determine various ranges of "confidence" that the results are correct as a function of the number of samples. This determining of "confidence limits" will not be discussed because of its relatively complicated mathematics. If the mathematics becomes complicated, a constructor will likely shun away from the use of the technique/model.

The productivity analysis of masonry operations is not limited to the use of the work sampling model. However, owing to its labor intense characteristic, the work sampling technique may prove an effective technique for estimating and citing improved mason operations.

EXERCISE 12.1

Determine the number of face brick to be purchased and placed for the drawing shown. Assume that all bricks are to be placed in a common bond and are standard size. Assume that the joint size is 3/8 of an inch.

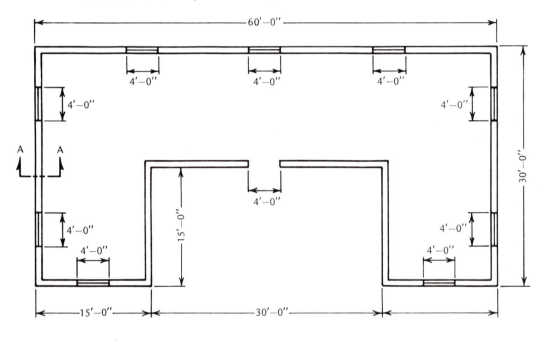

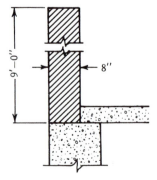

EXERCISE 12.2

Assume that a contractor analyzes a masonry work process with the objective of determining an estimating standard. He collects data by use of the Method Productivity Delay Model (MPDM) (discussed in Chapter 2). The data the contractor has obtained are in the data collection form on page 392.

Four masons on the crew are placing blocks simultaneously; 2 laborers assist them. Because the 4 masons work simultaneously, the contractor documents the placement of 4 blocks as a single cycle. Usually this entails each mason placing a block. Knowing this, and using the data shown, calculate the production equation that is part of the MPDM, including the determination of the average number of blocks placed per hour and the percentage of each expected delay.

EXERCISE 12.3

Mason productivity for placing bricks or concrete blocks is dependent on the location of the placement and the number and size of the blockouts (e.g., window or door openings in a straight wall). More labor time is required to place block on a second floor than on a first floor because the workers must operate from a suspended scaffold. Similarly, a window or door opening "interrupts" a mason's productivity when placing blocks in a straight wall.

Assume that a contractor has collected the following mason productivity for placement of concrete block for walls:

BLOCKS/MASON HOUR	FLOOR	NUMBER OF BLOCKOUTS	SQUARE FEET OF BLOCKOUTS
30	1	3	150
25	2	8	450
35	1	0	0
15	3	2	200
25	1	6	400
20	1	5	750

Assume that the contractor/estimator is preparing a mason estimate which requires that blocks be placed on the first and second floors. An equal number of blocks will be placed on each floor. The first floor has 3 blockouts with 200 square feet of openings and the second floor has 5 blockouts with 500 square feet of openings. Based on the collected data and the fact that 1,000 blocks have to be placed on each floor, estimate the number of mason hours required to perform the work.

PRODUCTION CYCLE DELAY SAMPLING

Page 1 of 1   Unit: Second
Method: CRANE BUCKET CONCRETE POUR   Production Unit: Concrete Drop

Produc-tion Cycle	Produc-tion Cycle Time (sec.)	Envir-onment Delay	Equip-ment Delay	Labor Delay	Mater-ial De-lay	Manage-ment Delay	Notes	Minus Mean Non-delay Time
1	120			✓				27
2	126		✓					33
3	98						✓	5
4	112		✓					19
5	108						✓	15
6	1122					✓	crane move	1029
7	116		✓					23
8	214		✓					121
9	92						✓	1
10	88						✓	5
11	100						✓	7
12	312		✓					219
13	110			✓				17
14	666		90%	10%				573
15	146		25%	75%				53
16	120		✓					27
17	138		✓					45
18	144		20%	80%				51
19	598	✓					crane slip	505
20	118		✓					25
21	138		✓					45
22	108						✓	15
23	98						✓	5
24	120			✓				27
25	116			✓				23
26	368		✓					275
27	118		✓					25
28	140		✓					47
29	136			✓				43
30	138			✓				45

Exercise 12.2

13

Structural Steel

*Structural Steel and Metals in a Building Project * Structural Steel Work Items * Taking Off Structural Steel * Pricing Structural Steel Work * Productivity Analysis of Steel Work*

STRUCTURAL STEEL AND METALS IN A BUILDING PROJECT

Buildings are designed to consist of varying amounts of structural steel and other metals. Some buildings derive their strength almost exclusively from the concrete or wood framing. These buildings may be totally absent of steel members. On the other hand, a building may be designed to consist of numerous steel members, including beams, girders, columns, and various other supporting members. This type of design is commonly referred to as a **steel skeleton building**. An example steel skeleton design is illustrated in Figure 13.1. Obviously, the amount of steel estimating take-off and pricing for the building shown in Figure 13.1 is extensive.

Many buildings are designed to include a combination of concrete or wood structural members and steel members. For such buildings estimating take-off may entail less effort than that required for the building illustrated in Figure 13.1.

Within the work classification of structural steel and metals is the inclusion of several different types of metals. Steel members and plates are by far the most common metal included in a given building design. However, other metals, including aluminum, and various alloys, including bronze plates, might be part of a building design. Because steel is by far the predominant kind of metal used in building, this chapter focuses almost exclusively on the estimating of structural steel.

Structural steel differs from reinforcing steel. Reinforcing steel is embedded in concrete and is taken off and estimated as part of the concrete take-off. Structural steel includes massive steel members in the shape of angles, channels, W-shaped sections, and solid tubing shapes. Also included in the structural steel estimate are various-sized steel plates that are part of the connections for the above-noted steel members.

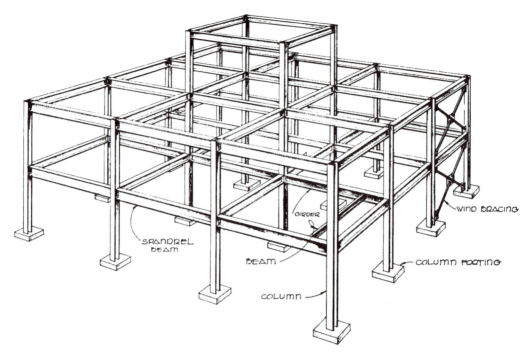

Figure 13.1   *Example of a Steel Skeleton Design*

Much as in the case of earthwork or sitework, the estimating of structural steelwork entails the recognition of considerable equipment costs related to placement. Partly owing to its heavy weight (steel weighs approximately 490 pounds per cubic foot), various types of cranes or similar equipment are utilized to put steel members in place.

Moreover, because of the relatively high cost of structural steel members, the material purchase cost of structural steel is large compared to the labor placement cost. Unlike other general contracting items such as concrete, masonry, and carpentry work, the material cost for structural steel may be significantly greater than the labor placement cost.

Structural steel is normally taken off and priced in units of weight. This is because a contractor purchases structural steel for a price per weight quantity. While weight is indicative of the material purchase cost for steel, this may not always be true in the case of labor or equipment cost. These costs may be better related to the *number* of steel members or steel connections to be fabricated. The potential problem of having two different units of measure will be discussed in a later part of this chapter.

STRUCTURAL STEEL WORK ITEMS

Estimating structural steel work essentially entails the possible take-off and pricing of 3 different types of work. The most obvious type involves the structural steel members themselves. These members include various angles, channels, W-sections, and the like that are designed to serve as structural beams and columns for the proposed building. Such pieces are available in standardized sizes. Steel members are seldom custom-designed or fabricated for a specific building; instead, the designer usually chooses from commonly available sizes.

The available sizes and shapes are widely publicized by steel fabricators; most are listed in the *American Society of Steel Associations Steel Manual.* A sample page from the manual is illustrated in Figure 13.2. Observe that the weight of each member per linear foot of length is listed, as well as the cross-sectional area and other design information. The listing of the weight per foot enables the estimator to convert taken-off linear feet of steel members into a weight quantity, which is the measure used for pricing the work.

Seldom does an estimator itemize the size or shape of each and every steel member as a separate work item. Instead, he groups the different sizes and shapes into a single summed weight quantity, and considers this a single work item. This relates to the fact that the contractor is most dependent on the total weight of the members and not their individual sizes or shapes.

Not all structural steel members in a building design are "standard." Perhaps the best examples of non-standard members are various fabricated joists that are designed to serve as roof members, floor support members, or purlins (steel members' joists supporting a slab). These are usually custom-designed for a specific use by fabricating several steel members to include various angles. An example steel joist/truss is illustrated in Figure 13.3.

Steel joists/trusses are often custom-fabricated by a steel fabricating vendor for a specific project. Their cost is dependent on the length of truss/joist itself, and the size, weight, and number of the individual members that make up the truss/joist. It is usually quoted on a price-per-linear-foot basis for a specific type of design. In other words, unlike "standard" structural steel members, which are taken off and priced in a measure of weight, fabricated steel joist/trusses are taken off and priced in a linear-foot measure.

Because each type or size of truss/joist has an individual price, the estimator should treat each different size or type as a separate work item for estimating. This approach is not as cumbersome as it may seem, because the number of different sizes or types of truss/joists that are part of a given building design is limited. Seldom are there more than 5 different truss/joist types on a single project.

A second type of structural steelwork that is part of a building project is an assortment of various steel plates that primarily serve the purpose of

$F_y = 36$ ksi	BEAMS W shapes							W 33

Allowable uniform loads in kips
for beams laterally supported
For beams laterally unsupported, see page **2** - 84

Designation		W33				W33			
Weight per Foot		240	220	200	152	141	130	118	Deflection Inches
Flange Width		15⅞	15¾	15¾	11⅝	11½	11½	11½	
L_c		16.7	16.7	16.6	12.2	12.2	12.1	11.9	
L_u		30.7	28.1	25.4	16.8	15.4	13.8	12.7	
	10							528	0.08
	11						557	522	0.09
	12				617	584	541	479	0.11
	13				599	551	500	442	0.13
	14				557	512	464	410	0.15
	15		747	684	519	478	433	383	0.17
	16	806	742	671	487	448	406	359	0.19
	17	765	698	632	458	422	382	338	0.22
	18	723	660	596	433	398	361	319	0.24
	19	685	625	565	410	377	342	302	0.27
	20	650	594	537	390	358	325	287	0.30
	21	619	565	511	371	341	309	274	0.33
	22	591	540	488	354	326	295	261	0.36
	23	566	516	467	339	312	282	250	0.40
	24	542	495	447	325	299	271	239	0.43
	25	520	475	429	312	287	260	230	0.47
	26	500	457	413	300	276	250	221	0.51
	28	465	424	383	278	256	232	205	0.59
	30	434	396	358	260	239	217	191	0.68
	32	407	371	336	244	224	203	180	0.77
	34	383	349	316	229	211	191	169	0.87
	36	361	330	298	216	199	180	160	0.98
	38	342	312	283	205	189	171	151	1.09
	40	325	297	268	195	179	162	144	1.20
	42	310	283	256	186	171	155	137	1.33
	44	296	270	244	177	163	148	131	1.46
	46	283	258	233	169	156	141	125	1.59
	48	271	247	224	162	149	135	120	1.73
	50	260	237	215	156	143	130	115	1.88
	52	250	228	206	150	138	125	110	2.03
	54	241	220	199	144	133	120	106	2.19
	56	232	212	192	139	128	116	103	2.36
	58	224	205	185	134	124	112	99	2.53
	60	217	198	179	130	119	108	96	2.71
	62	210	191	173	126	116	105	93	2.89
	64	203	186	168	122	112	102	90	3.08
	66	197	180	163	118	109	98	87	3.28

(Left margin: $F_y = 36$ ksi; Span in Feet)

Properties and Reaction Values

	240	220	200	152	141	130	118	
S in.³	813	742	671	487	448	406	359	For explanation of deflection see page 2 - 23
V kips	403	374	342	308	292	278	264	
R kips	133	122	110	92	86	81	76	
R_1 kips	22.4	20.9	19.3	17.1	16.3	15.7	15.0	
N_1 in.	15.6	15.5	15.5	16.1	16.1	16.1	16.1	

Load above heavy line is limited by maximum allowable web shear.

Figure 13.2 *Example Structural Steel Sections*

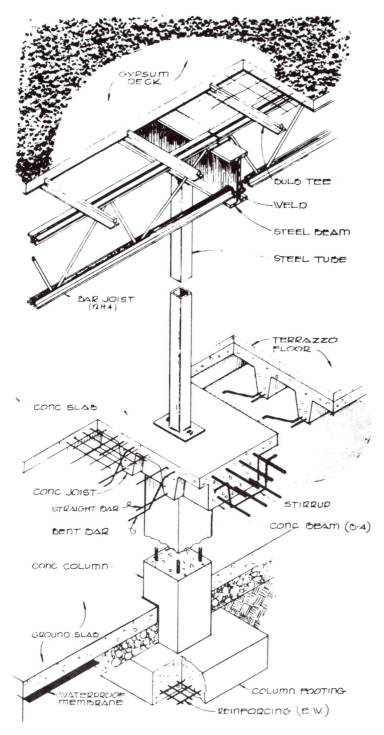

GYPSUM DECK

BULB TEE

WELD

STEEL BEAM

STEEL TUBE

BAR JOIST (12H4)

TERRAZZO FLOOR

CONC SLAB

CONC. JOIST

STRAIGHT BAR

BENT BAR

CONC COLUMN

GROUND SLAB

WATERPROOF MEMBRANE

STIRRUP

CONC. BEAM (B-4)

COLUMN FOOTING

REINFORCING (E.W.)

Figure 13.3   *Example of Steel Joist System*

397

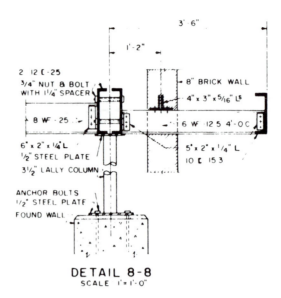

Figure 13.4 *Use of Plates for Connections*

aiding in the connection of two larger steel members (e.g., connecting a W-section beam and a W-section column). Figure 13.4 illustrates steel plates used for this purpose.

The estimator can take one of two different approaches in estimating the cost of structural steel plates. Many include the material and labor cost. The theory behind this method is that for standard building designs there is a defined relationship between the number of steel plates that are required for given weights of steel members. However, for more uncommon designs, the relationship may vary. When this is the case, the estimator should take off the steel plates as a separate work item. In this case, the amount of measure is usually in a *weight* quantity—i.e., pounds or tons. As was noted earlier, steel plates within a given building design vary in size. It would not be feasible or advantageous to consider each plate as a separate work item. Instead, the estimator converts each plate to a weight quantity, and sums them to yield a single work-item quantity. This "lumping" is justified because the purchase cost is most dependent on the total weight of plates; and the labor cost, while not dependent on the plate size, is somewhat dependent on the weight of plates. (Note: the labor cost is likely more dependent on the number of plates.)

Tables are available for converting some steel plates into a weight quantity. One of these is illustrated in Figure 13.5. However, some sizes may not be listed in such tables, so the estimator must be able to convert such plates into a weight measure. Most grades of steel weigh approximately 490 pounds per

WEIGHT OF RECTANGULAR SECTIONS
Pounds per linear foot

Width In.	3/16	1/4	5/16	3/8	7/16	1/2	9/16	5/8	11/16	3/4	13/16	7/8	15/16	1
1/4	.16	.21	.27	.32	.37	.43	.48	.53	.58	.64	.69	.74	.80	.85
1/2	.32	.43	.53	.64	.74	.85	.96	1.06	1.17	1.28	1.38	1.49	1.59	1.70
3/4	.48	.64	.80	.96	1.12	1.28	1.43	1.59	1.75	1.91	2.07	2.23	2.39	2.55
1	.64	.85	1.06	1.28	1.49	1.70	1.91	2.13	2.34	2.55	2.76	2.98	3.19	3.40
1 1/4	.80	1.06	1.33	1.59	1.86	2.13	2.39	2.66	2.92	3.19	3.45	3.72	3.98	4.25
1 1/2	.96	1.28	1.59	1.91	2.23	2.55	2.87	3.19	3.51	3.83	4.14	4.46	4.78	5.10
1 3/4	1.12	1.49	1.86	2.23	2.60	2.98	3.35	3.72	4.09	4.46	4.83	5.21	5.58	5.95
2	1.28	1.70	2.13	2.55	2.98	3.40	3.83	4.25	4.68	5.10	5.53	5.95	6.38	6.80
2 1/4	1.43	1.91	2.39	2.87	3.35	3.83	4.30	4.78	5.26	5.74	6.22	6.69	7.17	7.65
2 1/2	1.59	2.13	2.66	3.19	3.72	4.25	4.78	5.31	5.84	6.38	6.91	7.44	7.97	8.50
2 3/4	1.75	2.34	2.92	3.51	4.09	4.68	5.26	5.84	6.43	7.01	7.60	8.18	8.77	9.35
3	1.91	2.55	3.19	3.83	4.46	5.10	5.74	6.38	7.01	7.65	8.29	8.93	9.56	10.2
3 1/4	2.07	2.76	3.45	4.14	4.83	5.53	6.22	6.91	7.60	8.29	8.98	9.67	10.4	11.1
3 1/2	2.23	2.98	3.72	4.46	5.21	5.95	6.69	7.44	8.18	8.93	9.67	10.4	11.2	11.9
3 3/4	2.39	3.19	3.98	4.78	5.58	6.38	7.17	7.97	8.77	9.56	10.4	11.2	12.0	12.8
4	2.55	3.40	4.25	5.10	5.95	6.80	7.65	8.50	9.35	10.2	11.1	11.9	12.8	13.6
4 1/4	2.71	3.61	4.52	5.42	6.32	7.23	8.13	9.03	9.93	10.8	11.7	12.6	13.6	14.5
4 1/2	2.87	3.83	4.78	5.74	6.69	7.65	8.61	9.56	10.5	11.5	12.4	13.4	14.3	15.3
4 3/4	3.03	4.04	5.05	6.06	7.07	8.08	9.08	10.1	11.1	12.1	13.1	14.1	15.1	16.2
5	3.19	4.25	5.31	6.38	7.44	8.50	9.56	10.6	11.7	12.8	13.8	14.9	15.9	17.0
5 1/4	3.35	4.46	5.58	6.69	7.81	8.93	10.0	11.2	12.3	13.4	14.5	15.6	16.7	17.9
5 1/2	3.51	4.68	5.84	7.01	8.18	9.35	10.5	11.7	12.9	14.0	15.2	16.4	17.5	18.7
5 3/4	3.67	4.89	6.11	7.33	8.55	9.78	11.0	12.2	13.4	14.7	15.9	17.1	18.3	19.6
6	3.83	5.10	6.38	7.65	8.93	10.2	11.5	12.8	14.0	15.3	16.6	17.9	19.1	20.4
6 1/4	3.98	5.31	6.64	7.97	9.30	10.6	12.0	13.3	14.6	15.9	17.3	18.6	19.9	21.3
6 1/2	4.14	5.53	6.91	8.29	9.67	11.1	12.4	13.8	15.2	16.6	18.0	19.3	20.7	22.1
6 3/4	4.30	5.74	7.17	8.61	10.0	11.5	12.9	14.3	15.8	17.2	18.7	20.1	21.5	23.0
7	4.46	5.95	7.44	8.93	10.4	11.9	13.4	14.9	16.4	17.9	19.3	20.8	22.3	23.8
7 1/4	4.62	6.16	7.70	9.24	10.8	12.3	13.9	15.4	17.0	18.5	20.0	21.6	23.1	24.7
7 1/2	4.78	6.38	7.97	9.56	11.2	12.8	14.3	15.9	17.5	19.1	20.7	22.3	23.9	25.5
7 3/4	4.94	6.59	8.23	9.88	11.5	13.2	14.8	16.5	18.1	19.8	21.4	23.1	24.7	26.4
8	5.10	6.80	8.50	10.2	11.9	13.6	15.3	17.0	18.7	20.4	22.1	23.8	25.5	27.2
8 1/4	5.26	7.01	8.77	10.5	12.3	14.0	15.8	17.5	19.3	21.0	22.8	24.5	26.3	28.1
8 1/2	5.42	7.23	9.03	10.8	12.6	14.5	16.3	18.1	19.9	21.7	23.5	25.3	27.1	28.9
8 3/4	5.58	7.44	9.30	11.2	13.0	14.9	16.7	18.6	20.5	22.3	24.2	26.0	27.9	29.8
9	5.74	7.65	9.56	11.5	13.4	15.3	17.2	19.1	21.0	23.0	24.9	26.8	28.7	30.6
9 1/4	5.90	7.86	9.83	11.8	13.8	15.7	17.7	19.7	21.6	23.6	25.6	27.5	29.5	31.5
9 1/2	6.06	8.08	10.1	12.1	14.1	16.2	18.2	20.2	22.2	24.2	26.2	28.3	30.3	32.3
9 3/4	6.22	8.29	10.4	12.4	14.5	16.6	18.7	20.7	22.8	24.9	26.9	29.0	31.1	33.2
10	6.38	8.50	10.6	12.8	14.9	17.0	19.1	21.3	23.4	25.5	27.6	29.8	31.9	34.0

Figure 13.5 *Converting Sizes to Weights*

cubic foot. This knowledge enables the estimator to easily convert the dimensions of a plate to a volume measure, and in turn convert this to a weight quantity.

A third type of steel work item involves the material and labor cost related to the use of rivets, bolts, and welds that are used to connect steel members. While bolts are the most commonly used "connectors," still the use of welds and to a lesser degree rivets may be part of a specific building design. Two of these types of connections are illustrated in Figure 13.6.

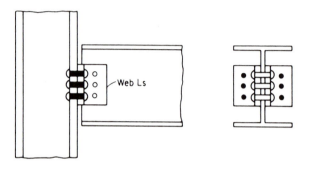

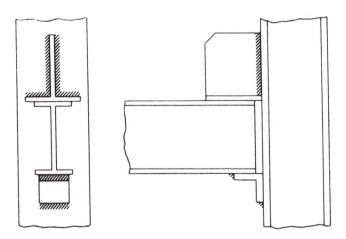

Figure 13.6 *Example Steel Connectors*

As in the case of steel plates, an estimator has two alternatives when he is determining the cost of bolts, welds, or rivets. He *could* count each and every bolt, weld, or rivet and sum them to determine a number of "connectors" to be placed. But the designer does not always delineate all of these on the working drawings. Even when he does, this is a very time-consuming method.

Perhaps equally important, concerning why the estimator may not take off each connector is the fact that the effort may prove wasteful. This is because there usually is a strong relationship between the weight of steel that needs to be put in place and the number of connectors and related work effort required. In other words, the erection of a determined weight of structural steel usually results in the need to use a defined amount of connectors.

Thus it is common for the estimator to include the cost of bolts, welds, and rivets in the cost of the structural steel itself. The result is that connectors are seldom defined as a work item in the steel estimate.

In summary, a structural steel estimate for a building construction project usually entails the take-off of the work items listed in Figure 13.7. As we have noted, the joist/truss work item actually entails several work items, one for each size or type. The taking-off of the work items noted in Figure 13.7 is discussed in the following section.

Structural Steel Framing
Lightweight Member Framing (Metal Building)
Steel Joists/Trusses
Steel Plates
Metal Decking

Figure 13.7 *Example Work Items for Structural Steel*

TAKING-OFF STRUCTURAL STEEL

The quantity take-off of structural steel essentially entails the orderly counting of steel members and plates and the subsequent efficient conversion of the members into a *weight* quantity. Because there may be many members to be identified, many of which are of a common size, the opportunity and need for efficient take-off procedures exists.

The relatively high material and labor cost for each individual steel member draws attention to the estimator's need to make sure he doesn't miss a member. Doing so is likely to result in a much more significant error in the total project cost estimate than if he inadvertently forgot an excavation, concrete, or masonry work quantity or member.

One aid the estimator should use in making sure each steel member is identified is to check off each one on the drawing as it is taken off and put on the estimating forms. The use of a colored pencil might enhance this procedure. Placing a check mark by each member, or "covering" the members with a colored pencil helps prevent counting a member twice or omitting one.

As we have said, structural steel members are available in standard sizes. In the take-off process, all the members are converted into a single weight quantity—that is, pounds or tons of structural steel. Given the fact that a given building design normally includes a relatively small number of different member sizes, an efficient structural steel take-off is accomplished by searching for and taking off members in an order of standard size. For example, the estimator would scan the drawings and locate each W-section, each channel section, and so forth. Consider the drawing illustrated in Figure 13.8. Each W-section labeled W14 × 43 is listed on the take-off sheet shown in Figure 13.9. Note that the members may be of differing lengths. Each of these varying lengths is specified via a line entry on the take-off sheet shown. Once all the members of the size listed are identified, the estimator can sum the total length of the members and convert this figure into a weight quantity by multiplying by the weight per foot of the member. As can be seen from Figure 13.9, this process continues until each category of member size and shape are taken off.

The take-off procedure shown in Figure 13.9 has two advantages. First, it provides the estimator an efficient means of taking off the members and determining a total weight quantity so that the work can be priced. The estimating process shown in Figure 13.5 has a second advantage. The take-off sheet in that illustration sets out the specific number of each member of a specific size and length. This information can be used by the construction firm's purchasing department in *ordering* the steel members.

The fact that steel members are ordered exactly to size means that there is little or no wastage of material in placing the steel members. Thus, no wastage factor need be added to the take-off quantities.

The take-off of steel truss/joists entails the identification of each size or type of truss/joist as a separate work item. Each size/type is measured in linear feet. The take-off process is relatively simple. Each member is taken off via an order of size or shape. All members of a specific size or type are taken off. The estimator marks off each member on the drawings. After all the members of a given size or shape are taken off, the estimator sums and then prices the linear feet of the member size/shape. As with the standard steel members, no wastage factor is necessary.

The most tedious structural steel take-off process involves the steel plates. As we said earlier the estimator may choose to recognize their existence by including them in the price of the related steel members, or may delineate and price the steel plates independent of the related steel members.

When in fact the steel plates are taken off as individual members, they are considered as a single work item and measured in a weight quantity.

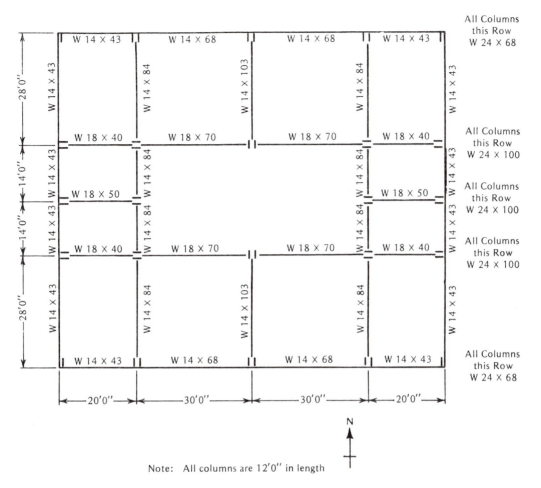

Figure 13.8 *Example Drawing for Structural Steel Take-Off*

Because plates vary significantly in size per project, a take-off process by order of size may not prove efficient. It is usually more efficient to take off the steel plates in the same order in which the steel plates or truss/joists they support are taken off. For example, the steel plates relating to the 14W42 steel member identified in Figure 13.8 would be taken off, then the plates relating to the next member, and so forth. The result of this take-off process is illustrated in Figure 13.9. No wastage factor is normally added to the steel plate take-off.

Steel connectors—including bolts, rivets, and welds—are seldom taken off as a separate work item. Instead, they are included in the estimate by adjustment of the cost of the related steel members for the connector costs.

PRACTICAL
FORM 346

QUANTITY SHEET

PROJECT	STRUCTURAL STEEL	ESTIMATOR	JKA	ESTIMATE NO	101
LOCATION	ANYWHERE	EXTENSIONS	DJA	SHEET NO	
ARCHITECT ENGINEER	JJA	CHECKED	SLL	DATE	9-12-X0

CLASSIFICATION STRUCTURAL STEEL (METALS)

DESCRIPTION	NO.	DIMENSIONS Length	TOTAL Length	WT (lbs) FF	(LBS) WEIGHT		ESTIMATED QUANTITY	UNIT
_05101_								
STRUCTURAL FRAMING								
BEAMS/GIRDERS								
E/W W 14×43	4	20.0'	80.0'	43.0	3,440			
W 14×68	4	30.0'	120.0'	68.0	8,160			
W 18×40	4	20.0'	80.0'	40.0	3,200			
W 18×70	4	30.0'	120.0'	70.0	8,400			
W 18×50	2	20.0'	40.0'	50.0	2,000			
N/S								
W 14×43	4	28.0'	112.0'	43.0	4,816			
W 14×43	4	14.0'	56.0'	43.0	2,408			
W 18×84	4	28.0'	112.0'	84.0	9,408			
W 18×84	4	14.0'	56.0'	84.0	4,704			
W 14×103	2	28.0'	56.0'	103.0	5,768			
COLUMNS								
W 24×68	10	12.0'	120.0'	68.0	8,160			
W 24×100	14	12.0'	168.0'	100.0	16,800			
TOTAL WEIGHT					77,264 ÷			
						2000 lbs/ton	38.6 TONS	

MFD. IN U.S.A.   FRANK R. WALKER CO., PUBLISHERS, CHICAGO

Figure 13.9 Structural Steel Take-Off

PRICING STRUCTURAL STEEL WORK

As is true of any type of work, structural steel is priced in the same units in which it is taken off. This presents some difficulty concerning the pricing, whether the units are structured members, steel joists/trusses, or steel plates. While the material cost for structural steelwork is likely most dependent on the weight of steel members or plates or the length of joist/trusses, the labor installation cost may be more dependent on the *number* of members to be placed. For example, approximately twice the labor effort is required to put in place 2 steel columns as that required for 1 column, even if the columns are of a different size.

Independent of the relationship between the labor cost and the number of steel members to be placed, both the labor and material cost are usually determined by means of the same cost basis. In the case of steel members or steel plates, the cost basis is usually one of *weight*. Tons or pounds of steel are the units of weight utilized. Structural steel joists/trusses are usually priced in units of *length*.

Fortunately, there usually is a strong relationship between the number of steel members or truss/joints and the weights of the steel members or the length of the steel truss/joists. So the use of the weight and length measures usually closely correlate to the *number of members*. The result is that the weight and length measures provide an appropriate pricing basis for steel members and truss/joists, respectively.

The pricing of the material cost for steel members or plates is relatively simple for the estimator. The price for the raw material is well publicized by steel manufacturers/fabricators. Usually the contractor is quoted a standard price per ton and then pays various "extras" for an unusually long transportation distance, a large number of different-member sizes, or perhaps a special or high-quality type of steel. Painting is another example of a cost that would be added to the base price. Unlike the case for other construction materials, the material cost for structural steel members or plates often exceeds the labor cost.

The labor effort required to place steel members or plates entails several somewhat separate steps or procedures. For one thing, the steel may have to be lifted off vehicles driven to the job site. This work is the contractor's responsibility because as part of the base price the steel is usually delivered only to the job site. A second labor-cost component entails the lifting and setting of the steel members, plates, or truss/joists into their location in the building. This "lifting" is usually achieved through the use of equipment such as cranes. The third labor-cost component involves making the steel connections, including the placing of bolts, rivets, or welds.

Average labor productivity figures for the first 2 labor efforts (lifting the members and putting them in place) for various types of structural steel members are illustrated in Figure 13.10. These figures are summaries of data collected by the author. Knowing the existing wage rates of the involved

Type of Work	Productivity
Erect Structural Steel	9 Hours/Ton
Erect Lightweight Steel Members	12 Hours/Ton
Metal Joists	2 Hours/1000 Pounds
Metal Decking	1 Hour/100 Square Feet

Figure 13.10 *Example Productivities for Placing Structural Steel*

workers (usually ironworkers), the estimator can multiply the productivities listed in Figure 13.10 by the relevant labor rates to yield a labor cost estimate for the structural steel-work.

Having established the raw material structural steel costs and the labor lifting and setting costs, the estimator has two other costs to determine before he finalizes the structural steel estimate. These two costs are the related "connector" costs and the cost of equipment utilized in the moving of the steel.

As was discussed earlier, the estimator seldom takes off the steel connectors individually. Instead, he usually includes material and labor cost in the material and labor cost for the related steel members, or calculates them as a percentage of these costs. For example, it is common for an estimator to consider that the material and labor costs for the connectors is approximately 2 to 3% of the material and labor costs for the related steel members. This approach is illustrated in Figure 13.9.

Owing to the weight of structural steel, one or more pieces of large construction equipment are usually needed. For example, a crane is often utilized to lift a steel member off of the delivery truck and into its location of placement.

The need for large equipment to place structural steel results in two costs. One is a "set-up" or **mobilization cost**. Owing to the size and sophistication of several types of equipment, this cost may be several thousand dollars or more.

A second equipment cost is the cost of *using* the equipment, whether it is a rental cost or an hourly cost of owning the equipment. The determination of an hourly rate for a piece of construction equipment and the design and use of an equipment costing system for estimating were discussed in Chapter 8. The reader would be wise to review that chapter because the material there is relevant to putting together an accurate structural steel estimate.

PRODUCTIVITY ANALYSIS OF STEEL WORK

To determine an estimating standard for the placement of structural steel members, a contractor can utilize the time-study model discussed in Chapter

2. Assume the structural steelwork entails the placement of steel beams and columns for a steel skeleton building. Assume each of the steel members are approximately the same size and so the productivity for placing each member is essentially equal. Each member weighs approximately 1,000 pounds.

In using the time study model, the construction method—in this case the placement of the structural members—is broken into a series of individual work steps. In studying the placement of the structural steel members, the contractor/estimator has broken the work process into the following 6 work steps:

1. Load steel member from truck onto ground.
2. Lift steel members from ground to position of placement.
3. Ironworkers position member into position.
4. Ironworkers make initial connection.
5. Ironworkers level and align member.
6. Ironworkers tighten connection.

In the time study cited, 6 cycles of the process of placing a steel member were analyzed. During each of these cycles, 4 ironworkers were employed at performing the work. The result of this time study analysis is illustrated in Figure 13.11. Note that the results of the collection of the times and their summation indicates an average cycle time of 3,786 seconds. When using the time study model, this time is referred to as the **select time**.

In the time study method, the select time is converted to the "normal" time by multiplication of the select time by a rating factor that reflects the fact that the workers are likely affected by the presence of an onlooker collecting the data. This presence may in fact positively or negatively affect the worker's productivity. Let us assume that in this case the collector of the data felt that the workers "speeded up" approximately 10% from what might be considered the normal rate. Given this fact, the normal time for the work process is calculated as follows:

Normal time = Select time × rating factor

Normal time = 3,786 × 1.10 = 4,164.60 sec.

Once the normal time is determined, the standard time for the work process can be calculated. This is accomplished by the recognition of an allowance time for personal breaks, rest, and the like. Often an allowance factor on the order of 10 to 15% is assumed. Let us assume a 15% allowance factor in our example. The standard time is therefore calculated as follows:

Standard time = Normal time + (Normal time) (Allowance factor)

Standard time = 4,164.6 + 4,164.6 (.15)

Standard time = 4,164.6 + 624.7

Standard time = 4,789.3 sec.

Given the crew size and given the weight for each member placed, this standard time per placement of structural steel member can be converted into a manhour per unit of weight productivity. As we noted earlier, each member placed in the 6 cycles listed in Figure 13.11 weighed approximately 1,000 pounds. The crew consisted of 4 ironworkers who work for 4,789.3 seconds each. This can be converted into manhours as follows:

$$(4,789.3 \text{ sec.}) \frac{(1 \text{ hr.})}{(3,600 \text{ sec.})} (4 \text{ men}) = 5.32 \text{ manhours}$$

Given a weight of 1,000 pounds per member, a standard productivity is calculated as follows:

$$\text{Standard time productivity} = \frac{5.32 \text{ manhours}}{\frac{1,000 \text{ lbs.}}{2,000 \text{ tons}}} = 10.64 \frac{\text{manhours}}{\text{ton}}$$

This determined productivity standard can then be used by the estimator for future projects. Naturally, the number should be modified to reflect the characteristics of each specific project.

One might ask why the work process described in Figure 13.11 was broken into 6 distinct steps when in fact the calculation process just illustrated only utilized the total cycle times. The main reason for this breakdown into individual work steps is that it enables the estimator or contractor to analyze the work process in detail, with the objective of revising the work process if necessary to bring about an increase in productivity. By breaking down the individual work steps to include the time required to perform each of the steps, the estimator or contractor may be able to identify work improvement potential which, if attained, can increase the productivity standard.

To analyze a construction method via the time study process, let us assume that the estimator analyzes the steel erection work process illustrated in Figure 13.11. The standard time was determined to be 4,789.3 seconds and this was in turn converted into a standard productivity of 10.64 manhours per ton of steel.

Inspection of the work process indicates that a considerable amount of time is required to unload the steel members from the truck onto the ground, and then in turn to lift them to their position of placement. The first work step takes an average of 197 seconds or 5.2% of the total work process; the later work step takes an average of 276 seconds or 7.3% of the total work process. In an attempt to improve the delivery and lifting methods, the estimator determines that the first work step, the unloading of the steel members from the truck onto the ground, can be eliminated. In effect, he assumes that the steel members can be lifted directly from the truck to the location of placement.

Work Step	Time in Seconds Cycle						Total	Average
	1	2	3	4	5	6		
1. Load steel members from truck onto ground	166	186	214	204	224	188	1,182	197
2. Lift steel members from ground to position of placement	270	226	322	284	302	250	1,654	276
3. Ironworkers position members	546	642	486	502	488	524	3,188	531
4. Ironworkers make initial connection	684	624	648	596	586	620	3,758	626
5. Ironworkers level and align members	1,240	1,080	1,114	1,108	986	1,112	6,640	1,107
6. Ironworker tighten connection	1,300	946	1,024	986	1,114	924	6,294	1,049
Total time	4,206	3,704	3,808	3,680	3,700	3,618	22,716	3,786

Figure 13.11

Work Step	Time in Seconds Cycle						Total	Average
	1	2	3	4	5	6		
1. Lift steel members from ground to position of placement	270	226	322	284	302	250	1,654	276
2. Ironworkers position members	546	642	486	502	488	524	3,188	531
3. Ironworkers make initial connection	684	624	648	596	586	620	3,758	626
4. Ironworkers level and align members	1,240	1,080	1,114	1,108	986	1,112	6,640	1,107
5. Ironworkers tighten connections	1,300	946	1,024	986	1,114	924	6,294	1,049
Total time	4,040	3,518	3,594	3,476	3,476	3,430	21,534	3,589

Figure 13.12

This change would result in the revised work procedure shown in Figure 13.12. The revised select time is as follows:

Select time = 3,589 sec.

Adjusting this select time for the rating factor results in the following normal time:

Normal time = 3,589 sec. (1.1) = 3,947.9 sec.

Finally, the standard time is determined by recognition of the 15% allowance factor:

Standard time = 3,947.9 + 3,947.9 (.15) = 3,947.9 + 592.2 = 4,540.1 sec.

This standard time of 4,540.1 seconds can in turn be converted into a productivity standard as follows:

$$(4540.1 \text{ sec.}) \frac{(1 \text{ sec.})}{(3,600 \text{ hr.})} (4 \text{ min.}) \frac{(2,000 \text{ ton})}{(1,000)} = 10.09 \frac{\text{manhours}}{\text{ton}}$$

This revised productivity standard can then be used for estimating future projects. Which productivity standard should be used for this purpose depends on whether or not the contractor will be required to unload the steel onto the ground or can both unload and place the members in one operation.

EXERCISE 13.1

For the construction drawings illustrated, take off the structural steel members and convert the taken-off members into a single weight in pounds of structural steel.

EXERCISE 13.2

Assume that a general contractor in attempting to help his structural steel subcontractor with productivity advice, analyzes the subcontractor's means of placing steel siding panels. Each panel is two foot wide and the heights vary from fourteen to eighteen foot high. The construction entails the placement of two panels, one a liner panel, and then a face panel.

The contractor analyzed six different crews working. The results of this analysis were as follows:

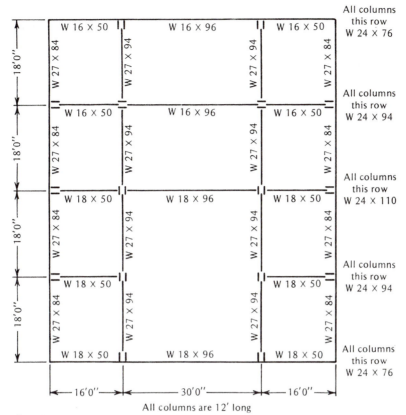

Exercise 13.1

	CREW 1	CREW 2	CREW 3	CREW 4	CREW 5	CREW 6
1. Number of crew members	4	3	4	3	4	3
2. Method of placement						
Both panels together	X	X			X	
One panel at a time			X	X		X
3. Type of scaffolding						
On wheels	X	X		X		
Hung			X		X	X
4. Number of panels placed per crew hour	25	22	20	19	22	18

Which method results in the highest productivity—i.e., number of panels placed per manhour?

If you had 7 men available and 1 of each of the two types of scaffolding, how would you allocate the men to the two different scaffolding systems and how should they place the panels—i.e., both panels (liner and face) together, or one panel at a time?

EXERCISE 13.3

The high cost of erecting structural steel includes the placement of steel connections. The connections might be bolts, rivets, or welds. The number of bolts, rivets, or length of weld required to fabricate two adjacent members varies depending on the load the members carry and the particular design. It is important for the contractor/estimator to establish the number of connectors in order to purchase them for a project. It is also important that they know the cost of placing connectors because this cost must be included in the labor-cost estimate for the structural steelwork.

Assume a contractor has collected the following data from past projects:

TYPE OF CONNECTIONS	NUMBER OF CONNECTIONS	TONS OF STEEL PLACED	TIME REQUIRED TO PLACE CONNECTIONS
Bolts	2,000	50	95
Bolts	4,500	90	165
Bolts	12,000	150	300
Bolts	3,000	60	110
Bolts	5,500	80	160
Bolts	10,000	120	280

What is the average number of bolts required per ton of steel placed?

What is the average number of manhours (expressed in minutes) required to place a single bolt?

What is the average number of manhours required to place bolts per ton of steel placed?

From the data illustrated, does it appear that the workers placing the bolts "learn" as the number of bolts increases on a job? In other words, is less time required to place a bolt when many bolts have to be placed relative to the time required when fewer bolts are placed? If your answer is yes, calculate the time required to place a bolt when fewer than 5,000 bolts are required versus the time required to place a bolt when more than 5,000 bolts are to be placed.

14

Carpentry Work

The Varied Nature of Carpentry Work in a Building *
Types of Carpentry Work * *Quantity Take-off of*
Carpentry * *Pricing Carpentry Work* * *Productivity*
Analysis of Carpentry Work

THE VARIED NATURE OF CARPENTRY WORK IN A BUILDING

Carpentry for a building project usually involves 2 somewhat different types of work—rough carpentry work and finish carpentry work. (A third type of wood-related work for a building project, the fabrication of wood forms for supporting cast-in-place concrete, is considered as part of the concrete work.)

Rough carpentry work entails wall framing, wood headers around windows and doors, and beams and joists that are part of framing ceilings and roofs. Rough carpentry wood members often serve a structural purpose in a building—that is, they carry live and dead loads. Often, in a finished building, rough carpentry wood members are "covered up" with interior or exterior finish materials such as drywall, roof shingles, and the like.

Finish carpentry work is a catch-all of such types of work as interior trim, wood cabinets, finished wood floors, and exterior finishes. Like rough carpentry work, finish carpentry work is performed by carpenters with or without the help of laborers. However, unlike rough carpentry work, the work processes for placing finish carpentry are non-repetitious and more variable concerning the effort required. Thus the take-off and pricing calculations for finish carpentry work is usually more difficult and more subject to error. Because of this, and also because of the specialized nature of the work, a general contractor is more apt to subcontract finish carpentry than he is rough carpentry. This ability to subcontract one aspect of the carpentry work without subcontracting the other is enhanced by the fact that there is a gap between the time the rough carpentry work is performed and the finish carpentry work is needed.

The amount of carpentry work—especially the rough work—on a building project may vary from none to a great deal of high-cost work for the entire building. For a light commercial building or a residential unit, most if not all of the structural members are wood. These members include the interior and exterior walls, roof trusses or joists, and interior carpentry trim and cabinets. The exterior of these types of projects might be wood siding. The result is that carpentry work is the predominant cost material for a residential unit.

At the other extreme, there are occasions when there is no carpentry at all within a building design. For example, some industrial or institutional types of buildings may be totally without carpentry work. However, most commercial projects contain some carpentry work in the form of interior finish trim. Both rough and finish carpentry work are discussed in this chapter.

TYPES OF CARPENTRY WORK

Rough carpentry work consists of fabricating several different sizes of wood members into stud walls, beams, columns, joists, and rafters. Wood braces between these types of members are also part of rough carpentry work.

Rough carpentry: Rough carpentry is measured in units or multiples of *board feet*. One board-foot measure, for example, is a wood member 1 inch thick by 12 inches deep by 1 foot long. A board foot is illustrated in Figure 14.1. It is possible to convert large-size boards into board feet by multiplying by the width and thickness. For example, a 1-foot-board 2 by 4—a common board size used in wall framing—is two-thirds of 1 board-foot measure. This is determined by multiplying the 2-inch side by the 4-inch side and dividing by 12, or 1 inch times 12 inches. Similarly, a 1-foot 2 by 6 board, commonly used for beams or rafters, is 1 board-foot measure. A board-foot measure is commonly abbreviated by the symbol **bfm**.

The purchase price of lumber or boards is dependent on the size of the boards and the quantity of board-foot measure. A 2 by 4 board of a given length may cost approximately the same as a 2 by 6 board of the same length, even though the 2 by 4 is only two-thirds as large as the 2 by 6 in regard to board-feet measure. In other words, the purchase price is not always totally proportional to the board-foot measure. On the other hand, when considering one size board only, e.g., a 2 by 4, the purchase price is linearly proportional to its number of board feet.

Given the fact that each board size may have a different purchase cost per running foot, it is advantageous to define each size as a separate work item for the quantity take-off function. In other words, all 2 by 4s would be identified as a rough carpentry work item, all 2 by 6s as another rough

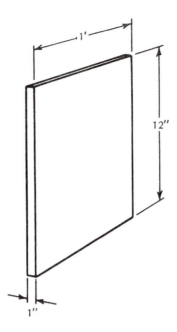

Figure 14.1 *One Board Foot Measure*

carpentry work item, and so on. This is illustrated in Figure 14.2. Each of these work items would be taken off in units of board-foot measure.

The identification of individual board sizes as work items presents somewhat of a problem in regard to the labor-cost component. As was true of structural steel members, the labor cost of placing wood framing members is in fact more dependent on the *number* of members to be placed than on their *size*. Thus, one might argue that rough carpentry material should be taken off using member size and labor in number of members. While this might be practical for a labor-cost consideration, the number of members would not be indicative of the material purchase cost.

2″ × 4″ Framing Members
1″ × 6″ Framing Members
2″ × 6″ Framing Members
2″ × 10″ Framing Members
Joists
Trusses
Rafters
Floor Sheathing
Finish Carpentry (e.g., Trim, Cabinets, etc.)

Figure 14.2 *Example Work Items for Carpentry*

This difficulty is lessened by the fact that there usually is a strong relationship between the number of members and the total amount of board-foot measure. This relationship exists because of the somewhat uniform length of wood members used for framing. For example, most stud walls consist of 8-foot-long 2 by 4 members.

In addition to various boards that are used for structural framing, on occasion a building is designed to include various finished wood beams, posts, or columns. For example, the interior of churches often include exposed wood members forming arches of the ceiling. Usually these large members are custom-designed for the specific building. Although they are in fact exposed and there might be considerable finish carpentry needed because of their bulkiness and the fact that they serve a structural purpose, the members are usually considered as part of the rough carpentry work. These types of wood members could be considered as one work item—*exposed structural wood framing.*

Finish carpentry: As we discussed earlier, finish carpentry may include a vast number of different types of work. Specialized wood finishes, hand-carved railings, and numerous types and qualities of wood wall panels or exterior siding are just a few examples of finish carpentry. Wood baseboard trim and cabinets are others. Because of the numerous possible types of finish carpentry work, it is difficult to define a standard set of finish carpentry work items. To some degree, every finish carpentry type of work might be defined as a separate work item. While this may seem tedious, it may be the only feasible approach. Fortunately, the number of distinct types of finish carpentry work for a specific project is often limited, and thus the definition of separate work items for every item does not present extreme take-off time difficulties. More typical types of finish carpentry work that are often part of a specific building design are illustrated as work items in Figure 14.2.

QUANTITY TAKE-OFF OF CARPENTRY

Rough carpentry: There are two somewhat distinct alternative take-off approaches that might be used by the estimator in regard to the rough carpentry work. For one, the estimator might delineate each board on the drawings, group these by work-item member size, and convert the number and length of the members into a board-foot measure. In effect, the estimator "counts" each member, including every wood stud, rafter, truss member, and so forth.

While this approach may be appropriate for some specialized framing sections or areas of the building design, it is usually not necessary or even possible for much of rough carpentry framing. It may not be possible because the designer might not even delineate every single member; instead, he might merely indicate the location and spacing for a specific type of framing.

A more practical approach is based on the fact that the spacing between members is often of a standard size. For example, building codes usually require wood wall studs to be placed at 16-inch centers. Roof trusses and rafters are also commonly spaced at 16-inch centers.

Even when the framing spacing is not standardized, it is usually uniform and designated on the drawings. For example, a designer commonly designates the wood framing work in a manner similar to that illustrated in Figure 14.3.

Perhaps the most important aspect of the quantity take-off relevant to efficiency is the *order* of the take-off. The large number of wood framing members that may be part of some building designs, especially residential and light commercial projects, results in a need to locate and delineate the wood framing efficiently.

As was noted earlier, all lumber sizes for wood framing are usually identified as separate work items. Thus it is common to take off the framing in an order dictated by lumber sizes—e.g., all 2 by 4s, all 2 by 6s, and so forth. To lend a specific order to the process, the estimator might progress from the smallest size to the largest.

In addition to determining the total board feet (or some multiple) for each lumber size, the quantity take-off process should also yield the total *length* required of each board size (i.e., work item), and the number of each member. This is necessary so the contractor can purchase the lumber.

The estimator has to take special care to make sure that all members are taken off efficiently, given the large number of members. To accomplish this, he should develop a systematic means of going around the drawings. Identifying members by purpose and location often serves as an efficient process. For example, all 2 by 4s in the first-floor wall framing might be delineated, then all second-floor wall framing. After having taken off the walls, the 2 by 4s in the floors might be taken off, then all the 2 by 4s in the roof system, and so on. After taking off all the 2 by 4s, the estimator can then proceed to take off the 2 by 6s, and so on. This format for the quantity take-off is illustrated in Figure 14.4.

The quantity take-off of special types of rough carpentry, such as exposed beams and columns sometimes found in church designs, entails merely delineating each specific member, its size, and the type of wood to be used. These special types of carpentry work are usually priced per member and not converted into board feet. Because they are usually custom-designed and thus very expensive, it is especially important not to miss any members in the take-off process. This could lead to a substantial error in the overall estimate. To reduce the chance that he might overlook a member, the estimator should scrutinize every structural and architectural drawing in detail.

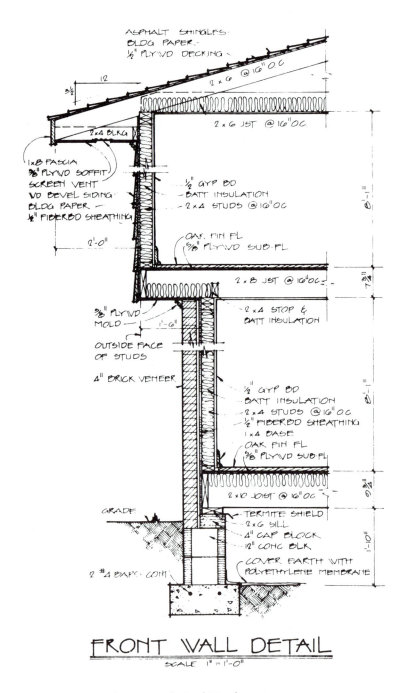

ASPHALT SHINGLES
BLDG PAPER
½" PLYWD DECKING

12
3½

2 × 6 @ 16" O.C

2 × 4 BLKG

2 × 6 JST @ 16" O.C.

1 × 8 FASCIA
⅜" PLYWD SOFFIT
SCREEN VENT
VO BEVEL SIDING
BLDG PAPER
½" FIBERBD SHEATHING

½" GYP BD
BATT INSULATION
2 × 4 STUDS @ 16" O.C

2'-0"

OAK FIN FL
⅜" PLYWD SUB-FL

2 × 8 JST @ 16" O.C

7¾"

⅜" PLYWD
MOLD

1'-6"

2 × 4 STOP &
BATT INSULATION

OUTSIDE FACE
OF STUDS

4" BRICK VENEER

½" GYP BD
BATT INSULATION
2 × 4 STUDS @ 16" O.C
½" FIBERBD SHEATHING
1 × 4 BASE
OAK FIN FL
⅜" PLYWD SUB-FL

2 × 10 JOIST @ 16" O.C

6½"

GRADE

TERMITE SHIELD
2 × 6 SILL
4" CAP BLOCK
12" CONC BLK

COVER EARTH WITH
POLYETHYLENE MEMBRANE

1'-10"

2 #4 BARS CONT

FRONT WALL DETAIL
SCALE 1" = 1'-0"

Figure 14.3 *Designation of Spacing of Wood Members*

PRACTICAL FORM 110			QUANTITY SHEET									

PROJECT CARPENTRY PROJECT **ESTIMATOR** JKA **ESTIMATE NO** 161

LOCATION ANYWHERE **EXTENSIONS** DJA **SHEET NO**

ARCHITECT ENGINEER JJA **CHECKED** SLL **DATE** 9-12-X0

CLASSIFICATION CARPENTRY

DESCRIPTION	NO.	DIMENSIONS Length			LENGTH' 2X4's	Length' 2×6's	Length' 2x		ESTIMATED QUANTITY	UNIT
CARPENTRY FRM.										
2x4's										
WALLS	120	8.0'			960					
	12	4.0'			48					
	8	5.0			40					
FLOORS	80	10.0'			800					
	30	8.0'			240					
	12	4.0'			48					
ROOF	6	8.0'			48					
2X6's										
WALLS	8	4.0'				32				
	2	6.0'				12				
FLOORS	80	8.0'				640				
	20	6.0'				120				
ROOF	60	8.0'				480				
	12	6.0'				72				
	4	4.0'				16				
2x10's										
FLOORS	12	8.0'					96			
	4	4.0'					16			
ROOF	32	8.0'					256			
	4	6.0'					24			
TOTAL FEET					2184	1,372	392			
Board Feet / Foot					0.667	1.000	1.667			
Board Feet					1,457	1,372	653		3,482	B.M.

Figure 14.4 *Carpentry Take-Off*

Finish carpentry: Finish carpentry work is considerably more varied than rough carpentry work. Thus it is more difficult to establish a specific take-off approach. Most finish carpentry members are priced individually. For example, cabinets are priced separately from trim work or wood paneling. In addition, each of these types of work may be priced in different units. The units typically used for somewhat common finish carpentry types of work are listed in Figure 14.5.

Type of Work	Units
Trim Work (e.g., Corner Boards)	Linear Feet
Soffits	Square Feet
Shutters	Per Pane
Wood Beams	Linear Feet
Countertops	Square Feet
Paneling	Square Feet

Figure 14.5 *Units for Various Carpentry Items*

Because finished carpentry varies significantly from one project to the next, it is essential that the estimator be attentitive to locating all the separate work items, and also be careful not to count work twice. The best means of identifying all the work and not counting work twice is to use a checklist of possible types of work and to develop an orderly means of going around the drawings. Using a colored pencil to check off the taken-off items on the drawings also aids this effort.

PRICING CARPENTRY WORK

Placement of carpentry members entails the use of carpenters and to a lesser degree, laborers. The work effort involves cutting boards and fastening them with nails.

The labor cost of the work is very dependent on the *number* of boards that have to be cut. If the carpentry work entails the fabrication of all 8-foot stud walls, few if any of the boards need to be cut. Therefore the carpentry crew can erect and fabricate a large number of board feet of lumber per manhour. On the other hand, if the work entails the fabricating of boards around windows and doors, the amount of board feet placed per manhour would be relatively low.

In addition to the number of cuts, the labor cost for fabricating carpentry work is dependent on the *size* of the members to be fabricated. As we discussed earlier, the labor effort required to place different-sized lumber is approximately equal, independent of the size of the member. For example, approximately the same labor effort is required to place an 8-foot 2 by 4 as that required to place a 2 by 6 board. Given the fact that there are more board feet in a 1-foot 2 by 6 than in a 1-foot 2 by 4, the labor cost for placing one board foot of these two different sizes of boards would vary.

Carpentry labor productivity, like most other types of general building work, is effected by weather, including temperature and precipitation. However, the impact of these factors on productivity is usually not as detrimental as on concrete or masonry work.

Sample carpentry labor productivities are illustrated in Figure 14.6. These figures assume "average" working conditions, including average weather and an average number of cuts required. Productivities are listed only for what is considered "typical" carpentry work.

Type of Work	Productivity
Wall Framing	2.5 Hours/100 Ft. B.M.
Placing Sills/Plates	7.0 Hours/100 Lin. Ft.
Framing Floor Joists	3.5 Hours/100 Ft. B.M.
Framing Rafters	3.0 Hours/100 Ft. B.M.
Placing Subflooring	1.0 Hours/100 S.F.

Figure 14.6   *Example Productivities for Carpentry Work*

Finish carpentry work, as we know, varies significantly and so it is next to impossible to list productivity figures for it. This "non-availability" of labor productivity data for every type of finish carpentry work results in an added estimating risk. It follows that the more types of work a contractor performs, the more the risk or uncertainty the firm undertakes.

The material cost for carpentry work is most dependent on the types and number of members required and the wastage factor that results when carpentry members are fabricated at the job site. Lumber prices for rough carpentry work varies depending on the type of wood and the size of the members. The type of wood is usually not of too much concern because most framing members are usually a Douglas fir type of wood. However, the sizes of members do vary in any building design. For example, it is common for a building design to include 2 by 4, 2 by 6, and perhaps 2 by 10 members. The purchase cost per foot of these members varies, as does the amount of board feet in a 1-foot length. And there is *not* a one-to-one ratio between either the length of members or the board feet and the purchase cost. In other words, there may be a price difference in a board-foot measure of a 2 by 4 versus a 2 by 6. Knowing this, the estimator is best served by first determining the amount of board feet of any member size (or type of wood) and then using this quantity as the basis for pricing. Rough carpentry wood members are purchased in a running-foot or board-foot measure.

The purchase cost of rough carpentry, and to a lesser degree finish carpentry, members depends on the *quantity* of wood purchased. As is true regarding the purchase of steel members, the contractor usually receives a discount for buying a large quantity of wood members. In effect, the more members of a specific size that are required, the more apt the purchase price per member is to be less. Depending on the quantity of material purchased, the purchase price per member, per foot, or per board foot can vary as much as 25%.

While the purchase cost from the vendor most dictates the carpentry material cost, the material wastage factor also must be considered. Wood members, including rough work and finish carpentry, are often purchased in somewhat standard lengths. For example, boards used as baseboards for interior trim work are often purchased in 8-foot lengths. Depending on the sizes of interior rooms, these may need to be cut to fit the room sizes. Such cutting adds labor time and results in some material waste. For example, assume that a room is of a dimension such that an 8-foot board must be cut into a 7-foot length. The remaining foot cut member may not be usable—representing about 12.5% of the total length, it is wasted. This adds to the material cost.

The material wastage percentage is dependent on the care taken by the on-site workers as well as the dimensions of the building design. By using an on-site material inventory, workers can minimize the amount of cutting required and the wastage. Without knowledge of the specific project design, and assuming adequate care taken by the on-site field personnel, estimators commonly use a 5% wastage factor in taking off and pricing carpentry work.

Various inventory purchasing models have on occasion been utilized to analyze the carpentry purchasing process. These models address the "ideal" or *least-cost purchase quantity* and *time of purchase* as a function of a quantity purchase cost, a demand for the material, and a storage cost for purchasing more material than is needed at any point in time. While these inventory models may sometimes be relevant to construction, especially when a large quantity of a certain type of material is needed, their application to construction is limited, because usually a contractor purchases material as needed—in other words, he does not stockpile material inventory. Therefore inventory purchasing models will not be discussed.

PRODUCTIVITY ANALYSIS OF CARPENTRY WORK

Let us assume that a contractor is analyzing the erection of carpentry framing to include the wall framing. Let us further assume that the firm has performed this operation numerous times in the past with varying sizes of work crews. Usually a combination of two labor crafts have been employed— carpenters and laborers. However, the number of each of the crafts used has varied.

Assume now that the contractor wants to determine a relationship between the output in board feet of lumber erected and the size of the crew employed, including carpenters and laborers. His objective in seeking to determine this relationship is to find the optimal crew size and to provide a basis for the carpentry cost on the project he is currently estimating.

Assume that, based on the collected accounting records from past projects, including timecards and daily output reports, the contractor has collected the data illustrated in Figure 14.7. The data illustrated indicates the daily output in board feet as a function of the number of carpenters and laborers on the crew. Assume that the data illustrated have all been collected in the geographic location of the work being estimated; therefore, the contractor judges it to be relevant to the analysis. Further assume that the contractor is working in a labor market in which there are no constraints on the size or combination of carpenters and laborers for a crew. In other words, he judges it feasible to use any size or crew composition he wants. Given an objective of determining the most productive, cost-effective crew, the contractor would like to determine crew output as a function of crew size and composition.

Board Feet of Framing Per Day	Number of Carpenters	Number of Laborers
0	0	4
0	2	0
400	1	1
650	2	1
700	2	2
1000	3	1
1100	3	2
1200	4	2
1300	5	2
1350	5	3

Figure 14.7 *Data Collected at Job-Site*

The reader is referred to the presentation of the production function model discussed in Chapter 2. This productivity model/technique was proposed as a means of modeling relatively complex production systems. The general form of the production function is

$$Z = a + bX^c Y^d$$

Where

a, b, c, and d are constants to be determined

Z is the production output

X and Y are the quantity of the production inputs

In attempting to model the carpentry framing operation, let us assume that Z equals the board feet of framing placed per day, and that X and Y equal the number of carpenters and laborers on the crew, respectively.

In trying to solve the 4 variables in the production function equation— i.e., a, b, c, and d—we can observe from the data illustrated in Figure 14.7 that whenever either carpenters or laborers are not included in the crew, no work is performed. In other words, it is determined that carpenters cannot work without the laborers and vice versa. Given the fact that when X or Y in the above production function is zero the output of board feet is zero, it follows that the variable a is equal to zero. Therefore the production function becomes the following:

$$Z = b X^c Y^d$$

In order to solve for the variable b, the fact is used that the number 1 raised to any power equals 1. Inspection of the data in Figure 14.7 indicates that when one carpenter and one laborer are used, the production output is 400 board feet. Therefore the variable b is equal to 400.

$$Z = 400 X^c Y^d$$

To solve for the two variables c and d, two of the data sets in Figure 14.7 can be used. For example, letting X equal 3 and Y equal 1, and letting X equal 4 and Y equal 2 results in a solution of $c = .5$, and $d = .3$.

The production function is now complete. It is as follows:

$$Z = 400 X^{.5} Y^{.3}$$

Having determined this equation, the contractor can utilize it to calculate the expected output in board feet as a function of the number of carpenters and laborers on the crew. Most importantly, outputs for crew combinations other than those illustrated in Figure 14.7 can be predicted.

The derived production function can also be utilized to determine the most cost-effective crew. With the knowledge of individual craft wage rates and the mathematical process of taking derivatives as described in Chapter 2, the lowest-unit-cost crew can be determined.

Note that it is not necessary to be able to model the data in Figure 14.7 via the production function if one wants to limit oneself to only the crew combinations listed in the data. The lowest-unit-cost crew can also be determined via the mathematical process shown in Figure 14.8. A carpenter hourly rate of $18 per hour and a laborer rate of $14 per hour were assumed in the

Board Feet of Framing Per Day	Number of Carpenters	Number of Laborers	Carpenter Cost Per Day*	Laborer Cost Per Day**	Crew Cost Per Day	Cost Per Board Feet
0	0	4	$ 0	$448.00	$ 448.00	—
0	2	0	288.00	0	288.00	—
400	1	1	144.00	112.00	256.00	$0.64
650	2	1	288.00	112.00	400.00	0.62
700	2	2	288.00	224.00	512.00	0.73
1000	3	1	432.00	112.00	544.00	0.54
1100	3	2	432.00	224.00	656.00	0.60
1200	4	2	576.00	224.00	800.00	0.67
1300	5	2	720.00	224.00	944.00	0.73
1350	5	3	720.00	336.00	1056.00	0.78

*Assumes Carpenter Hourly Rate of $18.00 = $144.00/Day
**Assumes Laborer Hourly Rate of $14.00 = $112.00/Day

Figure 14.8 *Cost Analysis of Crew Size*

analysis. However, without modeling the data via the production function, other crew combination outputs cannot be predicted or studied in terms of evaluating unit cost. The production function serves this purpose.

EXERCISE 14.1

For the carpentry construction shown in the illustrations on pages 426-27, calculate the total board feet of framing to be constructed—i.e., determine the amount of board feet for each lumber size.

EXERCISE 14.2

Convert the following list of lumber into a total quantity of foot board measure—i.e. board feet:

```
 28   2 x 4s    each   12 feet long
126   2 x 4s    each    8 feet long
 62   2 x 6s    each   18 feet long
 28   1 x 4s    each    8 feet long
 18   1 x 6s    each   12 feet long
 12   2 x 10s   each   12 feet long
  2   4 x 4s    each   10 feet long
  1   4 x 6             10 feet long
```

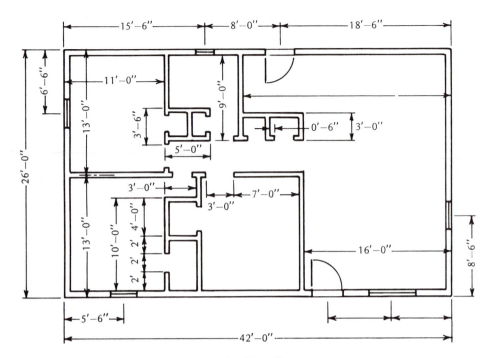

First Floor Plan

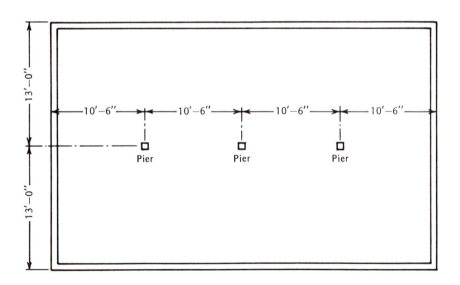

Foundation Plan

Exercise 14.1

426

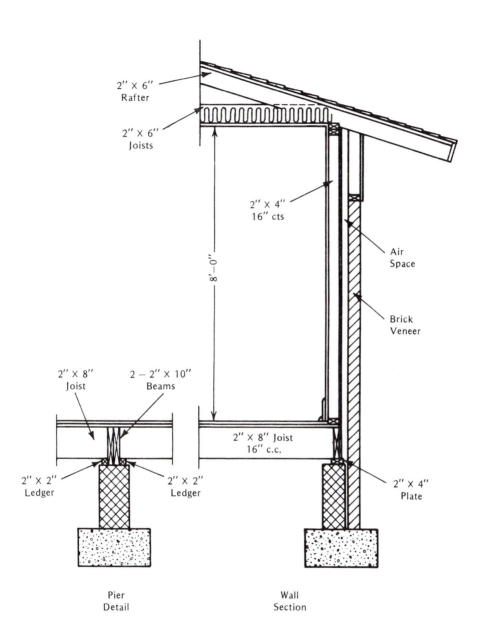

2″ × 6″
Rafter

2″ × 6″
Joists

2″ × 4″
16″ cts

8′–0″

Air
Space

Brick
Veneer

2″ × 8″
Joist

2 – 2″ × 10″
Beams

2″ × 8″ Joist
16″ c.c.

2″ × 2″
Ledger

2″ × 2″
Ledger

2″ × 4″
Plate

Pier
Detail

Wall
Section

Exercise 14.1 *Continued*

EXERCISE 14.3

The cost of carpentry work entails a considerable labor cost for cutting boards to various lengths required. In addition to requiring a labor time and cost, the cutting of wood members often results in considerable material waste. For example, a contractor may have to cut an 8-foot-long 2 by 4 board to a 7-foot length. The 1-foot section that remains will likely be wasted.

As a means of reducing the labor cost and material wastage associated with cutting members, a contractor/estimator may choose to purchase pre-cut members (members cut to the length required). Because these must be pre-cut by the supplier, the cost per board foot is higher than if the members are delivered in standard lengths.

The determination of whether or not wood members should be purchased precut involves consideration of the additional cost per board foot relative to the reduction in cost affiliated with the reduced labor cutting cost and material wastage.

Assume a contractor estimates that 36,000 board feet of carpentry are required for a proposed project. This includes approximately 6,000 boards. If they are purchased in standard sizes, the contractor estimates that the following number of cuts will be required:

NUMBER OF MEMBERS	NUMBER OF CUTS REQUIRED
3,000	0
2,000	1
1,000	2

Assume that the contractor estimates that each cut requires .25 manhours. Because of the cutting, the contractor estimates that a 4% wastage factor will result if the members are cut at the job site. Assume that a carpenter will make all required cuts and that his wage per hour is $15.

The contractor/estimator has obtained a material purchase cost for both the purchase of pre-cut and uncut members. The cost per board foot for uncut members is quoted at $.35. Pre-cut members can be purchased for an 8% higher cost.

Should the contractor purchase pre-cut or uncut members? Show calculations.

15

Miscellaneous Work and Specialty Work

WORK TYPICALLY PERFORMED BY SUBCONTRACTORS

A single contractor usually does not perform all the work for a single building. To construct a total building, even a relatively simple structure, many types of technical skills are required. To construct a functioning building, much work besides that discussed in the last 5 chapters (excavation, concrete, masonry, carpentry, and structural steel work) has to be performed. The skills required to perform these additional types of work are especially technical. Whereas many laypersons might be able to work with concrete, fewer would have the knowledge or skills to perform electrical work, install mechanical systems, and so forth. Because of their relatively complicated and technical nature, electrical, plumbing, and mechanical work are commonly referred to as **specialty work**.

A general contractor seldom performs specialty work—he usually has neither the skills and knowledge to estimate the work, nor the ability to subsequently perform it. Thus, he typically subcontracts specialty work. A single specialty contractor is normally engaged to do the electrical work, another to do the plumbing work, and yet another to do the mechanical work (which includes heating, ventilating, and air conditioning).

The general contractor usually solicits bids from several specialty contractors to do the work set out on the drawings. Often, more than one specialty contractor will be asked to submit a bid for a designated type of specialty work. Having received these bids, the general contractor usually selects the lowest qualified bidder, who then signs a lump-sum contract with the firm.

This process usually takes place while the general contractor is putting together his own estimate/bid. Obtaining firm bids for the specialty-type work enables the general contractor to calculate these fixed subcontracting costs within his own estimate/bid for the project.

Because the general contractor can typically obtain fixed bids from specialty contractors and in turn engage these firms as subcontractors, it might be argued that he does not need estimating skills relevant to specialty type of work. While this is in part true, he certainly must be able to evaluate the accuracy and competitiveness of the subcontractor/specialty bids. As we will see in a later section of this chapter, one way the contractor can evaluate sub-bid accuracy and competitiveness is to compare competing bids for the same work. However, this by itself may not be sufficient. The general contractor needs some estimating guidelines. This is not to say that he needs all-encompassing or detailed specialty-work estimating skills. If he had them, it might seem as if he should then also be able to perform the work. The fact is that the general contractor does *not* perform the work, and without the special expertise needed to do specialty work, he cannot therefore be expected to have amassed the historical data necessary to perform a detailed estimate for that work.

While a general contractor need not prepare a detailed specialty work estimate, he does need some knowledge of the specialty work to be able to coordinate it, and also some approximate estimating techniques so he can evaluate the specialty contractor bids. In this chapter we address the estimating of specialty work including means the general contractor can utilize to evaluate specialty contractor bids. We will not address a detailed take-off or pricing of this work.

It should be noted here there are other kinds of specialty work that may be sub-bid. For example, a specialty contractor may be engaged to "well-point" the job site to lower the water table; another may be engaged to damp-proof the building foundation. Numerous other specialty contractors might also be engaged. But because the work performed by such specialty subcontractors does not occur as frequently as electrical, plumbing, and mechanical work, and because some of that work may in fact be performed by the general contractor, less emphasis will be given in this chapter to these items. The same estimating techniques that are relevant to electrical, plumbing, and mechanical work would be applied to any other specialty work performed by a subcontractor.

MISCELLANEOUS SPECIALTY WORK

The previous 5 chapters addressed the "standard" types of work for a building construction project: excavation, concrete, masonry, carpentry, and structural steel work. The sum of the cost of these 5 work classifications usually comprises a large percentage of the total estimated cost. Add to these the cost of the necessary electrical, plumbing, and mechanical work and the total is a significant portion of the total building's cost. The fact remains, however, that there usually is additional work to be performed to complete a project. Such work might be referred to as "miscellaneous," because it consists of relatively low-cost but varied types of work.

Miscellaneous work includes work that may or may not be part of every building design—for example, insulation, doors, windows, glazing, lathing, and plastering, painting, construction of partition walls, elevators, and so forth. Most of these types of work are performed by specialist subcontractors.

The types of work considered in this miscellaneous work category can be categorized in as many as 8 different divisions of the CSI (Construction Specifications Institute) work. Reference was made to the 16 divisions of work for a building construction project in Chapter 9. Chapters 10 through 14 addressed work than can be classified as follows:

CSI DIVISION	WORK DESCRIPTION
2	Excavation
3	Concrete
4	Masonry
5	Metals (Structural Steel)
6	Carpentry

The fact that each of these work classifications in itself consists of various types of work, and the fact that the cost of any single work classification is usually significant, dictates the assigning a division number for each. But the division numbers for miscellaneous work serve somewhat as catch-alls for various types of work.

On a broad basis, one might correlate various miscellaneous work for a project according to the following classifications:

CSI DIVISION	WORK CLASSIFICATION
7	Thermal and Moisture Protection
8	Doors and Windows
9	Finishes
10, 11, 12, 13	Specialties
14	Conveying Systems

Some of the items of work that might be included in the above "miscellaneous" work categories were discussed as part of Chapters 10 through 14. In particular, work that might be performed by the general contractor, including waterproofing or dampproofing (this work is part of CSI Division 7), was discussed as part of Chapter 10, concerning excavations. Similarly, the estimating of carpentry, windows, and doors was in part addressed in Chapter 14, which dealt with carpentry.

However, most of the miscellaneous work—that is, work that is included in CSI Divisions 7 to 14—is subcontracted. The result is that the general contractor would probably not prepare a detailed estimate for this work. Instead, he would seek multiple bids from subcontractors and normally award the work to the low bidder. He would greatly rely on the premise that the receipt of multiple bids and the selection of the low bid would in itself assure the firm that the bid was reasonable and representative of an accurate cost estimate. Even if this does not prove to be correct, and the general contractor does not obtain what might be considered the most competitive estimate/bid, the relative cost of any one miscellaneous work item or contract would likely be small relative to the total project cost; and so would result in only a small error in the total project estimate. In other words, an error in any single miscellaneous item will likely prove less critical than an estimating error in major work categories.

Because most miscellaneous work is not performed by the general contractor, and also because most of the work varies widely from one project to the next, the taking-off and pricing of the miscellaneous types of work will not be addressed in this chapter. For the most part, the estimating of the various miscellaneous work for a project simply entails locating and counting the number of the items (e.g., specific types of metal windows) and applying a unit price that reflects the material and installation costs. Usually, only a firm specializing in the work in question has accurate pricing data for estimating the work. These data come from past records, and also from the specialty contractor's frequent contacts with vendor/suppliers of the specific items. In effect, estimating miscellaneous work is strongly dependent on a contractor's ability to seek quotes from vendor/suppliers.

IMPACT OF PROJECT ORGANIZATIONAL STRUCTURE

We have learned that not all specialty-type construction work has to be done by a subcontractor. It is also true that not all specialty contractors work as subcontractors. In recent years numerous alternative project organizational structures have evolved. For example, in addition to the traditional general contracting process, **project delivery organizational structures** include the design build process, turnkey construction, and the construction management process (CM). The general contracting process and each of these three project delivery systems is illustrated in Figure 15.1.

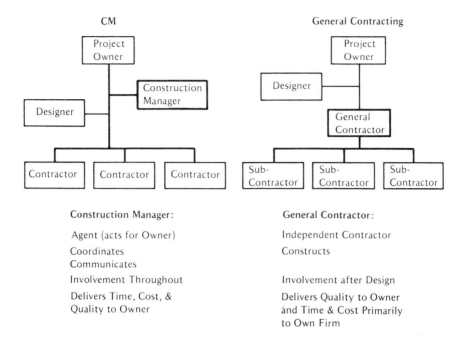

Construction Manager:

Agent (acts for Owner)

Coordinates
Communicates

Involvement Throughout

Delivers Time, Cost, &
Quality to Owner

General Contractor:

Independent Contractor

Constructs

Involvement after Design

Delivers Quality to Owner
and Time & Cost Primarily
to Own Firm

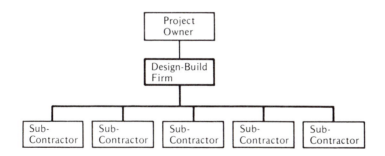

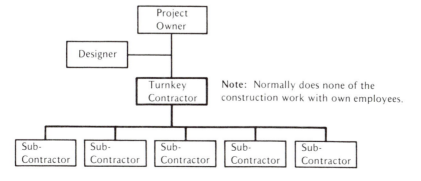

Note: Normally does none of the
construction work with own employees.

Figure 15.1 *Alternative Project Delivery Systems*

The CM process illustrated in Figure 15.1 has become an especially popular alternative to the general contracting process. Inspection of the CM organizational structure indicates that the contractors, including specialty contractors, are engaged directly by the project owner as ***prime contractors***. There is no general contractor role. Instead, the project owner engages a construction manager to oversee the individual prime contractors.

Often the construction manager engaged is a general-contractor–based firm. (*Note*: the construction manager may alternatively be a design-based firm.) In fact, the same firm may serve as a construction manager on some projects and a general contractor on other projects.

A construction management firm on a project does not necessarily perform any construction. Instead, it serves as a coordinator and communicator for the contractors. The estimating responsibility of the construction manager (or general-contractor–based firm) in regard to the individual prime contractors is similar to that of the general contractor as related to reviewing and evaluating subcontractor bids.

The construction manager is likely not in a position, or knowledgeable enough, to prepare or evaluate each prime contractor's estimate in detail. However, serving as an agent of the project owner, the construction manager must be able to evaluate the competitiveness and reasonable accuracy of the estimates. In order to accomplish this objective, the construction manager must implement approximate estimating techniques similar to those used by a general contractor evaluating estimates/bids from electrical, plumbing, or mechanical contractors. Perhaps the only difference is that because the construction manager might not do any actual construction work on the project, each of the engaged prime contractors in effect becomes a specialty contractor in regard to the construction manager evaluating his estimates/bids.

The estimating process, including the resulting estimate of individual specialty contractors is for the most part unchanged whether they are engaged as subcontractors or as prime contractors. The only real difference is that in the general contracting process, the contract agreement established by the estimate/bid would be between the specialty contractor and the general contractor, whereas in the typical CM process it would be between the specialty contractor and the project owner.

EVALUATING SPECIALTY CONTRACTOR BIDS

Because the general contractor cannot be expected to evaluate specialty contractors' estimates in detail, there is a need for alternative means of evaluating the estimates/bids. Perhaps the best alternative available is to solicit and compare alternative bids from several firms for the same work package.

A somewhat standard procedure in the construction industry is to obtain a minimum of 3 bids for each contract to be awarded to a subcontractor (or prime contractor in the CM process). In fact, on many public construction projects, the law requires at least 3 bids for each contract to be awarded.

The concept of requiring and evaluating at least 3 bids is based on the premise that the reasonableness or competitiveness of any one bid can be determined by comparing its dollar amount with the others. Soliciting multiple bids also in part assures the general contractor that the individual specialty contractors will be forced to be competitive.

To illustrate the benefit of requesting and utilizing at least 3 bids for an individual work package, let us consider 4 different sets of bids that might be received for a certain package to be subcontracted. These alternative sets of bids are illustrated in Figure 15.2. Let us assume these are bids for an electrical subcontractor.

Bid Set #1	Bid Set #2
$1,250,000	$1,250,000
1,300,000	1,280,000
1,210,000	1,600,000

Bid Set #3	Bid Set #4
$1,250,000	$1,250,000
1,640,000	1,480,000
1,600,000	1,690,000

Figure 15.2 *Benefits of Obtaining Multiple Bids*

In the first set of bids, the 3 bids are all rather close; there is less than 5% spread between any two of the bids. Given the premise that each of these bids has been prepared independently, the general contractor can be somewhat assured that they represent a fair and reasonable estimate/bid for the project.

In the second set of bids, the lowest two are approximately equal, but the third bid is considerably higher. The conclusion the general contractor should probably reach is that the two low bids, being prepared independently and being approximately equal, are likely to be reasonable. However, because one bid is considerably higher, the contractor should utilize yet another alternative means of bid evaluation. For example, he could use cost books or some of the estimating guidelines discussed in the following sections.

The third set of bids illustrated in Figure 15.2 contains one that is considerably smaller that the other two, which are of the same approximate dollar amount. The possible implication here is that the low bidder may have made a mistake. If this is the case, both the specialty contractor and the entity agreeing to the bid amount (the general contractor) are well advised not to agree to that bid amount. It could be argued that the general contractor should sign the agreement with the low bidder because this bid represents a "bargain." However, if this bid is too low as a result of an estimating error, the specialty contractor will likely have financial difficulties and/or project performance problems that will in turn result in project time and cost delays. The general contractor should be suspicious when a single bid is significantly lower than all the other bids, and he should investigate the dollar-gap difference.

The fourth set of bids illustrated in Figure 15.2 indicates a wide spread between each of the bids. Perhaps this set should cause the general contractor the most concern. It is possible that each of the contractors has a different need for work and thus the difference is explainable. On the other hand, the fact that the independently prepared bids show significant dollar differences indicates possible problems related to the defined scope of work set out on the drawings. It may be that the drawings and specifications are vague or unclear, causing each of the competing contractors to make a different set of assumptions in preparing their estimates and bid amounts. This difference of interpretation is likely to cause the general contractor problems once the construction phase starts. Costly claims regarding the scope of work to be performed are likely to evolve in the construction phase. To prevent this from occurring, the general contractor should investigate and determine the reason for the wide spread of bids.

The stated inferences that have been discussed regarding the four bids are not necessarily true for each and every case. For example, given the numerous alternative construction technologies and techniques available, and given the fact that each firm has a different need for work, it is possible that a wide spread between bids can legitimately occur. However, the fact remains that the soliciting and evaluating of several estimate/bids enhances the general contractor's ability to evaluate the reasonableness of specialty contractors' estimates/bids.

ESTIMATES FOR ELECTRICAL WORK

A general contractor would not perform a detailed take-off and pricing for work he intended to subcontract. However, occasions arise that require him to know how to evaluate the reasonableness of an estimate—for example, in the case of a wide variation in the bids received.

There are several ways in which a general contractor can approximate the cost of the electrical work for a project and thus determine the reasonableness of a specialty contractor's bid. Perhaps the 3 most common ways are to:

1. Evaluate the cost of electrical work as a percentage of total project cost.
2. Determine the approximate cost as a function of the floor space of the building.
3. Determine the approximate cost as a function of the number of fixtures.

Method 1: These methods are listed in an order of increasing potential of accuracy associated with the method. The first method calls for the general contractor to calculate the cost of an electrical contractor's bid as a *percentage of the total estimate project cost*. Based on his past experience and records, or based on the use of industrywide published cost books, the contractor can determine whether a specific bid is a much higher (or for that matter lower) percentage cost relative to the total project cost than should be expected. Example electrical cost percentages as a function of total project cost for various types of projects are illustrated in Figure 15.3. These cost percentages are based on data collected by the author.

Type of Project	% of Electrical Cost
Factories	11%
Fire Stations	10%
Commercial Garage	9%
Parking Garage	5%
Gymnasium	9%
Hospital	14%
Housing	10%
Library	11%
Medical Office	9%
Offices	10%
Post Office	11%
Restaurant	11%
Retail Store	10%
Schools	10%
Vocational School	13%
Theater	10%
Warehouse	10%

Figure 15.3 *Electrical Cost as a Function of Total Cost*

It should be noted that the percentages listed in Figure 15.3 can at best be utilized as guidelines. The fact of the matter is that the percentage of electrical cost on a building relative to the total project cost can vary significantly from one project to another depending on the type and quality of construction, the type of energy systems utilized, and the shape and height of the building. Nonetheless, the use of the data illustrated in the figure can be helpful in a first appraisal of an electrical contractor's estimate/bid.

Method 2: Electrical work in a building is designed to serve several standard functions, including supporting lighting, heating, and the provision of electricity to equipment in the building. Given a specific-purpose building— e.g., an office building—the amount of lighting, heating, and so forth is strongly related to the square feet of *floor space* in the building. The result is that there is usually a relationship between the square feet of building area for a specific type project and the cost of the electrical work. For example, past records obtained from prior projects might indicate that the electrical cost is $2.50 per square foot of finished floor area for a commercial project. Given the average unit cost accumulated from previous projects, the general contractor can use the unit cost as a basis for appraising an electrical contractor's bid for a proposed project. For example, let us assume that the project in question is an office building that has approximately 80,000 square feet of floor space. Given the general contractor's knowledge of the historical electrical cost of $2.50 per square foot of floor area, he can determine an approximate electrical cost for the proposed project of $200,000. If the electrical contractor's estimate/bid varies significantly from this amount, the contractor should likely investigate the reason. However, just as is true of any approximate estimating procedure, the unit cost per square foot approach is at best a guideline.

Method 3: The third approach to the evaluation of the electrical contractor's estimate/bid is to use a *cost-per-fixture approach*. Electrical work consists of roughwork and finish work. The finish work consists of the purchase and installation of electrical fixtures, including lighting fixtures, outlets, heating devices, and so on. Electrical roughwork consists primarily of the purchase and installation of conduit and electrical wiring to connect the electrical fixtures to the electrical energy supply. The amount of electrical roughwork that is needed to support a fixture is somewhat standard. The result is that it is possible to determine an approximation of the electrical roughwork for a building as a function of the number and type of electrical fixtures.

It is relatively simple and not very time-consuming for the general contractor to take off the number and type of electrical fixtures, such as outlets, lights, electrical devices, and so forth. These are set out on the electrical drawings for a project.

Cost books are available that state the approximate cost for an electrical fixture, including electrical roughwork. It should be noted that such costs are subject to inflation over time. In addition, at best, these are approximate costs in that the actual cost depends in part on the quality of the fixtures used, the actual shape and size of the building and the rooms, and the use of the designed space. However, because the estimating procedure relating cost to number of fixtures is more detailed than the prior two methods discussed, this method often provides the most potential for accuracy.

ESTIMATES FOR PLUMBING WORK

As is true for electrical work, the general contractor must be able to appraise the reasonableness of plumbing estimates. He can use essentially the same three approaches discussed for evaluating the electrical estimate/bid.

Method 1: Given a specific-use type of building, the plumbing cost for a building typically is approximately the same percentage of the total project cost. Sample plumbing costs as a percentage of total project costs for various types of buildings are illustrated in Figure 15.4. As in the case of electrical work, the actual plumbing cost for a specific building design can vary as a function of the quality of the specific design. For example, a designer may specify different qualities of plumbing pipe or plumbing fixtures, or the amount of fixtures may in fact vary.

Type of Project	% of Plumbing Cost
Factories	7%
Fire Stations	2%
Commercial Garage	9%
Parking Garage	3%
Gymnasium	7%
Hospital	10%
Housing	10%
Library	5%
Medical Office	7%
Offices	5%
Post Office	6%
Restaurant	9%
Retail Store	5%
Schools	7%
Vocational School	8%
Theater	13%
Warehouse	5%

Figure 15.4 *Plumbing Cost as a Function of Total Cost*

Method 2: It is also possible to approximate the plumbing cost for a project by means of calculating the cost as a function of the total square feet of floor area. As is true of electrical work, given a specific-use building, there is a relationship between the amount and cost of plumbing work and the floor area of a building. Such a relationship can be established from historical records of previous projects. Knowing this historical plumbing cost per square foot of floor area, the general contractor can multiply this number times the floor area for the proposed project and determine an approximation of the plumbing cost for the proposed project.

Method 3: Often the most accurate means of approximating the plumbing cost is to determine the cost as a function of the number and type of plumbing fixtures. For a specific building design, there is usually a defined relationship between the amount of plumbing roughwork—that is, the purchase and placement of piping and the number and type of plumbing fixtures required is somewhat standard. For example, for a commercial construction project, the number of water closets is dictated by the size and purpose of the building.

Given the fact that there are not a relatively large number of different types of plumbing fixtures in any one building design, it is a relatively simple process for the general contractor/estimator to determine both the number and type of plumbing fixtures from the drawings. With this information, the estimator can then utilize tables that set out the approximate cost per fixture, the dollar amount to include the related plumbing roughwork for the fixtures.

The estimator should view the use of this type of data as a means of only approximating the plumbing cost for a specific building. This type of historical cost data does not acknowledge the fact that plumbing fixtures are available in varying qualities. Also, this type of historical cost data may not always recognize the actual cost of the plumbing roughwork because a given building design may be somewhat unique and consist of an unusual ratio of plumbing fixture costs to plumbing roughwork. However, while such an estimate is approximate, it can serve as a means of evaluating the reasonableness of a plumbing contractor's estimate/bid.

ESTIMATE OF MECHANICAL WORK

Mechanical work for a building construction project consists of the purchasing and installation of certain relatively expensive mechanical equipment, including such items as furnace, an air conditioning unit, and possibly generators, heat pumps, air compressors, and so forth. This type of work is also sometimes referred to as **HVAC work** (heating, ventilating, and air conditioning work).

The cost of each mechanical piece of equipment or fixture is usually relatively expensive compared to any single item in electrical or plumbing work. Mechanical work is not characterized as consisting of the installation of what is considered roughwork (i.e., the placement of electrical conduit and wiring and the placement of plumbing pipe).

Because mechanical work consists primarily of the purchase and installation of equipment that is easily identified on the drawings, it is a relatively simple process to take off the mechanical work from the drawings. It is realistic to assume that the general contractor can perform this task. However, given the specialized aspect of mechanical work, it is unlikely that he would have the necessary past data or the knowledge to price the taken-off work. So an approximate estimate becomes necessary.

The objective of a general contractor's approximate estimate of the mechanical work is usually to evaluate the reasonableness of the specialty contractor's estimate/bid, as is true for electrical and plumbing work.

Estimating the mechanical work via an evaluation of the percent of its cost as a function of the total building cost, or as a function of the total floor area are usually reliable methods. Given the unique types of mechanical equipment that may be part of a specific building, there may be a wide variation in the cost of the mechanical work relative to the total building cost or to the square feet of floor area.

Use of historical data: The most reliable means of approximating the cost of the mechanical work in a building is to take off the major pieces of mechanical equipment and use available published data for approximating their cost. It is true that the purchase cost, and to a lesser degree the installation cost, for various types of mechanical equipment varies significantly. The cost of a furnace depends on its type, size, and quality of the unit. For example, a furnace may be a gas, electric, or steam system, and may be purchased with or without an accompanying air conditioning unit. Any one furnace may be available in different energy-producing capacities or sizes. However, a contractor can obtain approximate prices from widely circulated cost books for "standardized" equipment. Any historical costs are obviously subject to inflation factors and such things as possible discounts, sales, and so forth. A contractor needs to adjust historical data for these considerations.

By utilizing historical data, the general contractor can determine an approximate cost for each major piece of equipment he has taken off the drawings. Having done this, he would then merely add all the approximate equipment costs to obtain the total mechanical cost.

The resulting estimate should be viewed as approximate. It lacks the needed pricing accuracy that can only be obtained by a specialty contractor's knowledge of purchase costs of specific types of equipment. However, the estimate does serve as a means of evaluating the more sophisticated mechanical contractor's estimate/bid.

USE OF HISTORICAL COST BOOKS

Previous chapters emphasized the need for the contractor to collect and use his own cost data from previous projects as a basis for estimating a future project. In effect, there is no substitute for such data. However, a contractor may not always have cost records for all the work to be performed on any single proposed project. This means that on occasion he must utilize cost data collected by other firms or a publishing company. This type of historical cost data is available in published cost books. The data in these books represent an accumulation of cost data from several firms and/or projects.

While the use of cost books is advantageous if a contractor does not have his own data, he should be aware of several shortcomings inherent in the use of these books. For example:

1. Cost books represent an accumulation of data from several firms. Thus data are postulated for "average" firms. The fact is that no one firm is likely to be "average."
2. Many of the cost books do not take into consideration the exact geographic location of the project being estimated. In effect, the data in the cost books represent the average cost in total for a number of cities in various geographic locations. However, material, labor and equipment costs vary significantly from one city to another, and from one geographic location to another.
3. Likewise, the productivity of job-site workers also varies from one location to the next. Owing to several factors, including the variable demand–supply for labor and the training provided to labor crafts at various locations, labor productivity might vary by 30% or more from one city to another. The data in some cost books may not reflect this fact.
4. The data in widely circulated cost books represent averages accumulated over periods of time. In some parts of the country, productivity is greatly a function of the weather and thus the time of the year. For example, the productivity for certain defined work is likely to be less in the winter months in Chicago than if it were performed in the spring or summer. In effect, published cost data average the productivity over several months to include very hot and cold temperatures. If a project is to be built in a specific time period, the actual productivity and cost may vary from the "average" published data.

In addition to these factors, the actual costs for a project may vary from the published averages for other reasons. A contractor's costs also relate to the firm's location relative to the project site, and the size of the work. For example, a cost book may indicate a $3-per-cubic-yard cost for excavating work. However, if the contractor only has a few cubic yards to do, the unit cost will probably be higher. On the other hand, a contractor performing 10,000 cubic yards of excavation can likely perform the work at a lower unit cost than the published historical average indicates.

Some widely circulated cost books provide the user with a means of adjusting the published "average" cost to reflect some of the above-noted factors. For example, price indexes may be included that reflect needed adjustments up or down from the averages for specific cities. While these adjustment factors prove useful, the fact remains that there is no substitute for collecting and using one's own historical cost data. It follows that the more work a contractor subcontracts—the premise being that the contractor lacks his own cost data for this work—the more uncertain he is that the estimates are the best measure of anticipated actual costs. This does not mean that a contractor should not subcontract work; subcontracting is a necessary part of the construction process.

EXERCISE 15.1

On occasion, the mechanical and electrical work for a building construction project might be subcontracted to one entity, rather than two separate ones. Assume that a general contractor is considering this option. As part of the process of obtaining subcontractor bid quotes, the firm obtains the following bids from different specialty contractors:

Bids on electrical work only

Contractor A	$851,000
Contractor B	880,000
Contractor C	884,000
Contractor D	910,000

Bids on mechanical work only

Contractor B	$745,000
Contractor C	960,450
Contractor E	755,000
Contractor F	890,000

Bids on combined electrical and mechanical contract

Contractor A	$1,810,000
Contractor C	1,798,000
Contractor E	1,950,000
Contractor G	1,800,000

Based on the objective of minimizing the cost of the work to be subcontracted, how should the general contractor award the electrical and mechanical work (i.e., the number of contracts and the dollar amount of the contracts)? What other additional factors should he consider besides the lowest cost? If there are additional factors, would you recommend that a different contract situation be used than you indicated in your first answer? If so, what contracts should be awarded?

EXERCISE 15.2

One of the arguments than can be made for multiple prime contracting is that there will be less contractor markup or adding profit to a subcontractor's bid price. In other words, the general contractor often obtains bids from subcontractors who have marked up their bids for overhead and profit; he in turn, has marked up the subcontractors' bids to reflect his own overhead and profit; and thus there is a double markup. The awarding of multiple contracts is common when the construction management (CM) process is utilized.

Based on the assumption that a general contractor who obtains a $1,000,000 project subcontracts 40% of the $800,000 earmarked for direct-cost work, and that the subcontractors have marked up their work 20% for overhead and profit, and that the general contractor in turn has marked up the sub-bids an additional 10%, determine the amount of contractor markup on markup that the project owner might realize if he awards separate but equal multiple contracts.

EXERCISE 15.3

Various specialty contractors engaged by the general contractor as subcontractors are required to perform their work at a time well after they submitted their bid/ estimate and well after the start of the construction work for the project. The fact that this delayed work is subject to inflating material and labor costs means that subcontractors must include an estimate of the inflation cost in their bids/estimates. The determination of this inflation cost is made more complex by the fact that the actual start of subcontractors' work might be delayed because of the poor performance of the general contractor or prior subcontractors.

Let us assume that an electrical contractor is required to submit a bid on January 1, 19X0, for work on a project. Independent of any consideration of inflation on material and labor costs, he estimates his material cost to be $200,000 and the labor cost $150,000.

The electrical contractor estimates that the material costs will increase at a rate of 1.5% a month over the next 2 years, and the labor cost will increase at a rate of 1.0% a month over the next 2 years.

Based on a study of the proposed project plan the electrical contractor has estimated various probabilities associated with starting and completing his work in various months as follows:

START WORK	COMPLETE WORK	PROBABILITY
March 1, 19X1	May 1, 19X1	.2
April 1, 19X1	June 1, 19X1	.5
May 1, 19X1	July 1, 19X1	.3

Assuming that the material is to be purchased on the first day of starting work, and the labor cost will be expended linearly over the duration of the work performed, calculate the material inflation cost and the labor inflation cost that the electrical contractor should include in the estimate.

16

Project and Company Overhead

*Importance of Overhead Costs * Types of Job Overhead Costs * Construction Bonds * Insurances * Project Finance Cost * Other Job Overhead Expenses * Company Overhead*

IMPORTANCE OF OVERHEAD COSTS

In addition to direct labor, material, and equipment costs, a construction firm typically incurs various costs referred to as **overhead costs** when it constructs a project. Accurate determination of these costs is as important in the preparation of an estimate as is true for the labor, material, and equipment costs. In fact, it has been the author's experience that an estimate more often proves inaccurate because of an error in determining overhead costs than an error in direct labor, material, or equipment estimates. Nevertheless, the accuracy of the overall estimate is only as good as the weakest link. The estimator may obtain a near-100% accurate determination of a proposed project's direct labor, material, and equipment costs, but if he misses the mark for his overhead estimate, the effect could be disastrous.

There are 2 broad categories of overhead costs that a contractor must include in its construction estimate. Some overhead costs are easily traceable to the performance of a specific project. These costs are commonly referred to as **project or job overhead costs**. A contractor also incurs costs that pertain to his maintaining an office and a firm; such costs are not easily identified with a specific project. These are commonly referred to as **company overhead costs**, or as *general and administrative overhead costs*.

The importance of classifying overhead costs into 2 separate types is emphasized by the fact that the estimator must use a different approach for calculating the 2 different types of overhead costs when he is preparing a construction estimate. Job overhead costs—including such costs as job trailors, on-site supervision, insurances, and project finance costs—are directly traceable to a project and should be itemized by an estimator when preparing an estimate. In other words, each job-cost overhead item should be singled out and budgeted individually for the detailed contractor estimate.

446

Company overhead costs are determined for a specific project by means of an *allocation process*. Company overhead costs are not easily traceable to a specific project. For example, such items as officer salaries, office secretarial wages, and utility costs associated with maintaining a company office would be tedious if not impossible to trace directly to a single project. It is more practical to allocate these costs to individual projects by means of an accounting application process.

The most difficult task the construction estimator has when he is estimating the job overhead costs for a project is to single out each item (e.g., specific types of insurance required, expected cost of the individual items). On the other hand, the difficulty in establishing the dollar amount of company overhead costs for a project centers around the establishment of a correct overhead basis and the calculation of the cost to be applied to a specific project. Types of job and company overhead costs and procedures for determining each of these types of overhead for a project are discussed in following sections.

TYPES OF JOB OVERHEAD COSTS

Job overhead costs, as we have said, are those costs that are directly traceable to a specific project. Given this direct traceability, it can be argued that these costs are direct costs rather than overhead costs.

As we saw in Chapter 3, which addressed the topic of cost accounting, an indirect cost is often referred to as an *overhead cost*. However, the concept of whether a cost is considered a direct or indirect cost really depends on the *definition of the cost object*. Given a defined cost object, a cost that is directly traced to the cost object is referred to as a direct cost, whereas if it is not traced directly to the cost object, it is referred to as an indirect cost.

Job overhead costs are seldom traced directly to individual pieces of a job (i.e., work items); therefore in regard to individual project work items, job overhead costs are "indirect" and thus the term "overhead costs" is fitting.

The type and dollar amount of job overhead that must be estimated for each project is somewhat unique for each one. Perhaps the most common 3 types of job overhead costs, and often the most significant in regard to dollar amounts are *bond, interest,* and *project interest costs*. We will discuss each of these.

CONSTRUCTION BONDS

When submitting a bid for building a construction project, the contractor is required to purchase several types of bonds from a bonding or surety

Project and Company Overhead

company. The cost of the bonds must be identified and included in the overall estimate of a project.

The project owner, by means of his bidding instructions and contract documents, states the types of bonds the contractor must purchase.

Bid bond: A bid is an offer to contract. Theoretically, since it is only an offer, the contractor could withdraw it. Thus, to protect himself from such an event, the project owner will usually require the contractor to submit a **bid bond** with his bid. The bid bond protects the owner, since it assures him that the contractor will build the owner's project for the amount stated in the bid proposal. If a bid proposal contains an obvious contractor mistake (e.g., a gross multiplication error), the courts may rule that the contractor is not subject to the conditions of his bid bond. In such a case, the courts will usually rule that because of the obvious mistake, there has been no contract agreement.

The contractor purchases a bid bond from a bonding/surety company in the same manner in which he purchases insurance from an insurance company. Typically, he may have to submit a bid bond for about 10% of his bid.

Certified check: An alternative to the bid bond is a **certified check** from the potential project contractor to the project owner. The amount of the check would be stipulated by the owner. Rather than cashing the check, the owner merely retains it until the contractor has performed the work, and then returns it. The checks submitted by contractors who do not win the project contract are returned.

Performance bond: Another bond very common to the construction process is a **performance bond**. A performance bond guarantees the project owner that the contractor will build the project owner's project according to specifications. On most public construction projects, a contractor performance bond may be required by law. And many, if not most owners of private construction projects also require a contractor to obtain a performance bond.

The bonding/surety company that issues a performance bond to a contractor risks the possibility that the contractor will fail financially, meaning the bonding/surety company will have the obligation to complete the project and pay the bills for the labor and materials. Because of this risk, the bonding/surety company evaluates a contractor's financial strength, its reputation for completing projects, and the risk conditions of the project; these factors determine whether or not the company will initiate a performance bond and the cost of the bond. Performance bonds may cost from about $10 per $1,000 of contract amount to as little as $4 per $1,000, depending on the above considerations.

Other bonds: While a bid bond and a performance bond are perhaps the 2 most common bonds a contractor has to obtain and itemize in its estimate, additional less common bonds may also be required. A **supplier and/or subcontractor bond** may be required. These are guarantees from a material supplier and/or a subcontractor regarding material supplied or workmanship provided. A project owner may choose to "duplicate" its guarantee by requiring both its general contractor and the general contractor's subcontractors to bond the same work.

A *maintenance bond* is a bond that guarantees the project owner that the general contractor agrees to correct all defects of materials and work for a specified time period following project completion. It is similar to a warranty on the work performed.

A *license or permit bond* guarantees the project owner that the contractor is in compliance with local codes and ordinances. This protects the owner from a secondary liability that may occur because of a contractor's non-compliance with a code or ordinance. It is also possible that a project owner may require the contractor to obtain some other additional bonds to serve a specific purpose. Every bond cost must be included in the contractor's estimate.

INSURANCES

Another job overhead cost the contractor must include in his estimate is the sum of his project insurance costs. There are often numerous types of contractor insurances associated with building a project. The contractor will find it advantageous to have a checklist of insurances available when he is estimating the cost of a project. Such a checklist will help eliminate the possibility of overlooking a project insurance cost.

The premiums a contractor pays an insurance company for many of the project insurances are directly related to the contractor's reputation and his accident record. To keep insurance premiums to a minimum, a contractor should maintain a good safety program on his construction projects.

In discussing insurances that are required by a contractor in building a project, it is convenient to classify them according to 3 types: those that are (1) required by law, (2) required by the project owner, or (3) simply insurance the contractor purchases to cover his own risk.

A contractor is required by state or federal law to carry several types of insurance: workers' compensation, unemployment, old age, disability, and motor vehicle insurance are examples.

Workers' compensation insurance. This covers a contractor's employees in case of personal injury or death. It removes the problem of determining who is at fault, the contractor or the employee. This insurance provides the contractor's employees with the benefits specified by the various state and federal compensation laws in the event of injury or death arising out of the employment.

Most compensation acts contain at least some provisions specifying benefits for the most common occupational diseases.

Benefits vary widely among the several states. In those states where medical benefits are limited, it is advisable to carry full or extra legal medical coverage.

Certain compensation laws are restricted and provide benefits for only specified occupations. The workers' compensation policy may be broadened to voluntarily pay compensation to employees injured in other occupations.

In addition to providing the statutory benefits, the standard workers' compensation policy affords employers liability insurance against claims for bodily injury or disease arising out of employment in the states named in the policy. This coverage against liability not within the scope of the compensation law is subject to a standard $25,000 limit which can be increased for a moderate premium charge.

An injured employee (or his or her dependents), in the event of death, generally has the right of choosing the benefits of the state in which the accident occurred or where the employee was hired. In addition, a number of states impose penalties upon an employer who, although subject to the compensation act, has failed to provide insurance coverage. Accordingly, it is important that these points be discussed with one's insurance representative before entering a state not covered by a policy and before hiring anyone in such other state.

The general insurance rates should be used as a guide only. They are basic manual rates per $100 payroll, applicable to each job classification. Rates and experience modifications are subject to frequent and sometimes radical changes and should be verified with an insurance representative for each specific project.

Reference has been made herein to base or manual rates and to experience modification of the manual rates or premium. A program of retrospective rating is available on an optional basis. This rating plan, superimposed on experience rates premiums, is usually attractive and appropriate for larger contractors. Subject to certain state limitations as to combinations permissible, it is available on an intrastate or interstate basis for workers' compensation and the liability lines either separately or in combination.

An insurance representative can explain the details of this rating plan as well as any other features of coverage or rating on these various lines of insurance.

Contractors' public liability insurance: This policy protects against liability for bodily injury to a member of the public arising out of the contractor's operations and against property damage liability. It can be obtained under either a schedule general liability form policy or a comprehensive general liability policy form.

Under the schedule form, the several hazards to be insured are specifically named and the coverage may be applied either to specified locations or operations of the insured contractor. The comprehensive form is broader in nature and covers all the usual hazards of a general liability nature other than those specifically excluded.

Under either form, considerations should be given to at least the lines of coverage briefly outlined below:

Owners, landlords and tenants, premises exposure: This hazard includes such exposures as offices, warehouses, and fixed locations in contrast to specific contracting operations as such. The usual rating basis is a charge per 100 square feet of area and in the case of the usual contractor, the premium for this hazard is relatively minor compared to that for the operations hazard.

Products liability: This is completed operations coverage applying to accidents that occur after the contractor's operations have been completed. Rates per $1,000 of receipts or contract price vary over a wide range according to the operations being conducted. These rates are frequently the subject of individual account-rate determination rather than the application of any base manual rates.

Owners protective: This covers the liability of the owner for injuries to the public arising out of operations of independent contractors. Characteristic rates for general construction operations in most states are .018 bodily injury and .011 property damage per $100 of contract cost. These rates apply to the first $500,000 of cost on each specific project. Corresponding rates for the next $500,000 of contract price are .009 and .006; for contract cost over $1,000,000, the rates are .002 and .004. Rates in some states are somewhat higher.

Contractor's protective: Such policies cover liability of the contractor for injuries to the public arising out of operations of subcontractors. Rating basis and rate levels stated above for owner's protective are applicable under this line to sublet contract costs.

Additional features can be incorporated into the general liability policy either for application on an individual exposure basis or to be applied on an automatic or blanket basis.

Contractual coverage: This applies to the liability of others assumed by the contractor. Rates depend on the nature of exposure and degree of liability assumed.

Accident-defined coverage: In this case, policies broaden the usual "bodily injury" nature to include such personal injury as libel, slander, false arrest, malicious prosecution, and so on. Pricing varies with the degree of exposure and extent of coverage afforded.

Malpractice coverage: These policies cover a contractor for liability arising out of error of mistake in professional treatment in on-the-job hospital or medical facilities under his control. This line is individually rated.

Increased limits of liability: Here, the general liability rates are at standard limits of $5,000 per person, $10,000 per accident for bodily injury and at $5,000 per accident, $25,000 aggregate for property damage. The percentage increase factors for various higher limits may be secured from an insurance representative.

Automobile liability coverage: Bodily injury and property damage coverage for owned vehicles, non-owned vehicles, hired vehicles, or independent contractors' vehicles may be secured under a schedule form or a comprehensive form of automobile liability policy. Coverage can be selected either on the basis of a listing of owned automobiles or on an automatic basis whereby coverage applies to all owned vehicles. The rating basis for owned vehicles is based on size and weight, usage of the vehicle, and geographical territory involved.

Physical damage coverage (fire, theft, and collision) on owned vehicles may be afforded either under the same policy that affords automobile liability coverage or as a separate policy.

Property damage insurance: The contractor should make a careful study of what his responsibilities are in regard to damage to other property and to the work completed so his policies will mesh with those carried by the owner.

If provision is not made in the specifications that fire insurance will be carried by the owner during construction, it is important that the contractor protect his interests by carrying some type of builders' risk insurance. This is normally written on a standard fire policy together with an extended coverage endorsement that extends the fire and lightning provisions of the standard fire policy to include windstorm, hail, explosion, riot, smoke damage, and damage from aircraft and vehicles.

An additional endorsement, vandalism and malicious mischief, may be obtained. Some companies afford a broader coverage of an all-risk nature, although there are a number of exclusions.

There are two common builders' risk forms that may be written: (1) The *builders' risk reporting form* affords coverage for the contractor from the time material is placed on the job site until the building is completed. Under this form the contractor reports to the insurance company the actual value of the building at the end of each month as construction progresses and pays a premium each month on the additional values that go into the building. (2) the *completed value builders' risk form* is normally preferred by a contractor, because it gives him 100% coverage without the necessity of keeping records as to the value of the building as it progresses or the necessity to report at the end of each month.

This form is written for the completed value of the building. The normal rate charged is about 50% of the rate on the basis that at the beginning of the job there is little value and at the end there is full value. Normally, the premium under the two types of policies is remarkably close.

The interests of a general contractor can be covered and he may exclude or include the interests of subcontractors who also may be employed in the construction of the building.

Temporary structures, tools, and equipment belonging to the contractor and on the premises of the building site may also be included in the coverage for the hazards insured.

Costs of such insurance vary by location, type of construction, and available public fire protection and may be obtained from an insurance person locally.

Contractors' equipment floater: Many contractors possess many thousands of dollars worth of contracting machinery and equipment. To properly protect these values, such equipment and machinery is usually insured under a contractors' equipment floater policy.

Insurance under this form is quite flexible so that many situations may be insured to fit the contractor's particular needs. The insurance is normally written on a named-peril policy that includes loss or damage by fire or lightning, explosion, windstorm, flood, earthquake, collapse of bridges and culverts, theft, landslide, collision, and upset. Or if desired, a policy may be issued protecting from practically all risks of physical loss or damage. Such contracts usually write a deductible clause.

Installation floater: The contractor who specializes in plumbing, heating, air conditioning or electrical equipment may find an installation floater more suitable to his needs. This type of policy may be written for an individual job to cover all the jobs he may secure during a year's time.

Coverage may be on a so-called "named-peril" policy usually includes the hazards of transportation of material to the job site and cover fire and lightning, windstorm and hail, explosion, riot, and civil commotion, smoke damage, and aircraft and vehicle damage at the job site. It may also include coverage for vandalism and theft if desired.

PROJECT FINANCE COST

Another significant dollar-amount job overhead cost that an estimator needs to quantify and include in his estimate for a project is the finance cost associated with building a project. Often this cost can be significant because a contractor may have to pay his bills (e.g., for labor and material) before he in turn bills and collects for work performed for the project owner. The dollar amount of the finance cost incurred by a contractor may be further increased, because it is typical for a project owner to retain some of the amount due the contractor until the completion of the project. This retainage is often 10% of the amount due.

A project finance cost is a direct cost for a project. A contractor should not use an application process to determine the interest or finance cost for the estimate. Instead, he should calculate his forecast interest cost for a project and include it as a direct cost of building a project.

Fundamental to the calculation of a project finance or interest cost is the *preparation of a project plan*. It is necessary for a contractor to determine what work he is going to do in a specific time period if he is going to be able to determine his financing needs as a function of time. Thus, we will now discuss project planning. While this topic will be presented as a basis for determining the estimated finance cost for a project, there are actually many other applications of project planning.

Project Planning

Project planning entails defining activities necessary to complete a project and determining when each of the activities is to be performed. Project activities include elements of work such as placing concrete footings, stripping wall forms, and placing steel joists. Determining when activities are to be performed is constrained by the technical relationship between activities and the available resources of the firm. For example, the firm's stripping of concrete wall forms depends on the completion of placing and curing the concrete in the wall. Similarly, if a crane is required to perform two different activities and the construction firm is limited to using one crane, then one activity must be performed before the start-up of the other activity.

In practice, the project-planning effort of the construction firm varies from very detailed computer-generated project plans to virtually no plan at all. Unfortunately, more often than not, the latter is more characteristic of a planning effort. The absence of planning effort is no doubt one of the main reasons for the construction firm's lack of adequate project cash-flow analysis. Without a prepared plan, such an analysis and determination of a project finance cost is not possible.

Two project-planning management tools or models are in use by some construction firms: (1) the **bar chart model** (also referred to as a Gantt chart), and (2) the **critical path method** or model (CPM).

Bar chart model: A typical bar chart model for a project is shown in Figure 16.1. The defined activities of the project are listed vertically in the

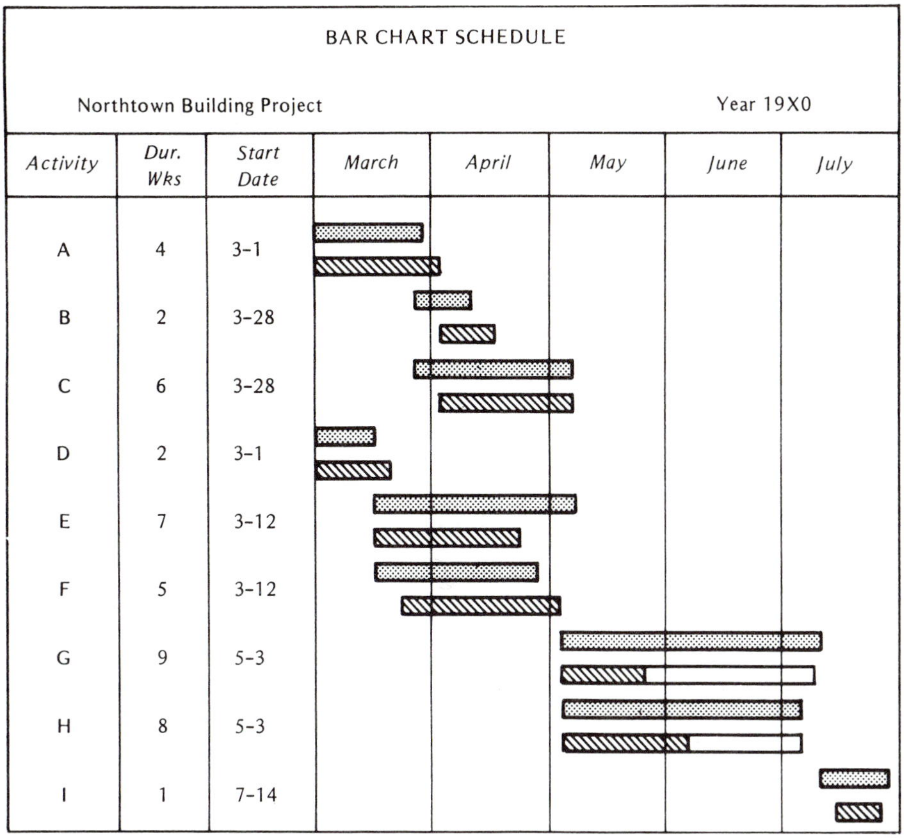

Figure 16.1 *Example Bar Chart*

bar chart. In addition to columns for the quantities of work to be performed for each activity and columns for the start and finish of each activity as determined by the activity logic and overall project plan, a time scale is illustrated on the diagram. The planned time for performing an activity is shown by a horizontal bar. The bar begins at the estimated starting date of the activity and ends at the estimated finish date. In addition to the horizontal bar, an additional bar can be drawn for each activity to indicate its actual progress. The progress bar is often cross-hatched to differentiate between the indication for planned progress and that illustrating actual progress.

The bar chart helps the construction firm plan and control a project. In addition, it can be used to assist in scheduling labor, material, equipment, and cash. (The application of planning models to the scheduling of cash is discussed in following sections of this chapter.)

Owing to its visual effectiveness and the simplicity of its preparation, several construction firms prefer using the bar chart over CPM. However, bar charts have several limitations. They do not show the interdependencies of all project activities as well as does the CPM. In addition, CPM offers more potential for scheduling techniques. In defense of the bar chart, its use by the construction firm enhances cash-flow analysis.

Critical path method: Like the bar chart model, a CPM plan for a project can serve as the basis for preparing a cash-flow analysis. Because of its increased sophistication, the CPM offers the user more potential applications than the bar chart model.

The CPM technique is an application of a science referred to as *network theory*. When a CPM network model is being constructed for a project, an arrow-notation network or a circle-notation network can be prepared. Arrow-notation CPM diagrams are in greater use and are more advantageous as visual tools for analyzing cash flow for a construction project. For this reason, our discussion of CPM is limited arrow notation.

In an arrow-notation CPM diagram, the arrows represent activities. The relation, or logic, between project activities is represented by connecting the tail of the arrow representing one activity to the head of the arrow representing another activity. This is illustrated in Figure 16.2. The diagram represents the fact that Activity B cannot start until Activity A is completed. This is all the diagram represents; it does not indicate that Activity B starts as soon as Activity A is completed—it may or may not start as soon as A is completed. Note that A and B are merely abbreviations for a more detailed description of the activities as indicated in Figure 16.2.

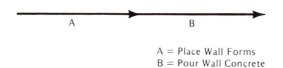

A = Place Wall Forms
B = Pour Wall Concrete

Figure 16.2 *Dependent CPM Activities*

If two activities have no relationship to each other, they can be shown several different ways on a CPM arrow-notation diagram. Figure 16.3 illustrates 3 different ways of indicating that Activities C and D have no dependence on one another.

Our major interest in CPM in this chapter is its use as a basis for the analysis of a project's cash flow. However, CPM is also used for assisting in several other management functions. Fundamental to the use of CPM for any application is an understanding of basic CPM calculations, which are aimed at satisfying 3 objectives:

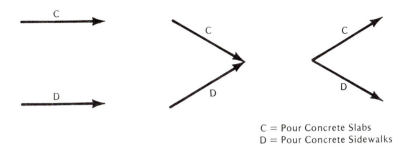

C = Pour Concrete Slabs
D = Pour Concrete Sidewalks

Figure 16.3 *Independent CPM Activities*

1. Determining the minimum project completion time or duration.
2. Determining which project activities dictate or control the minimum project completion time. These activities are called the *critical activities* and form a path through the CPM diagram (hence the term "critical path method").
3. Determining how much time each project activity can be delayed without affecting the minimum project completion time.

In order to accomplish these objectives, it is necessary to perform a series of calculations on a CPM diagram. One set of calculations is determined from a forward pass through the CPM network—also referred to as the *earliest start schedule*, since the calculations performed will yield information pertaining to the earliest time that the project activities can possibly start and finish. The second set of calculations is called the *backward pass*—also referred to as the latest start time schedule, since the calculations will yield information pertaining to the latest time activities can possibly start and finish without delaying the calculated minimum project completion time.

In order to illustrate how the 3 objectives of basic CPM calculations are satisfied by the 2 described schedules, we will perform calculations on the CPM project plan shown in Figure 16.4. The planned duration of each CPM activity is placed alongside the corresponding activity. (*Note:* in the construction of the CPM diagram there is no intended relationship between the length of an arrow representing an activity and the duration of an activity.)

In performing the sets of calculations on the CPM diagram, we will generate 5 different pieces of information about each activity:

Earliest start time for an activity (EST)
Earliest finish time for an activity (EFT)
Latest start time for an activity (LST)
Latest finish time for an activity (LFT)
Total float for an activity (TF)

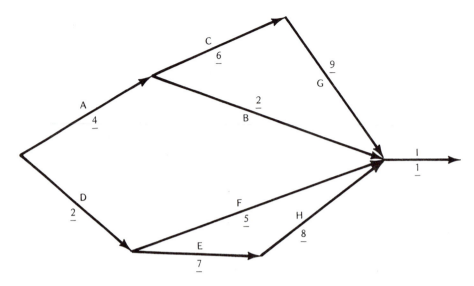

Figure 16.4 *Example CPM Diagram*

Again, the earliest start time (EST) for an activity is defined as the earliest possible time the activity can start. Assuming a starting date for the project illustrated in Figure 16.4 as the end of week 0 (which is equivalent to the beginning of week 1), the earliest start time for Activities A and D are calculated as 0. (Note: CPM activity durations might be stated in days, weeks, or months.) Nothing constrains the start of Activities A and D. The EST answers, along with the answers from other calculations for the CPM diagram in Figure 16.4, are illustrated in Figure 16.5.

The earliest finish time (EFT) for an activity is calculated as an activity's earliest start time plus the activity's duration. In other words, if an activity

Activity	Duration	EST	EFT	LST	LFT	TF
A	4	0	4	0	4	0
B	2	4	6	17	19	13
C	6	4	10	4	10	0
D	2	0	2	2	4	2
E	7	2	9	4	11	2
F	5	2	7	14	19	12
G	9	10	19	10	19	0
H	8	9	17	11	19	2
I	1	19	20	19	20	0

Figure 16.5 *Answers to CPM Calculations*

such as A can start on week 0 (the end of week 0) and it takes 4 weeks to complete, the soonest A can be completed is the end of the 4th week.

The EST of any other activity is calculated as the maximum EFT of activities immediately preceding the activity in question. In the case of Activity C, only one activity, Activity A, directly precedes it. Therefore, the EST of activity C is equal to the EFT of Activity A, or the end of 4 weeks.

Other activity ESTs and LFTs can be calculated in a similar manner. Let us consider Activity I shown in Figure 16.4. Its start is constrained by completing Activities B, F, G, and H. The EFTs of Activities B, F, G, and H are 6, 7, 19, and 17 weeks, respectively. Because Activity I cannot start until Activities B, F, G, and H are complete, Activity I's EST is the end of the 19th week. Similarly, Activity I's EFT is the end of the twentieth week. The ESTs and EFTs of all the activities shown in Figure 16.4 are shown in Figure 16.5.

Because Activity I is the terminating activity of the project shown in Figure 16.4, Activity I's EFT also represents the minimum completion time of the project. In other words, the minimum completion time of the project is the end of the 20th week. The first objective of the CPM calculations is achieved.

The latest start time (LST) and latest finish time (LFT) for each CPM activity are calculated from the backward pass through the CPM network. As a starting point for the calculations, the project duration is set equal to the minimum project duration (20 weeks in the example in Figure 16.4).

The LFT for an activity is defined as the latest possible time an activity can finish without delaying the predetermined project completion time. Because Activity I is a terminating activity in the CPM diagram shown in Figure 16.4, its LFT is therefore 20. This answer is also shown in Figure 16.5.

The LST of an activity is the activity's latest time (LFT) minus the activity's duration. Because Activity I has a duration of one week and must be completed by the end of the 20th week (its LFT), it must be started no later than the end of the 19th week (or 20 − 1 = 19).

The LFT of any other activity is equal to the minimum LST of the activities immediately following the activity in question. For example, only Activity I immediately follows Activity G. Therefore, the LFT of Activity G is equal to the LST of Activity I, or the end of the 19th week.

The LSTs and LFTs of the remaining activities shown in Figure 16.4 are calculated in a similar manner. Let us consider Activity A. It is followed immediately by Activities B and C. The latest start times of Activities B and C are the end of weeks 17 and 4, respectively. In order for each of these activities to start no later than their LSTs, Activity A must be done by the smaller of the two or the end of the fourth week.

The last CPM calculation is determining each activity's total float—which is the amount of time an activity can be delayed assuming no other activity is delayed, without affecting the minimum completion time of a project. Total float is a function of both the earliest and latest start-time schedules. Total float for an activity is the difference between the LST and EST for the activity or the difference between the LFT and EFT for the activity. Either of these calculations will yield the same total float value. The calculated total floats for the activities illustrated in Figure 16.4 are shown in Figure 16.5.

Inspection of Figure 16.5 shows that Activities A, C, G, and I have total float of 0. The LSTs and ESTs for these activities are equal, as are their LFTs and EFTs. In other words, these activities cannot be delayed. They dictate the minimum project duration and form a path through the CPM network. Therefore, the second basic CPM calculation objective, determining the critical activities, is complete. The total floats illustrated in Figure 16.5 for other activities represent the possible delay times for the non-critical activities and satisfy the third and final objective of the basic CPM calculations.

The calculated activity times and total float times shown in Figure 16.5 can be used as a basis of scheduling project activities and the resources used for each activity. Knowledge of possible delay times for the activities provides the planner a means of allocating scarce resources and leveling the demand for a resource such as cash.

Time-scale CPM: To better enable the prepared CPM diagram to analyze a project's cash flow, a time-scale CPM diagram or network may be prepared. In the CPM arrow-notation diagram just discussed, the lengths of individual arrows were not related to the duration of the activity represented by the arrow. However, in a time-scale CPM network, the arrow lengths used to represent activities are proportional in length to the various activity durations.

The time-scale CPM network is prepared by connecting activity arrows according to their defined logic. Each activity is represented by an arrow of a length dictated by the activity duration and the time scale that is placed on the horizontal axis of the time-scale CPM. Figure 16.6 illustrates a time-scale CPM network for the previously discussed CPM diagram shown in Figure 16.4.

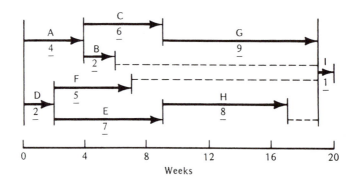

Figure 16.6 *Time Scale CPM Diagram*

Note that the time-scale CPM shown in Figure 16.6 corresponds to the earliest start-time schedule. It therefore indicates the exact timing of each activity and is less general than the CPM diagram shown in Figure 16.4.

The time-scale CPM network or diagram shown in Figure 16.6 looks very similar to the bar chart model shown in Figure 16.1. Both have an improved visual effect over the CPM diagram shown in Figure 16.6. More important, both the time-scale CPM and the bar chart model provide a means of charting cash inflows and outflows as a function of time. This will be illustrated in the following section. However, one should not lose sight of the additional information obtained from the CPM diagram or time-scale CPM diagram versus the bar chart model.

Project Planning and Determination of Finance Cost

The project plan, whether it is a bar chart plan or a CPM plan, serves as the basis for charting the cash inflows and outflows for building a construction project. Therefore, preparing such a plan is fundamental to determining project cash requirements, estimating finance cost, and managing the inflows and outflows of project cash.

To illustrate the dependence of determining estimated finance cost and analyzing cash flow on the project plan, we will consider the project represented by the CPM diagram in Figure 16.4 and the time-scale CPM diagram shown in Figure 16.6.

In addition to the activity durations that were determined, let us assume that the direct cost of performing each activity is as shown in Figure 16.7.

Each activity's duration is shown alongside each activity in the time-scale CPM diagram shown in Figure 16.8. The diagram represents the project's earliest start-time schedule.

	Direct Cost
Activity A	$80,000
Activity B	60,000
Activity C	90,000
Activity D	50,000
Activity E	70,000
Activity F	100,000
Activity G	45,000
Activity H	160,000
Activity I	40,000
Total direct cost	$695,000

Figure 16.7 *Calculation of Cost of Project*

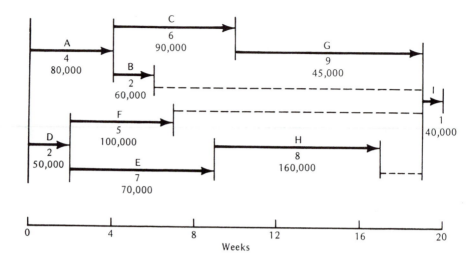

Figure 16.8 *Project Plan with Cash Requirement*

In preparing a cash-flow analysis for the project, let us further assume that the construction firm building the project has related overhead expenses of $2,500 per week and includes $30,000 of profit in its bid. The total contract price is calculated as shown in Figure 16.9.

Let us assume that the contract between the project owner and the construction firm states that a progress report is to be made every 4 weeks and billings for the work completed in that period are to be paid to the construction firm within 2 weeks thereafter. Let us further assume that the owner is to retain 10% of each progress payment, which will be paid as part of the final progress payment if the work is completed satisfactorily by the construction firm.

Charting the construction costs for each of the 4-week intervals is shown as the cash outflow in Figure 16.10. They are calculated as the sum of the expended direct costs for the activities worked on during the 4-week period

Direct costs	$695,000
Overhead costs	
(20 weeks times $2,500)	50,000
Profit	30,000
Total contract amount	$775,000

Figure 16.9 *Calculation of Total Project Cost*

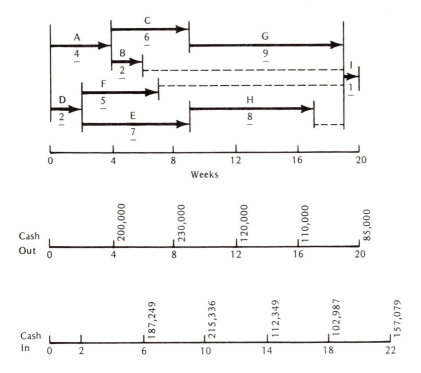

Figure 16.10 *Time Scale CPM with Costs Shown*

and the $2,500 per week or $10,000 per 4-week overhead cost. It is assumed that the costs expended on any activity during a time period are proportional to the percentage of work completed for the activity in the time period. For example, during the first 4-week time period, 2 weeks of a total of 5 weeks of duration for Activity F are completed. Therefore, the assumed cost spent for Activity F during the first 4-week time period is two-fifths of F's total direct cost of $100,000, or $40,000.

The total cash outflow for the first 4 weeks is calculated as shown in Figure 16.11. The cash outflow for the other 4-week intervals is calculated in a similar manner.

Admittedly, the construction firm may not have to pay all of its bills incurred in the same time period. For example, it may be possible to defer the payment of a material bill. However, this fact could be interjected into charting the timing of the cash outflows.

The cash inflows shown in Figure 16.10 are calculated by summing the costs incurred in the previous 4-week period, adding a proportionate share of the construction firm's profit that it will bill with the interim progress reports, and multiplying by 90% to reflect the fact that the owner will retain

Activity A:	4/4 times $80,000	=	$ 80,000
Activity D:	2/2 times $50,000	=	50,000
Activity E:	2/7 times $70,000	=	20,000
Activity F:	2/5 times $100,000	=	40,000
Overhead costs:	4 months times $2,500	=	10,000
	Total cash outflow		$200,000

Figure 16.11 *Calculation of Cash — First 4 Months*

10% as a retainer. The profit billed in a given period is assumed to be calculated by multiplication of total profit by the percent of total costs expended during the progress billing period. All cash inflows are assumed to be received 2 weeks after the progress report is made.

The calculation of the first cash inflow received at the end of the 6th week of the project is as shown in Figure 16.12. The other cash inflows shown can be calculated in the same manner. The cash inflow received at the end of the 22nd month includes the owner's payment of the accumulated retainer amounts.

Direct costs and overhead costs in previous period	$200,000
Profit billed ($30,000 times $200,000/$745,000)	8,054
Total billed	$208,054
Less retainer	20,805
Total cash received	$187,249

Figure 16.12 *Calculation of Cash Received — First 4 Months*

The cumulative cash outflows and inflows are shown in Figure 16.13; they reflect the assumption that the construction firm's cash outflows occur linearly throughout a progress-report period. The difference between the firm's cumulative cash outflows and inflows at any point in time represents the overdraft or surplus of cash from the project. An overdraft represents the amount of cash the construction firm will have to have available from its own funds or borrowed funds in order to perform the project as scheduled — in effect, preparing the cash flow for the project charts out the cash needs for

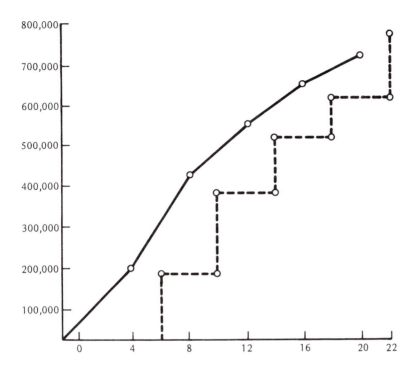

Figure 16.13 *Cumulative Cash In and Out*

the project. Knowing these cash needs ahead of time allows the firm to secure financing well in advance so that it does not get caught in a cash shortage that could lead to contract default and even bankruptcy. Effective cash
management defines financing needs well in advance of when the cash is actually needed.

The cash-flow analysis shown in Figures 16.10 and 16.13 corresponds to
an earliest start-time schedule. A latest start-time CPM schedule could also
be prepared along with a corresponding charting of each cash inflow and
outflow. The cash inflows and outflows from this schedule would differ from
those illustrated in Figures 16.10 and 16.13. In fact, numerous other schedules that combined parts of the earliest start and latest start time schedules
could be prepared. Each of these would also result in a different timing of
cash inflow, outflow, and the amount of cash overdrafts or surplus at any
one point. The construction firm's knowledge of its alternative schedules
and the availability of activity delay times in the form of total floats becomes
one of the tools of effective cash management.

Once the contractor decides on which schedule he is to implement, he can prepare a plot of his expected cash out and cash in. For example, let us assume the contractor decides he will use the earliest-start-time schedule illustrated in Figure 16.6. Having this schedule and the plot of the expected cash out and in as a function of time, the contractor can calculate his weekly or monthly cash surplus or shortage. Assuming a specific cost of capital or interest rate, he could then calculate the estimated finance cost to build the project according to the proposed schedule. It is this finance or interest cost that should be included in the project estimate.

OTHER JOB OVERHEAD EXPENSES

In addition to the previously discussed bond costs, insurance costs, and project finance costs, a contractor incurs other types of job overhead costs that must be included in his project estimate. Each of these costs should be identified and itemized in the project estimate. Some of the typical job overhead costs a contractor can expect to incur during project construction are the following:

Supervision: This job overhead expense should include the wages and fringe benefits of the job superintendent, any assistant superintendents, project engineers, timekeepers, and additional job-site employees carried on the contractor's payroll.

Temporary office: Often a contractor will use a trailer located at a job site to coordinate personnel and store temporary equipment and/or material. In addition to the cost or rental of the trailer, the contractor should include in his estimate the cost of maintaining the trailer, including the telephone and utility expenses.

Temporary toilet: Legally, almost every construction project must have at least one temporary toilet at the job site. On large project sites, many may be required. Costs of these facilities and their maintenance are job overhead costs incurred by a contractor.

Temporary water and utilities: On most projects, the contractor has to provide temporary water, light and power, and heat for the project. The cost of providing any one of these can be significant if the site is undeveloped. The contractor needs to do a survey of the project before the bid is made in order to determine his expected cost. (Occasionally, some of this work and the accompanying cost is covered in the work specifications of one of the subcontractors. For example, some of the temporary electrical work may be included in the electrical subcontract.)

Protecting and repairing work: Because of the sequencing of construction work, certain types of work performed, such as windows or stone work, are subject to damage as the work proceeds. It is usually the contractor's responsibility to put in place any materials required either to protect the finish work or to repair it in case of damage. These expected expenses must be included in the contractor's estimate.

Cleanup: During construction of a project, a certain amount of scrap material and rubbish gets scattered throughout a project site. Usually the contractor is responsible for interim cleanup of the project site and the final cleanup at the completion of the project.

Photographs: It is becoming quite common for a project owner or his designer to require the project contractor to take photographs of the project site as the work progresses. These photographs are in turn used by the project owner or designer to evaluate percentage of completion and the equity of interim contractors' project payment requests.

Surveys: Often the project owner's design team is responsible for performing surveys. However, on occasion some of the surveying may be the responsibility of the contractor. This would be specified in the contract documents.

Building permits: Federal, state, or municipal laws may dictate numerous building permits for a project, including permits for the building itself, as well as for elevators, ventilating systems, and electrical units. A single permit may cost as little as a few dollars or as much as $100 or more. The contractor is usually responsible for purchasing these permits as part of his contract responsibilities; thus he should maintain a permit checklist.

These listed job overhead costs are not meant to be considered exhaustive; to a certain degree, each project is unique in regard to the job overhead costs. It is for this reason that the construction estimator has to be especially attentive to making sure he has included each and every job overhead cost for a project.

COMPANY OVERHEAD

Costs incurred by the firm that support the production process but are not *directly* related need to be allocated to a project. These costs are referred to as *general and administrative costs or expenses* or *company overhead costs*. They are also commonly referred to simply as overhead costs, or as office overhead. Examples of these costs include the costs of maintaining the home office, entertainment expenses, and marketing expenses.

Two somewhat separate questions must be addressed when applying general and administrative (G&A) costs or company overhead to specific projects. First, the construction firm must decide on the basis for the overhead allocation process. Second, once an overhead basis is chosen, there still is the question of the mechanics of determining the amount to be applied to a specific project.

The objective of any allocation of expense is to attempt to apply costs to the cost object in an amount that matches the total costs incurred because of the existence of the cost object. In regard to project estimating, the cost object is the construction project.

Too often *total direct cost*, including labor and material costs (and equipment, if it is treated as a direct cost), is taken for granted as the best basis for applying G&A costs to projects. While this may prove an accurate basis for some firms, other allocation bases may prove more accurate for other firms.

Some construction firms find that they attain a better matching of applied and actual cost if they apply G&A costs to specific projects as a function of only the *direct labor cost* of those projects. This is the second most common method in use.

One might also make a good argument for applying G&A costs to specific projects as a function of the *expected duration* of the projects. Most G&A costs are fixed in nature. The total dollar amount of these types of costs is often a function of time only. For example, many of the costs of maintaining a company office are fixed. The cost of maintaining the office for 2 months is approximately twice the cost of maintaining it for 1 month.

It can be argued that the longer a project takes to complete, the more G&A costs are incurred and the more should be applied to the project. This is a defensible argument and while not in wide use, many construction firms could probably get a better matching of actual and applied overhead if they would recognize duration in their G&A allocation process.

The mechanics of overhead allocation are relatively simple. However, in spite of this, construction firms continue to ignore the process in favor of an antiquated process. Too often, firms use the same percentage for overhead allocation year in and year out without any periodic review or analysis of the percentage. For example, a construction firm might add 15% of total direct costs to each estimate for a number of years without ever analyzing the overhead percentage. The danger of this process is that a firm's overhead costs, volume, and gross margin change yearly, each at a different rate.

The objective of the allocation or application of G&A costs to projects is to apply every dollar of G&A costs, no more and no less, to all projects over a defined period of time. If the firm uses an allocation basis that falls short of applying $50,000 of G&A, a profit of $30,000 is actually a loss of $20,000. If the firm falls short of applying some of its overhead, the unapplied amount is referred to as *underapplied overhead*.

Given the basis by which overhead costs are to be applied, the process involved requires 2 estimates to determine the application rate: (1) an estimate of the G&A costs for the time period in question and (2) an estimate of the dollar amount of the overhead basis must be estimated.

An example: As a minimum, the construction firm should analyze its G&A application rate annually. To illustrate the mechanics of overhead allocation, let us assume a construction firm allocates its G&A overhead costs to projects on the basis of the sum of direct labor and material costs, and that the firm reviews and establishes its G&A application rate annually. The example being considered is illustrated in Figure 16.14.

The first step of the allocation process illustrated in Figure 16.14 is determining the estimate of the coming year's G&A costs. It is these costs that are to be applied in total to projects constructed during the year. The estimate of a coming year's G&A costs or expenses is not as difficult as it might seem. Usually the components that make up the total G&A costs do not vary significantly from one year to the next. Therefore, determining the estimate of the coming year's G&A costs can often be attained by merely multiplying the prior year's figures by an inflation factor. Obviously, any major increase or decrease of G&A cost stemming from a planned expansion or contraction would also have to be acknowledged in the estimate.

Step 1: Estimate of Annual Overhead
(General and Administrative Expense)

Last Year's G and A	$270,000
10% Inflation	27,000
Firm Growth	23,000
Estimated G and A	$320,000

Step 2: Estimate of $ of Cost Basis for Allocation

Estimated Volume	$4,000,000
Gross Margin	20% = $ 800,000
Labor and Matl $	$3,200,000

Step 3: Overhead Allocation Rate

$$\frac{\text{Overhead costs estimated (G \$ A)}}{\text{Labor and material estimate}} \quad \frac{320,000}{3,200,000} = 10\%$$

Step 4: Cost to Apply to a Specific Project

Estimated Labor and Matl. Costs	= $500,000
Overhead to Apply	
Labor and Matl. Costs 500,000 times allocation rate 10%	50,000
Costs	$550,000

Figure 16.14 *Overhead Allocation Process*

Typically company overhead costs would include the following:

Principals' salaries	Taxes and social security
Rent	Dues
Fees paid to consultants	Books and periodicals
Duplication	Contributions
Postage	Insurance
Travel and entertainment	Depreciation
Automobile expenses	Amortization
Utilities	Hospitalization
Offices supplies	Advertising
Telephone and telegraph	Miscellaneous
Employees' salaries	

As a means of determining an estimate of the firm's direct labor and material costs for the coming year, volume is projected along with the firm's gross profit margin. These 2 variables are then used to estimate the coming year's direct labor and material costs. This "backing into" the estimate of direct labor and material costs as a function of volume is commonplace because the amount of the direct labor and material costs is in great part determined by the dollar volume of work performed.

Once the volume is projected ($4,000,000 in Figure 16.14), the gross margin is used to calculate the direct labor and material costs. In Figure 16.14 the total cost of sales is assumed to consist only of the direct labor and material costs. If equipment or other costs were assumed to be part of the cost of sales, these costs would have to be subtracted from the expected volume to yield the expected direct labor and material costs.

The calculations shown in Figure 16.14 yield an application rate of 10%. Therefore in the year in question, each time a project is estimated, an overhead cost is applied to the project equal in amount to 10% times the project's estimated direct labor and material costs. For example, for the $500,000 project in Figure 16.14, the applied G&A cost is 10% multiplied by $500,000 or $50,000.

It is advantageous for the construction firm to review the calculated overhead rate at various times during the year. If it is apparent after 4 months into the year that the firm's estimate of annual volume is too high, an upward adjustment in the overhead rate is probably justified.

The constraint to good overhead allocation is the competitive nature of the construction industry. Apart from a firm's accurately calculated overhead rate is the fact that the construction firm still has to be low enough in its bid to secure the project contract. It is possible that the firm's allocation process results in project bids that are non-competitive. However, it does not follow that the firm should make a bid that contains an allocation rate less than its calculated rate. What more likely follows is that the firm needs to lessen its total overhead costs to a point where it is again competitive.

EXERCISE 16.1

Assume a contractor keeps records regarding his yearly costs. Part of this information is as follows:

Yearly direct labor costs	$600,000
Yearly material costs	$300,000
Yearly equipment costs	$400,000
Yearly company overhead costs	$150,000

On an upcoming project, the contractor estimates his project costs as follows:

Direct labor costs	$150,000
Material costs	$ 50,000
Equipment costs	$200,000

The contractor also estimates that the duration of the project in question will be 5 months, and that he will have to use all of his company resources throughout the project. On the average, the contractor builds 3 construction projects per year. Determine the company overhead cost to be assigned (estimated) to the project in question, by calculating it as a function of each of the following:

Direct labor costs
Material costs
Equipment costs
Total project direct costs (i.e., direct labor costs, material costs, and equipment
    costs)
Duration of project
Number of projects

Discuss the advantage or disadvantage of using each of these various means of calculating a project's company overhead cost.

EXERCISE 16.2

A contractor must prepare a project plan for a project that consists of placing a large culvert. He divides the project into the following project activities:

Activity	Activity Identification
Dig hole	A
Obtain sub-base material	B
Obtain pipe	C
Move pipe to site	D
Fine-grade soil in hole	E
Place sub-base material	F
Compact sub-base material	G
Place pipe and level it	H
Remove excess sub-base material from site	I
Backfill with soil	J

(Observe that each of the activities is assigned a letter. This is done only to shorten the identification of the activities for the CPM diagram.)

The contractor next identifies the technological logic that exists between the activities and recognizing his somewhat limited resources for the project, establishes activity durations. The defined logic and the activity durations are as follows:

Activity A can be done initially and takes 8 working days.
Activity B can be done initially and takes 16 working days.
Activity C can be done initially and takes 24 working days.
Activity D can be done after C and takes 2 working days.
Activity E can be done after A and takes 5 working days.
Activity F can be done after B and E and takes 4 working days.
Activity G can be done after F and takes 8 working days.
Activity H can be done after D and G and takes 6 working days.
Activity I can be done after F and takes 2 working days.
Activity J can be done after H and takes 5 working days.

Help the contractor by preparing an arrow-notation CPM diagram for the above-described project.

Given the CPM diagram shown as the answer to Part 1 of the question perform CPM calculations for the diagram to determine the earliest start time, earliest finish time, latest start time, latest finish time, and total float for each of the activities. Use the table format shown below:

ACTIVITY	DURATION	EST	EFT	LST	LFT	TF
A	8					
B	16					
C	24					
D	2					
E	5					
F	4					
G	8					
H	6					
I	2					
J	5					

Assume the cost of each activity is 1000 times its duration. Assume the contractor is paid for work in 4 week increments; each payment being one period late.

For the project illustrated in this exercise, draw a time-scale CPM diagram illustrating the latest start-time schedule.

For the project schedule illustrated in the foregoing part of Exercise 16.2 calculate the expected cash outflow for the second 4-week period.

For the project schedule illustrated in the foregoing sections of Exercise 16.2, calculate the expected cash inflow for the second 4-week period.

EXERCISE 16.3

The amount of fixed overhead to apply to projects should be based on the accounting principle that any allocation method used should result in the allocation "matching" the actual incurred costs. Estimators often struggle with the determination of the proper basis or bases to be used in applying overhead. The "best" bases may be dependent on several factors. For example, some would argue that the amount of fixed overhead that is incurred for a project is dependent on the dollar size of the project, its duration, the relative amount of labor to material or equipment cost, and so on.

In order to determine a proper overhead allocation basis, assume that during the last year a contractor/estimator has collected the following data:

MONTHS	JOBS COMPLETED	COST OF PROJECT	LABOR COST	MATERIAL COST
January–March	School	$3,000,000	$1,000,000	$1,000,000
	Hospital	2,000,000	1,000,000	600,000
April–June	School	1,000,000	600,000	300,000
	Office Building	500,000	200,000	200,000
July–September	Office Building	2,000,000	1,200,000	600,000
October–December	Hospital	5,000,000	3,000,000	1,500,000
	Office Building	3,000,000	2,000,000	700,000

Assume that during the next time period—i.e., January through March—the contractor is anticipating performing $4,000,000 of office building work. It will consist of approximately $1,800,000 of labor cost and $1,800,000 of material cost. How much overhead cost would you recommend the contractor include in the bid? (Assume that this is the only project the firm will be doing during this 3-month period.)

17

Determination of Contractor Profit

*Definition of Profit * Profit Margin of Contractors *
Profit: a Breakeven Analysis Approach * Contractor
Profit Considerations * Bid Strategy*

DEFINITION OF PROFIT

Profit can be defined as the *return to innovators or entrepreneurs* for taking risk. Profit is linked with risk and uncertainty. Typically, the more risk there is in a business venture or industry, the higher the potential profit. Given the many possible uncertainties that surround building construction projects, the contractor can be characterized as a risk-taker.

Because the contractor is asked to take a significant risk, it follows that his profit margin should be relatively high. However, this potential is tempered by the relatively fierce competition that exists between contractors. As an industry approaches perfect competition, the potential profit margin of the industry diminishes.

If a business had no uncertainties, there would be little or no potential for profits. For example, a contractor could liquidate his firm and invest his equity in a secure bank account. However, the income derived would be relatively small relative to what he might receive for building a project. By the same token, the contractor also has a potential to lose money from a project.

Risk also results in a relatively wide fluctuation of profits. The result is that a contractor realizes widely varying profits (and losses) on a project. It is not unusual for a contractor to obtain a very large profit on one project and incur a substantial loss on yet another project.

In a free economic system, profits can be viewed as the reward to an individual or firm for being innovative or efficient. On the other hand, losses are the *penalty* for using inefficient methods or devoting resources to uses not desired by a consumer or client.

474

A firm that seeks profits should not be viewed as being greedy. Without profit-seekers, our economic system would fail, there would be no incentive for technology change, and human lifestyles would not improve.

It is true that a contractor has to be attentive to factors other than just profits in carrying out his business function. He has to be concerned with promoting good will with his owner–clients, and also with satisfying the needs of his employees. However, monetary profit is the driving force in carrying out his business.

PROFIT MARGIN OF CONTRACTORS

The profit a specific contractor can include in a project bid is dependent upon the risk he is expected to take and the amount or number of his competitors. A contractor performing cost-plus type of work (i.e., a process whereby he gets paid for his actual incurred costs plus a profit) takes little risk and can therefore be satisfied with a smaller potential profit than a contractor submitting a guaranteed maximum lump-sum contract.

Typically, the more specialized a contractor is, the less his competition. For example, an electrical contractor performing specialized work is likely to have less competition for his work than a general contractor. The result is that the electrical contractor likely can include a higher profit margin in his project estimate than can a general contractor. The profitability of some types of contractors as reported by a publisher of financial ratios is illustrated in Figure 17.1. These are *average* profit margins; a specific firm, because of its innovation, efficiency, or lack of competition, may be able to operate with a higher (or lower) profit rate than that shown in the figure. Independent of this fact, one might conclude from Figure 17.1 that contractor profits and profit margins are relatively small relative to those of other industries and given the risk they are asked to take. This is in great part the impact of almost perfect competition in the construction industry.

Type of Contractor	Net Profit Margin (Average)
Commercial Contractor $1 M Volume	3.2
Commercial Contractor $10 M Volume	2.7
Commercial Contractor $100 M Volume	2.3
Heavy and Highway Contractor	3.8
Residential Contractor	4.8

Figure 17.1 *Profit Margins of Average Contractors*

PROFIT: A BREAKEVEN ANALYSIS APPROACH

Cost-accounting concepts of breakeven analysis and contribution provide another means of looking at profits. The approach considers volume, gross margin, and overhead. Breakeven analysis and contribution also prove to be useful concepts when it comes to evaluating the feasibility of a construction firm bidding below its total costs to cover some of its overhead.

The breakeven analysis equation is as follows:

Profit = Revenue – Variable Expenses – Fixed Expenses

With this equation it is possible to calculate a construction firm's breakeven volume or revenue, commonly referred to as its **breakeven point**. The breakeven point or volume is the *level of volume at which the firm's revenue exactly equals its total costs*. At this volume, the construction firm neither makes money nor loses it.

Breakeven analysis is a useful way for the construction firm to determine what level of volume it needs in order to cover its planned overhead expenses. In order to perform this analysis, we usually assume that the construction firm's costs of sales are variable expenses and its operating or general administrative costs are fixed costs. While there are exceptions to these assumptions, most construction costs are variable, and most operating costs may be considered fixed within a given range of volume.

The breakeven-point calculation for the construction firm assumes that the firm has estimated its overhead or fixed expenses for the period in question. Let us assume that this amount is $240,000 for an upcoming year for a given construction firm. The analysis also assumes that the construction firm can predict its gross margin on work to be performed. In reality, this prediction is not difficult, given a review of the last year's performance. Without an innovative approach to decreasing construction field costs and securing more profitable projects, a construction firm's gross margin does not vary significantly from one year to the next. For purposes of example, let us assume a construction firm is able to attain a gross margin of 12%.

Given the estimated annual overhead expenses and the firm's gross margin, the necessary breakeven point or volume needed to cover the overhead is calculated as follows:

At breakeven point:

Revenue = Cost of Sales + Operating Expense

but

Gross Margin = (Revenue – Cost of Sales)/Revenue

therefore, at breakeven point,

Gross Margin = Operating Expenses/Revenue

or

Revenue = Operating Expenses/Gross Margin

In the example of the $240,000 overhead and 12 percent gross margin, the breakeven volume is calculated as follows:

$$\text{Breakeven Volume} = \frac{\$240,000}{.12} = \$2,000,000$$

This calculation yields the fact that the construction firm is going to have to secure at least $2 million of volume in the 1-year period to prevent a loss. Similarly, if the required annual volume is judged unfeasible, the calculation of the breakeven volume draws attention to the fact that the construction firm will probably have to reduce its overhead to prevent a loss.

Establishing a required volume needed to cover overhead in effect defines a minimum goal for the firm. Setting such a goal often motivates the firm to achieve it and also serves as the basis for justifying certain marketing plans and expenditures. Periodically, the breakeven point should be recalculated. This is in recognition of changes in overhead expenditures and modifications in the gross margin.

The concept of contribution is also useful in profit analysis. Contribution is defined as follows:

Contribution = Revenue – Variable Expense

A specific construction project yields the construction firm a positive contribution as long as the dollar contract amount is greater than the cost of construction (remember, we identified the cost of construction of sales as mainly a variable cost). As long as a project has positive contribution, it contributes to covering some of the firm's overhead. For example, let us assume that a construction firm contracts to build a project for $2,000,000 and subsequently its cost of construction (including direct and indirect costs) is $1,800,000. A $200,000 **contribution** results that helps to cover some of the firm's overhead.

Having introduced the concept of contribution, we can understand the practice some construction firms have of occasionally signing a contract to build a project for less than total costs. (By **total costs** is meant the costs of construction and an allocated percentage of the firm's overhead costs.) This practice is illustrated in Figure 17.2.

Obviously, a construction firm should never bid below its estimate of construction costs. Doing so would have a negative contribution in regard to both project profits and company profits. If a firm's total costs exceed a contract amount, it can obtain a profitable contribution as long as the costs of construction are less than the contract revenue. Even then, the firm has to avoid letting the practice of bidding below cost become habit forming. A

Assume firm's monthly overhead equals $40,000	Firm Considering Bidding Project of One Month Duration		
	Alternative 1	Alternative 2	Alternative 3
Direct costs of project equal $180,000	Bid High (Very little chance of getting contract)	Bid Low (Very good chance of getting contract)	Don't Bid
Bid	$230,000	$200,000	
Costs:			
Direct	180,000	180,000	
Fixed	40,000	40,000	40,000
Profit (Loss)	$ 10,000	($ 20,000)	($40,000)
Contribution to overhead and profit	$ 50,000	$ 20,000	0

Figure 17.2 *Process of Bidding Below Cost to Cover Overhead*

continual practice of bidding below total cost means that the firm is not covering its overhead costs. Eventually, the problem may compound to result in more severe financial results such as bankruptcy.

CONTRACTOR PROFIT CONSIDERATIONS

The profit a contractor adds to his proposal represents the amount of money in excess of his costs which he desires as a return for building a project. Disregarding profit, all contractors bidding on a single project may estimate the same cost for building it. Thus, the profit a contractor adds to his estimated project cost should represent the difference between his *winning or losing the project contract* in the competitive bidding procedure. In reality, various contractors bidding on a single project will probably have different estimates of project cost. This is because of the different structure of cost information, the different construction methods, and the different take-off procedures used by various contractors. However, even in the case of varying contractor project cost estimates, the profit a contractor adds to his bid often determines whether or not he will win the project contract.

Winning the project contract is not the only consideration when the estimator adds a profit to his project cost calculation. Consistent with the overall contracting objective, the contractor wants to maximize his profits or his projected rate of return. After submitting a bid proposal in a competitive bidding procedure, he receives the project contract if his bid is judged to be the *lowest responsible bid*. If he bids too high (not the lowest responsible bid), he receives no work and must absorb the cost of making the bid proposal. On the other hand, if he bids low to win the project contract, he risks losing money on the project owing to his small profit margin. Thus, when attempting to determine the optimal profit to add to a bid to be submitted in a competitive bidding procedure, the contractor is confronted with conflicting constraints.

Although he may have a little more freedom in determining his project profit in a negotiated contract environment, his task is not vastly different. To receive work from a project owner, his bid (and therefore his profit) must be competitive with that of other contractors.

When discussing a contractor's desired profit associated with building construction projects, it is meaningful to discuss *long-term profit*, and desired *short-term, or project profit*. Usually, a contractor (or any type of firm) has a defined desired long-term profit. However, due to the characteristics of a given construction project, or because of the characteristics of "environment" of the contractor at the time of bidding a project, the contractor may often have to adjust his desired long-term profit in favor of using a desired project, or short-term profit.

Various criteria are used by different contractors to determine their desired long-term profit rate associated with building projects. Probably the overriding factor in determining a long-term profit rate is the *profit earned by other contractors*. Due to the near-perfect competitive nature of the industry, a contractor's profit rate is somewhat shaped by the industry's profit rate; 3% is typical of many contractors. Naturally, such a profit rate is also a function of time. The point here is that a contractor's profit rate tends to be equal to the overall industry profit rate, because of the competitive nature of the industry.

The *profit rate required to internally finance contractor growth* is another factor that dictates the contractor's desired long-term profit rate. Thus, the profit rate that will enable the contractor to finance planned expansion from retained earnings and depreciation should be considered.

Another consideration is the *historical profit rate of the contractor*. By considering past profit rates, the contractor may be able to formulate his desired future profit rate.

If the contracting firm is a corporation, the profit rate necessary to *attract equity capital* must be considered. This profit rate must be high enough to enable the firm to sell new shares of stock at a price that will not dilute the ownership of the current stockholders of the firm.

Other considerations which may somewhat dictate the contractor's desired long-term profit rate include the *public concept of a fair industry profit,* and *labor's attitude* toward a fair profit rate. To a lesser degree, the contractor may also consider the *amount of new competition* that will enter the industry under various profit rates.

As previously indicated, particular conditions may result in a contractor's desired profit rate for a single project being different than his desired long-term profit rate. There are several conditions that would require a contractor to adjust his desired long-term profit rate.

One of these conditions is the contractor's *expected competition* on a project. It is helpful for the contractor to know how many contractors he is bidding against, who they are, and what their expected bids will be. Obviously, the contractor does not usually have such information available. The more he has, however, the easier it will be for him to determine an optimal profit. As will be shown in the discussion of bidding strategy models in the following section, the optimal profit rate decreases as the number of contractors bidding on a project increases.

A contractor's profit rate for a single project is also a function of the *contractor's need for work*. Certainly, if he needs work, he should be willing to accept a somewhat less than desired profit margin on a project.

The *duration* of a project and its *dollar value* of contract should also be considered in the determination of a profit rate. Given 2 projects of equal dollar-contract value but different expected duration, the project with the longer duration should have a higher total profit, due to the time value of money and opportunity costs. Similarly, if 2 projects have equal duration but different dollar values of contract, the contractor may be willing to accept a lower profit rate for the project with the larger dollar value of contract. This, of course, assumes all other conditions affecting the profit for the projects are equal for each project.

A contractor's desired project profit rate must account for the *risk of the project*—that is, there may be some risk associated with the contractor performing part or all of the work. For example, a contractor may be inexperienced in performing excavation work. If a project in question has a lot of required excavation work, the contractor may evaluate the project as risky, and therefore increase his profit rate for it. This risk cost, or premium, that a contractor associates with a project is often referred to as his **contingency cost**. Sometimes a contractor's project contingency cost is addressed independently of his profit calculations.

Probably the most important overall consideration in the contractor's determination of a profit rate for a project is his *expected project rate of return versus his opportunity rate of return*. The contractor's rate of return (profit as a function of his project investment) from a project should exceed the rate of return from alternative investment opportunities.

BID STRATEGY

As discussed in the previous section, the optimal profit that a contractor should add to his bid proposal is partly determined by the contractor competition for the project. This is particularly true in the competitive bidding procedure. Whether or not a contractor wins or loses a project contract often depends on how well he has formulated his information about his competitors. A contractor must always search for ways to gain an advantage over his competitors. This includes using new and cheaper construction methods and management practices.

Bidding Model: Immediate/Expected Profit

One of the ways a contractor might gain an advantage over his competitors is to formulate bidding information about their past performances. The contractor can formulate this information into some type of bid strategy. For our purpose, we will consider bid strategy as a *combination of various bidding rules a contractor follows for bidding, based on a formulation of information.* In our case, the information will be past bids of contractors. Formulating the information is often referred to as determining a bidding model.

An Example—Immediate Profit, Expected Profit: As discussed earlier, the profit in a contractor's bid for a particular project is the amount of money he intends to make on the project. For purposes of discussion, let us assume that the contractor's estimated cost of building a project is indeed accurate and is equal to his actual building cost. Therefore, on a particular project for which he submits a bid, he will either receive his desired profit (assuming he wins the bid), or he will receive zero profit (assuming he does not win the bid). It becomes clear that a contractor's long-term profit (average profit over a long period of time) will not only be a function of the profit within his bids, but a function of how many projects he receives from the number of projects for which he submits bids.

Owing to the possibility of there being 2 levels of profit (depending on whether a contractor wins or loses the contract for a project), it is necessary to define 2 different types of profit. A contractor's **immediate profit** on a project is defined as the difference between his bid price for the project and his actual cost of building it. The contractor's immediate profit for the project is also equal to the difference between his bid price and his estimated cost. If we let X represent the contractor's bid price on a particular project, and A equal the contractor's actual cost of building the project, then the contractor's immediate profit (IP) on the project is given by the following formula:

$$IP = X - A$$

If a contractor submits a high bid (a large included profit), his chance for receiving the contract in a competitive bidding environment is very small. As he reduces his profit, and therefore his bid, his chance for receiving the contract increases.

If we assign probabilities of receiving the contract to various bids the contractor considers feasible, we may calculate an **expected profit** for the various bids. The expected profit of a particular bid on a proposed project is defined as the immediate profit of the bid for the project multiplied by the probability of the bid winning the contract. In a competitive bidding procedure, winning the contract implies that the bid is the lowest responsible bid. If we let p represent the probability of a particular bid winning the contract for a project, then the expected profit (EP) of the bid is given by the following formula:

$$EP = p(X - A) = p(IP)$$

Assume that a contractor is interested in a certain project, called Project ABC. Assume that his estimated cost for the project is equal to the actual cost of the project, which is $20,000. The contractor has a choice of submitting 3 different bids for the project. These bids and their probabilities of winning the project contract are as follows:

BID NAME	AMOUNT	PROBABILITY OF WINNING CONTRACT
B_1	$30,000	0.1
B_2	$25,000	0.5
B_3	$22,000	0.8

The probabilities shown are estimated from the contractor's evaluation of the chance of being the lowest bidder. Of course, bid B_1 has the highest immediate profit ($30,000 − $20,000, or $10,000). However because of B_1's low probability of winning the contract, it may not be the best bid to make to maximize overall profits. The calculation of the various bids' expected profits is as follows:

BID NAME	PROBABILITY × IMMEDIATE PROFIT	EXPECTED PROFIT
B_1	0.1 (10,000)	$1,000
B_2	0.5 (5,000)	$2,500
B_3	0.8 (2,000)	$1,600

It is observed that bid B_2 has the highest expected profit. Expected profit may be conceived as representing the average profit a contractor can expect to make per project, if he were to submit the same bid to a large number of similar projects. Expected profit does not represent the actual profit the contractor expects to make on a project. In the problem described, the contractor would either make a profit of 0 or a profit of \$5,000 if he submitted bid B_2, whereas the expected profit is calculated to be \$2,500. Since immediate profit does not recognize the probability of a bid winning a contract, expected profit becomes the more informative profit. Therefore, because most contractors have the objective of maximizing total long-term profits, expected profit calculations are more meaningful than immediate profit calculations, and should be used to determine the optimal profit and bid. In the example described, therefore, the contractor should submit bid B_2.

Bidding Models: Knowledge of Competitors

By using the discussed expected profit concepts in addition to information concerning the past bids of his competition, a contractor can develop a bidding strategy which he may use to optimize his profits. At a competitive bid letting it is common practice to announce openly all the bids of the respective contractors. The intelligent contractor can record the bid prices of the contractors, along with his own bid and his own estimate of the project's cost. If it is possible for him to learn the actual cost of the project (either through building it himself or through information obtained from others), he should also record this information. The intelligent contractor should also take note of any special conditions, such as knowledge about a particular contractor's need for work. Having recorded this past bidding information about his competitors, he may formulate the information into a bidding strategy for future projects. Naturally, the more information the contractor has available and the more accurate his information, the better is his chance of having his strategy prove successful.

When a contractor is bidding on a project in a competitive bidding atmosphere, he generally finds himself in one of the following states regarding his competition. In the most deterministic or ideal state, the contractor knows who his competitors are; thus, he also knows the number of competitors. A somewhat infrequent situation occurs when the contractor knows how many competitors there are for the project, but does not know who they are. Since there is less information available in this case than in the case of known competitors, the bidding strategy will be less deterministic and, therefore, less reliable than the bidding strategy for known competitors. An even less deterministic and less desirable situation occurs when the contractor knows neither who the competition is nor how many competitors there are. The bidding strategy for this situation will be less reliable than the previous two cases, owing to the lack of more complete information.

Determination of Contractor Profit

Example—One Competitor Known: Consider the case in which the contractor knows who his competitors will be for a competitive bid project letting. In particular, let us assume that the contractor knows he is only going to be competing against one contractor (Contractor XYZ). Assume that the contractor has bid against Contractor XYZ many times in the past and has kept records of Contractor XYZ's bids. For each of the projects, the contractor has also recorded his own estimated cost. Having this information, the contractor can calculate the ratio of Contractor XYZ's bid price to the contractor's cost estimate for the various projects. His recorded information is summarized as follows:

CONTRACTOR XYZ'S BID/CONTRACTOR'S ESTIMATED JOB COST	FREQUENCY OF OCCURRENCE
0.8	1
0.9	2
1.0	7
1.1	12
1.2	21
1.3	18
1.4	7
1.5	2
	Total: 70

Using the frequency table of the various ratios, the contractor can calculate the probability of each bid ratio by dividing each bid's frequency of occurrence by the total number of measured bid occurrences. For example, the probability of a ratio of 1.0 is 7/70, or 0.10. Probabilities of the other ratios are as follows (the probabilities are rounded off to 2 decimal places):

CONTRACTOR XYZ'S BID/CONTRACTOR'S ESTIMATED JOB COST	PROBABILITY
0.8	0.01
0.9	0.03
1.0	0.10
1.1	0.17
1.2	0.30
1.3	0.26
1.4	0.10
1.5	0.03
	Total: 1.00

Having calculated the probabilities of the various ratios, the contractor can calculate the probability of his various bids being lower than contractor XYZ's bids. To eliminate theoretical bid ties, it will be assumed that the contractor will bid different ratios than the computed ratios for Contractor XYZ. For example, to be lower than Contractor XYZ's bid-to-cost ratio of 1.1, the contractor might make a bid with a bid-to-cost ratio of 1.05. Let us assume that the contractor decides upon the following bid-to-estimated-cost ratios as being feasible:

CONTRACTOR'S BID/ CONTRACTOR'S ESTIMATED JOB COST	PROBABILITY THAT CONTRACTOR'S BID IS LOWER THAN BID OF XYZ
0.75	1.00
0.85	0.99
0.95	0.96
1.05	0.86
1.15	0.69
1.25	0.39
1.35	0.13
1.45	0.03
1.55	0.00

The calculated probability of being the lowest bidder, or winning the contract, for any particular bid ratio is found by merely summing all the probabilities of Contractor XYZ's ratio being higher than the particular bid. For example, if the contractor is to make a bid with a bid-estimated-cost ratio of 1.35, the probability of winning would be the sum of 0.03 (the probability that the ratio of XYZ's bid to the contractor's estimated cost is 1.5), and 0.10 (the probability that the ratio of XYZ's bid to the contractor's estimated cost is 1.4). Thus the probability of contractor bid with a bid-to-estimated-cost ratio of 1.35 winning the contract is 0.13.

The contractor may now use this information to form a bidding strategy for bidding against Contractor XYZ. He may do this by calculating the expected profits of his possible feasible bids. *Expected profit* of a bid was defined as the *immediate profit of the bid multiplied by the bid's probability of winning the project contracted.* Immediate profit for a bid was defined as the bid price minus the actual cost of the project. Let us assume that the contractor's estimated cost of the project is equal to the actual cost. Ideally, the contractor would like his estimator to calculate the actual cost correctly, but this is not always the case. However, due to the lack of information about the actual cost of the project, let us assume that the estimated cost of the project is the actual cost. The immediate profit for each of the contractor's possible bids then becomes equal to the bid price minus the estimated cost of the project. The bid prices are given in terms of the estimated cost of the project. Letting c equal the estimated cost of the project, the immediate profit of the contractor's possible bids may be stated in terms of c. The immediate profit of the bids may be found by merely subtracting 1.0c (the estimated cost of the job) from the respective bids. For example, for a bid of 1.35c, the immediate profit is 1.35c – 1.0c, or 0.35c. The expected profit of the possible bids may then be found by multiplying their immediate profits by their respective probabilities of winning the project against contractor XYZ. For a bid of 1.35c, the probability of winning against Contractor XYZ was calculated as 0.13; therefore, the expected profit is 0.13 multiplied by 0.35c, or 0.0455c. The expected profits for the contractor's feasible bids are as follows (they are rounded off to 3 decimal places):

CONTRACTOR BID	EXPECTED PROFIT OF BID WHEN BIDDING AGAINST CONTRACTOR XYZ
0.75c	$1.00(-0.25c) = -0.250c$
0.85c	$0.99(-0.15c) = -0.149c$
0.95c	$0.96(-0.05c) = -0.048c$
1.05c	$0.86(+0.05c) = +0.043c$
1.15c	$0.69(+0.15c) = +0.104c$
1.25c	$0.39(+0.25c) = +0.098c$
1.35c	$0.13(+0.35c) = +0.046c$
1.45c	$0.03(+0.45c) = +0.014c$
1.55c	$0.00(+0.55c) = +0.000c$

Observe that the bid of 1.15 multiplied by the estimated cost of the project yields the maximum expected profit of 0.104c. This implies that when bidding against Contractor XYZ, over time it would be most profitable for the contractor to submit a bid with a bid-to-estimated-cost ratio of 1.15. For example, if the estimated project cost were $100,000, the bid proposal should be $115,000. Considering the possibility of not winning the contract, the contractor's expected profit for such a bid would be $10,400. Of course, the contractor should keep his bidding information about Contractor XYZ current. The best bid ratio for the contractor to use in future projects against Contractor XYZ may change, depending upon Contractor XYZ's bidding performances.

Example—Two Competitors Known: If a contractor were bidding against several known competitors rather than only Contractor XYZ, he could formulate his bidding strategy in a similar manner. Let us assume that the contractor knows he will be bidding against two known competitors on an upcoming job, Contractor XYZ and Contractor UVW. Assume the contractor's information about Contractor XYZ is the same as in the previous example. Further assume that the contractor has also gathered information about Contractor UVW's bidding performances, and has calculated the probability of his own bids being lower than Contractor UVW's. This information along with the probabilities of winning versus contractor XYZ are as follows:

CONTRACTOR'S BID/CONTRACTOR'S ESTIMATED JOB COST	PROBABILITY OF CONTRACTOR'S BID WINNING VERSUS:	
	XYZ	UVW
0.75	1.00	1.00
0.85	0.99	1.00
0.95	0.96	0.98
1.05	0.86	0.80
1.15	0.69	0.70
1.25	0.39	0.60
1.35	0.13	0.27
1.45	0.03	0.09
1.55	0.00	0.00

To calculate the expected profit of the feasible bids, the contractor must determine the probability that his bid is lower than both Contractor XYZ's and Contractor UVW's bids. Both of these events are independent. The probability of being lower than XYZ is independent of the probability of being lower than UVW. From probability theory, one may show that the probability of the occurrence of joint events which are independent is given by the product of their respective probabilities. For example, the probability that the contractor's bid of 1.15c wins (i.e., is lower than XYZ's and UVW's bids) is the product of 0.69 and 0.70, or 0.483. Knowing the bid's probability of winning the contract, we can calculate its expected profit as 0.483 multiplied by its immediate profit of 0.15c, resulting in an expected profit of 0.07245c. The expected profits for all the bids are calculated as follows (they are rounded off to 3 decimal places):

CONTRACTORS BID	EXPECTED PROFIT
0.75c	$1.00(1.00)(-0.25c) = -0.250c$
0.85c	$0.99(1.00)(-0.15c) = -0.149c$
0.95c	$0.96(0.98)(-0.05c) = -0.047c$
1.05c	$0.86(0.80)(+0.05c) = +0.034c$
1.15c	$0.69(0.70)(+0.15c) = +0.072c$
1.25c	$0.39(0.60)(+0.25c) = +0.059c$
1.35c	$0.13(0.27)(+0.35c) = +0.012c$
1.45c	$0.03(0.09)(+0.45c) = +0.001c$
1.55c	$0.00(0.00)(+0.55c) = +0.000c$

Note that the contractor should submit a bid that has a ratio of bid cost to estimated project cost of 1.15. Thus, the contractor should make the same bid he should have made when bidding against only Contractor XYZ. However, the expected profit of 0.072c, in bidding against the two contractors, is less then the expected profit of 0.104c, in bidding against the single contractor. This is because of the added competition. The more competition a contractor has, the less likely he is to receive the contract. The problem of more than two known competitors is handled in a similar manner. We should not conclude that the optimal bid remains unchanged with increasing competition. In general, the optimal bid will have a tendency to decrease with an increasing number of competitors.

Example: Number, not names, of competitors known: If the contractor knows the number of his competitors, but does not know who they are, he must make some adjustments to his bid strategy. Since he has a less deterministic problem than when he knew who his competitors were, his bidding strategy will be less reliable. The best the contractor can do when faced with the problem of a given number of unknown competitors is to assume that they are average. He collects information from all the contractors against whom he has bid and totals the information to derive a theoretical average competitor. Having this average-competitor information, the contractor can compute the probabilities that his feasible bids are lower than the average contractor's bids. Let us assume that a contractor has done this and has achieved the following results:

AVERAGE CONTRACTOR'S BID/CONTRACTOR'S ESTIMATED JOB COST	PROBABILITY THAT CONTRACTOR'S BID IS LOWER THAN BID OF AVERAGE COMPETITOR
0.75	1.00
0.85	0.98
0.95	0.95
1.05	0.85
1.15	0.60
1.25	0.40
1.35	0.20
1.45	0.05
1.55	0.00

Knowing the probabilities of winning against the average bidder and knowing the number of competitors for a particular project, the contractor may now determine his best bid. The probability of having a bid lower than several competitors is the product of the probabilities of the bid being lower than those of the individual average competitors. The probability of a bid being lower than n competitors' bids may be found by raising the probability of being lower than the average competitor to the nth power. For instance, in the described example, let us assume that the contractor is bidding against 5 unknown competitors. The probability of the contractor's bid of 1.15c being lower than those of 5 competitors is $(0.60)^5$, or approximately 0.078. The expected profit of the bid of 1.15c is, therefore, 0.078 multiplied by 0.15c, or 0.012c. The expected profits for the contractor's feasible bids against 5 unknown competitors are as follows (the calculations are rounded off to 3 decimal places):

CONTRACTOR'S BID	EXPECTED PROFIT OF CONTRACTOR'S BIDS VERSUS 5 UNKNOWN COMPETITORS
0.75c	$(1.00)^5(-0.25c) = -0.250c$
0.85c	$(0.98)^5(-0.15c) = -0.135c$
0.95c	$(0.95)^5(-0.05c) = -0.039c$
1.05c	$(0.85)^5(+0.05c) = +0.022c$
1.15c	$(0.60)^5(+0.15c) = +0.012c$
1.25c	$(0.40)^5(+0.25c) = +0.003c$
1.35c	$(0.20)^5(+0.35c) = +0.000c$
1.45c	$(0.05)^5(+0.45c) = +0.000c$
1.55c	$(0.00)^5(+0.55c) = +0.000c$

It is observed that the bid of 1.05c has the highest expected profit (0.022c). Obviously, as the number of competitors increases, the expected profit of a particular bid by the contractor decreases. Although it is not as obvious, the actual best bid often decreases with an increasing number of competitors. Using the information about an average bidder, the contractor's best bid, and his expected profit for various numbers of competitors, are as follows:

NUMBER OF COMPETITORS	CONTRACTOR'S BID WITH GREATEST EXPECTED PROFIT	EXPECTED PROFIT OF BID
1	1.25c	+0.100c
2	1.15c	+0.054c
3	1.15c	+0.032c
4	1.05c	+0.026c
5	1.05c	+0.022c
6	1.05c	+0.019c

Note that the expected profit of the best bid decreases as the number of competitors increases. Also note that the best bid ratio drops as the number of competitors increases. This happens because as the number of competitors increases, it becomes more difficult for the contractor to submit the low bid, and he is forced to lower his profit and, therefore, his ratio of bid price to estimated projected cost. As the number of competitors becomes very large, one would expect the contractor's expected profit on his optimal bid to approach zero. Also, his best bid ratio would approach 1.0c.

Example—Unknown number and name of competitors: A contractor is often confronted with the problem of bidding against an unknown number of unidentified competitors. Since his information is limited, the reliability of his bidding strategy will be limited. If possible, he should estimate the number of competitors and proceed as if they were all average. Using his information about an average bidder, the contractor may then determine his best bid. It should be observed that the best bid does not change rapidly with the number of competitors. The expected profit changes rapidly with the number of competitors; however, the best bid is somewhat stationary. Therefore, if the contractor can even approximate the number of competitors, he can probably determine the best bid. Another alternative is to use past bidding information to determine the probability of the various numbers of expected competitors. The mathematics of the bidding strategy model based on this added information becomes more complex.

Other Bidding Considerations

In the bidding strategy models discussed, it was assumed that a contractor considered only a finite number of bids as being feasible. In reality, the number of feasible bids is infinite. For example, when he is attempting to submit a bid lower than a competing contractor's bid-to-cost ratio of 1.15, the contractor could use a bid-to-cost ratio of 1.14, 1.135, 1.13, 1.2, 1.1, and so on. The bidding models discussed may be formulated to account for unlimited numbers of feasible contractor bids. However, to avoid the mathematical operation of integration that would be required to formulate the information, this section assumed only finite feasible bids. In view of the type of information that would be required for formulating the infinite feasible contractor bids (a probability function), it is more realistic to consider only finite feasible contractor bids.

In the bidding models discussed, it was assumed that the contractor's estimated cost was equal to his actual cost of building a project. However, this is seldom the case. Several attempts have been made to construct bidding models which account for this fact by assuming a probability distribution for the ratio of estimated to actual cost. Other variables, such as the expected number of competing contractors, have been handled in a similar manner. As the number of variables considered in the bidding model are increased, the mathematics becomes more complex. In fact, the complex nature of the mathematics and the detailed information required for such models often makes their use impractical. The usefulness of bid strategy models depends on cost, benefits, and reliability. The feasibility of any bid strategy model depends on the contractor's ability to gather the information it requires.

The bid strategy models discussed determined the optimal contractor bid for a project as a function of competition on the project. The resulting optimal bid did not reflect other projects and contractor characteristics which affect the determination of the contractor's optimal bid. The models discussed did not consider such things as competition's need for work, the dollar value of the contract, the contractor's need for work, and so on. If the contractor knows that a competitor definitely needs work, he must adjust his competitor's past performances. As with the use of any other construction management tool, a contractor cannot use bid strategy models blindly.

E X E R C I S E 17.1

The amount of profit a contractor should include in an estimate/bid should recognize the productivity risk inherent in performing construction methods. The amount of this risk is dependent on the type of construction work and the methods utilized to perform the work. Assume that absent the consideration of the risk of the construction methods to be used, an estimator has intentions of determining profit as 5% times the sum of the labor and material costs.

To simplify the problem, assume that the construction work to be performed only consists of 6 required types of construction work. From consideration of the type of work, and also past records, the contractor/estimator has determined that 2 different methods of construction are possible for each required 6 types of work. The average and expected dollar cost ranges of these two methods are as follows:

CONSTRUCTION WORK	METHOD #1			METHOD #2		
	LOW COST	AVG. COST	HIGH COST	LOW COST	AVG. COST	HIGH COST
1	$88,000	$90,000	$115,000	$94,000	$95,000	$96,000
2	52,000	54,000	56,000	56,000	56,000	56,000
3	65,000	68,000	72,000	69,000	69,000	69,000
4	98,000	102,000	105,000	104,000	106,000	106,000
5	86,000	86,000	90,000	86,000	86,000	86,000
6	90,000	94,000	98,000	90,000	92,000	108,000

Which construction methods would you recommend for the lowest average cost?

Which construction methods would you recommend to minimize the risk (i.e., to minimize the variation from the average or expected)?

Would you recommend that the previously noted 5% profit margin should be modified given the above ranges? If the answer is yes, suggest the profit margin to be used if all methods labeled #1 are used versus the profit margin when all methods labeled #2 are used.

EXERCISE 17.2

Assume that a construction contractor is expected to be bidding against 2 contractors on a proposed project. One of the expected bidders is Contractor XYZ. Assume that your contractor has kept data regarding XYZ's bidding and summarized the example data regarding XYZ that was set out in this chapter. Assume that the other expected bidder is Contractor UVW. The construction contractor has also bid against Contractor UVW in the past and, based on the collection of Contractor UVW's past bidding performances, the contractor has calculated his probability of being lower than Contractor UVW's bids by bidding the following:

CONTRACTOR'S BID/ CONTRACTOR'S ESTIMATED JOB COST	PROBABILITY OF CONTRACTOR'S BID WINNING VERSUS CONTRACTOR UVW
0.75	1.00
0.85	0.97
0.95	0.94
1.05	0.90
1.15	0.80
1.25	0.70
1.35	0.34
1.45	0.15
1.55	0.00

Assuming that your contractor client wants to submit a bid that has the highest expected profit, determine his best bid for bidding against contractors XYZ and UVW simultaneously. Also calculate the expected profit of this best bid.

EXERCISE 17.3

The term "contribution" is sometimes used relevant to the study of overhead and profit. The theory of contribution is based on the fact that if a firm covers its variable costs with a bid, it "contributes" to covering some if not all fixed overhead.

Given this theory, contractors/estimators have on occasion argued that a firm is better off bidding a job below total cost (assuming that the firm does not cover variable cost) than not bidding at all. In the first case, the firm is at least covering some of its overhead, whereas none is covered in the second option.

To illustrate the concept of contribution, assume that a contractor/estimator calculates the variable expense for a project to be $800,000. Based on a study of overhead costs, the firm estimates that during the project, $200,000 of fixed costs will be incurred. The estimator believes that the firm can bid no more than $900,000 to be competitive. Determine the amount of contribution for bidding versus no bidding. Does it follow that a firm is better off bidding below total cost versus not bidding at all? If it doesn't, state what is wrong with the "contribution" approach.

18

Computers and Estimating

*Computers and Construction * The Technology of Computers * Computer Hardware * Computer Software * Batch Processing, On-Line Processing, and Real-Time Systems * Three Types of Computer Software * Industry Programs * Applications of Computers to Estimating * Future Applications of Computers to Construction*

COMPUTERS AND CONSTRUCTION

There is much talk today about "computerized" estimating. This is a natural event, given the growth of computers and the criticality of the estimating function. The estimator is in search of any advantage he can obtain in preparing an estimate; thus his interest in computers.

Computers have been historically viewed by the construction industry as accounting machines. The early use of computers in construction and contracting were in applications concerning the payroll, accounts payable, and overall general ledger accounting functions. But today, computers play an increasing role in the contractor's project management functions, including estimating.

Computers will continue to have an increasing number and type of applications in contractor's business. Applications vary from accounting to calculations of cut and fill excavation problems, to CPM diagrams, to financial modeling, and so forth. The discussion of computers in this chapter will be limited to their application to the estimating function.

THE TECHNOLOGY OF COMPUTERS

Computers are a rather recent innovation; early development was as recent as the late 1940s. Modern computers can perform almost any quantitative function conceivable.

A computer is a high-speed electronic machine capable of performing mathematical operations and storing and executing instructions that enable it to perform a series of operations without human aid. The earliest computers used vacuum tubes; later machines used transistors and printed circuitry. These have been outdated by electronic circuitry and, more recently, by microelectronic circuitry and machines with small electronic chips. In addition to being faster than earlier machines in performing mathematical operations, modern computers have increased memory capacity, can handle data from remote locations, and can execute several different mathematical operations and jobs simultaneously.

A computer system may be divided into 2 basic components: software and hardware. Software consists of the computer programs along with programming languages and system documentation. In effect, the software tells the computer what to do.

Computer hardware consists of 4 basic components:

1. Central Processing unit (cpu).
2. Input devices.
3. Output devices.
4. Secondary storage devices.

The relationship between these components is illustrated in Figure 18.1. Each of the components of a computer, along with various computer software, is discussed in the following sections.

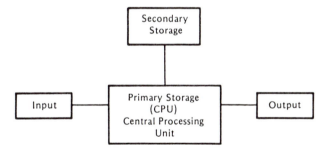

Figure 18.1 *Computer Hardware Relationship*

COMPUTER HARDWARE

The **central processing unit** (cpu) is the fundamental element of a computer system. It performs three somewhat independent functions:
1. Primary storage.
2. Arithmetic and logical operations.
3. Control.

Computer vendors offer various types and sizes of cpu's with their computer systems. The cost of the computer system, including all the hardware and software, is most dependent on the cpu. The performance and capability of a specific cpu is measured by the execution time of performing addition, multiplication, and so on; the access time, which is the time required to retrieve data from memory; and the memory capacity. Execution times vary widely from less than one to several hundred microseconds. Access time in modern computers is extremely fast, averaging around one microsecond. The memory size largely dictates computer hardware cost. Memory size is commonly measured in thousands of words of available storage; for example a computer having a capability of 16,000 words of data would be identified as a 16K machine.

Input devices serve as the media for providing data to the computer system. The actual input hardware used depends on the form of input, among which are:

1. Punched cards.
2. Magnetic tape.
3. Punched paper tape.
4. Magnetic ink character recognition (micr).
5. On-line data entry by terminal.
6. Diskettes (floppy disks).

In past years, punched card data input was the most common, and it remains popular although recent years have seen considerable increases in use of on-line data input devices. The most common form of punched card is the 80-column card. The hardware for reading data from the cards is referred to as a **punched card reader.** Card readers are usually characterized by the speed at which they can read cards. The range is from 200 to 1,500 cards per minute.

Magnetic tape input provides a fast means of data entry. Data are represented on magnetic tape by means of magnetic bits. The original recording of source data on magnetic tape may be done by a keyboard encoder similar to keypunch, or by the transfer of data from punched cards by a device called a converter.

Data on magnetic tape are read into the computer system using a hardware device called a magnetic tape drive unit. Such a unit can read magnetic tape data at speeds up to 200,000 characters per second.

Punched paper tape input is somewhat uncommon. The tape is normally prepared as the by-product of other devices such as output from the computer processing. Punched paper tape is read by a paper tape reader. This process is relatively slow.

Magnetic ink character recognition (micr) input is seldom used in the construction industry. With this type of input device, account numbers and dollar amounts on checks are encoded on documents using a special magnetic ink; micr readers can read these source documents directly with speeds averaging around 1,000 characters per second.

On-line or data terminals make possible direct input of data into the computer system. This is done by means of an on-line keyboard similar to a typewriter. The two most common forms of data terminal are the teleprinter and the cathode ray tube (CRT). On a teleprinter, input is typed into a paper copy as it is entered into the computer system. Output from the computer system is also printed via the teleprinter. Input and output on a crt terminal appear on a screen similar to a television screen. Several crt devices also produce a paper copy of input and output as a duplicate of that produced on the screen.

One of the distinct advantages of on-line or data terminals is that they make it possible to input data at numerous locations. All that is needed is a terminal that is connected to the computer system data bank. On-line terminals are often used in conjunction with a time-sharing computer system (discussed in a later section of this chapter).

The most recent innovation in computer input technology is the **diskette.** A diskette is similar in appearance to a phonograph record. The use of diskettes is increasing rapidly owing to the ease with which they access random information. In addition to improving the efficiency of data input, diskette input devices can be easily stored and handled. Diskettes are sometimes also referred to as floppy disks.

Many input devices such as teleprinters, crt terminals, and magnetic tape drives double as output devices. The most commonly used output hardware is the **printer.** In addition to printing reports, the printer is capable of preparing invoices, purchase orders, and paychecks. Typical printer speeds range from 300 to 2,000 lines per minute.

Card punches are also common output devices. Attached to the computer system, they can punch 200 to 500 cards per minute. Later they can serve as input to the computer system. The card punch and card reader are often combined as one device.

Punched paper tape output is prepared by a paper tape punch device. The tape can also serve as input for a later computer run. Usually punched paper tape is relatively inefficient because of the difficulty of storing the tape. Thus, this form of output is rarely used.

Secondary storage units, including magnetic disks and drums, serve the purpose of holding data that is not being manipulated or operated on in the cpu. When the data are stored in the secondary storage unit rather than in the cpu, the amount of data the computer system can handle is increased considerably. These units are commonly classified as **sequential-access storage, direct-access storage,** or **random-access storage.** A sequential access storage is one in which any single record stored on the device may be accessed only after all other records that precede it are read. A direct- or random-access storage device has the capability of having any word record stored in the device accessed directly without the necessity for reading any other record.

COMPUTER SOFTWARE

Computer programs comprise a significant portion of computer software. They are prepared by a process referred to as **programming,** which is preparing a set of computer instructions that provides the means of solving a specific problem.

Programming in machine language is very difficult; therefore, programming languages have been devised that translate the computer system user's language into machine language. This conversion is accomplished by means of a computer program referred to as an **assembler.** The language and program the computer user supplies the system are referred to as the **source program.** Source programs are written in one of several procedure-oriented languages, such as FORTRAN, RPG, and COBOL. COBOL is used extensively for business applications, including estimating. Other languages include BASIC and PL-1. Numerous languages have been developed to be compatible with specific industry needs. Because many of these languages are designed for specific industries, it becomes possible to utilize languages similar to the vocabulary used in the construction industry. For example, the computer word FORCE is part of a structural design language called STRESS. This use of symbols compatible with industry language makes it possible for an individual to easily learn the computer language without having detailed or broad knowledge about software.

Computer software is not limited to computer programs. Programs need to be documented as to their structure, components, and use, and this is considered part of software. Too often the importance of program documentation is underemphasized. Computer software also involves documenting the computer system itself. This includes an indication of equipment configuration, data record layouts, and program listing. Software documentation is essential to the overall computer system.

BATCH PROCESSING, ON-LINE PROCESSING, AND REAL-TIME SYSTEMS

The term **batch processing** refers to the process by which data or transactions are accumulated in batches and processed at designated points in time. In regard to punch card input, batch processing consists of accumulating a stock of punch cards and subsequently processing them.

Batch processing is characterized by two separate clerical operations: first, the data are prepared (punch cards, tape, etc.), then the resulting input source is fed into the computer system. This is contrasted to the single operation that is characteristic of an on-line system.

An alternative source of assembling transactions for batch processing is referred to as **remote batch processing**: this is used for batch processing at one or more locations removed from the central computer. The transaction data are accumulated at each location (e.g., cards are punched and accumulated). When a batch of transactions is ready for processing, it is read into a terminal and transmitted to the central computer by telephone lines. The data in turn are written on a tape or disk at the central computer site. In this way information is sorted and stored for processing.

On-line processing is becoming increasingly popular. This refers to the processing of individual data or transactions through the computer system as they occur and from the point of origin of the transaction. The advantages of on-line over batch processing are several. First, the on-line data entry and processing typically requires less clerical time because the clerical operation of transferring source documents is eliminated. Perhaps more important is that on-line processing is compatible with **real-time computer systems**. This combination provides the computer user quick and accessible input, processing, and output. On-line systems also possess the advantage of having inquiry capabilities.

On-line processing is not without its disadvantages. Compared to the hardware required for batch processing, on-line equipment is costly. Additional hardware, such as direct-access file storage and data terminals is needed for on-line systems. It is also less efficient than batch processing in terms of the cpu hardware.

The decision as to which to select – batch or on-line processing – must be made in terms of the individual application in question. Often a case can be made for on-line if the computer user needs inquiry capabilities and must obtain information quickly. This is often characteristic of the construction firm's need.

A real-time system is a particular type of on-line system. The characteristic that typifies the real-time system is quick response time; it is so short that the response itself is useful in controlling a physical process. The response time can usually be considered immediate.

It is sometimes difficult to distinguish between an on-line system and a real-time system. A real-time system is an on-line system; however an on-line system may not be a real-time system. Simply having on-line data entry and processing capability is not sufficient to qualify a system as real-time. It must also have the hardware and programs that enable it to process input and quickly generate a response.

The computer hardware necessary for a real-time computer system includes the following:

1. Central processing unit.
2. On-line file storage devices.
3. Data terminals located at points of desired data input and output.

4. Data communications network to connect the terminals with the central processor.

A multiplex may be another piece of hardware used in a real-time computer system. It is a device that controls communication between several terminals.

THREE TYPES OF COMPUTER SOFTWARE

Today there are literally thousands of computer software programs in use by the contractor for various applications. Each of these programs can be categorized as to one of three different types of programs.

User Written Software Programs
General Application Programs
Industry Programs

In early years of computer usage by business, almost all software programs were written by the computer user. Using computer software languages such as FORTRAN, COBOL and more recently PASCAL and BASIC, programs were written to perform specific defined tasks. An example of one of these user written software programs that allows the computer user to input various work item labor and material costs and proceeds to sum the costs is illustrated in Figure 18.2.

Even relatively simple custom designed programs such as that shown in Figure 18.2 can take an individual hours if not days to write. Even after taking the time to do this, the benefit to cost ratio is limited given one end user of the program.

In addition, unless the individual using his or her own application software program is very skilled software writer, the program will likely have flaws. The two most common flaws with user-written programs are 1) lack of error checks, and 2) lack of data entry editing controls. Lack of error checks means that if the software program user enters something other than the input data that is intended, the program may collapse or go into an error mode. Lack of data editing program routines prevents the program user from correcting or "backing up" in the program to change data that was input incorrectly.

The end result is that while it is possible for a computer user to write his or her own estimating program, such an effort often proves to be less than beneficial relative to the quality of the program written compared to the time and effort required. In addition, one must weigh this alternative compared to purchasing one of the many pre-designed programs that are now available at a relatively low cost. These pre-designed programs fall into one of two categories; general application programs and industry programs.

General application software programs are programs written for use by a wide range of businesses for a wide range of applications. Because the

```
100 CLS
110 PRINT "ENTER PROJECT NAME
120 INPUT NA$
130 PRINT "ENTER PROJECT LOCATION
140 INPUT LA$
150 CLS
191 LAT=0
192 MAT=0
193 TAT=0
200 FOR N=1 TO 50
210 PRINT "ENTER WORK ITEM NAME ";N
220 INPUT W$(N)
230 PRINT "ENTER LABOR AMOUNT "
240 INPUT LA(N)
250 PRINT "ENTER MATERIAL AMOUNT"
260 INPUT MA(N)
270 TA(N)=LA(N)+MA(N)
275 LAT=LAT +LA(N)
276 MAT=MAT + MA(N)
277 TAT=TAT + TA(N)
280 PRINT "DO YOU WANT TO ENTER ANOTHER WORK ITEM
290 PRINT "ANSWER YES OR NO
300 INPUT B$
310 IF (B$="NO") THEN 400
320 CLS
330 NEXT N
400 CLS
405 F$="$$#####,####.##"
410 PRINT TAB(32);NA$
420 PRINT TAB(32);LA$
430 PRINT ""
440 PRINT "WORK ITEM DESC.";TAB(20);"LABOR $";TAB(40);"MATERIAL $";TAB(60);"TOTA
L $"
450 PRINT "---------------";TAB(20);"---------------";TAB(40);"---------------";
TAB(60);"---------------"
460 FOR K= 1 TO N
470 PRINT W$(K);
475 PRINT TAB(20);
480 PRINT USING F$;LA(K);
485 PRINT TAB(40);
490 PRINT USING F$;MA(K);
495 PRINT TAB(60);
500 PRINT USING F$;TA(K)
510 NEXT K
520 PRINT "---------------";TAB(20);"---------------";TAB(40);"---------------";
TAB(60);"---------------"
530 PRINT "TOTALS";TAB(20);
540 PRINT USING F$;LAT;
545 PRINT TAB(40);
550 PRINT USING F$;MAT;
555 PRINT TAB(60);
560 PRINT USING F$;TAT
```

```
                         OFFICE BUILDING
                         PEORIA ILLINOIS

WORK ITEM DESC.    LABOR $            MATERIAL $           TOTAL $
---------------    ---------------    ---------------      ---------------
WALL FTGS          $34,534.00         $34,755.00           $69,289.00
PAD FOOTINGS       $93,434.00         $34,232.00           $127,666.00
WALLS              $94,594.00         $34,343.00           $128,937.00
BEAMS              $93,434.00         $34,232.00           $127,666.00
SLABS              $34,343.00         $9,343.00            $43,686.00
KEYWAY             $3,434.00          $5,454.00            $8,888.00
---------------    ---------------    ---------------      ---------------
TOTALS             $353,773.00        $152,359.00          $506,132.00
0
```

Figure 18.2   *Example "Write Your Own" Program*

software vendors that develop these programs sell these to a diverse and wide-ranging type of businesses, these programs are often referred to as the horizontal software market. Businesses to include a construction firm, an accounting firm, a retail firm, and a manufacturing firm would purchase the same general application program.

Two of the more common and widely sold general application programs are **spreadsheet programs** and **database programs.** A spreadsheet program is essentially a program that provides the computer user a large template of columns and rows. Through the use of easy-to-use computer commands, the program user can format the cells and rows in large templates to do complicated mathematical calculations almost instantaneously. For example, a template can be formatted to add a long list of labor, material, equipment, and subcontractor costs for defined work items. In addition to being able to have the template add all the costs almost instantaneously, the template can be used to enable the user to perform sensitivity or "what if" analysis. By changing any single number in the template, the user can study and evaluate the impact on the results of other cells. In effect, this enables the computer user to evaluate the sensitivity of any variable to a change in another variable. This ability to analyze a "what if" situation enables the user to study and evaluate a range of possible solutions or results. This enables a better business decision.

There are literally hundreds of applications of spreadsheet programs for the construction process. An example is illustrated in Figure 18.3. Other possible applications are listed in Figure 18.4. Just as there are many applications of spreadsheet programs for a wide range of businesses to include construction, there are also many computer software vendors.

The ability to do "what if" analysis with a spreadsheet program enables the construction estimator to prepare an estimate, subject it to various risk variables in the estimate, and evaluate a "range" of possible estimates or outcomes. For example, upon preparing an estimate template, the construction estimator can change expected labor productivity, a material price, or an environmental variable such as the expected number of rain days. Upon changing such a variable, the project being considered is "re-estimated" within seconds. The end result is that the use of a spreadsheet program enables the estimator to evaluate the real world of construction uncertainty. One might argue that in a world of uncertainty and unexpected results (the construction process) it is more meaningful to prepare and evaluate a range of estimates rather than a single estimate. LOTUS, EXCEL, and QUATTRO PRO are just a few of many spreadsheet programs available. The vendors that have designed and marketed these programs are normally relatively large software companies.

Given the far reaching and diverse number of businesses that purchase these programs, the supplying vendor usually has considerable sales, and a relatively large staff of employees. This enables the vendor to design a program that has been well tested and is relatively easy to use (user

```
PROJECT: MONMOUTH WTP
ARCHITECT: WELLS ENGINEERS
LOCATION: MONMOUTH, IL.
BID DATE: 4 FEB.
PROJECT DURATION (MOS.):       7.0
```

ESTIMATE SECTIONS	SUB.	MATERIAL	LABOR
1. GENERAL REQUIREMENTS		$17406.00	$36055.00
2A. SITE WORK-EXCAV., ETC.	Y	$28386.00	$135.00
2B. BIT. PAVING, LANDSCAPING		$0.00	$0.00
2C. DEMOLITION		$0.00	$0.00
3A. CONCRETE FOUNDATIONS	N	$180.00	$32429.00
3B. CONCRETE FLOORS, ETC.	N	$15378.00	$14681.00
3C. STRUCT. CONCRETE, PRECAST	N	$1016.00	$2096.00
3D. RE-STEEL, MISC. ADJ.	Y	$9398.00	$5355.00
4. MASONRY	Y	$7000.00	$0.00
5A. STRUCT. & MISC. STEEL	Y	$6985.00	$1866.00
5B. METAL DECKING		$0.00	$0.00
6A. ROUGHH CARPENTRY, HDW.	N	$1577.00	$2571.00
6B. FINISH CARPENTRY, ETC.		$0.00	$0.00
7A. WATERPROOFING, INSUL.		$0.00	$0.00
7B. PREFORMED METAL PANELS		$0.00	$0.00
7C. ROOFING	Y	$7480.00	$0.00
7D. SEALANTS	N	$60.00	$160.00
8A. HOLLOW METAL, FIN. HDW.	Y	$1348.00	$160.00
8B. SPECIALTY DOORS	Y	$625.00	$0.00
8C. GLASS & GLAZING, WINDOWS		$0.00	$0.00
9A. DRYWALL, PLASTER, ACOUST.		$0.00	$0.00
9C. PAINTING, WALLCOVERING		$0.00	$0.00
10. MISC. SPECIALTIES		$0.00	$0.00
11. EQUIPMENT	Y	$1250.00	$1120.00
12. FURNISHINGS		$0.00	$0.00
13. SPECIAL CONSTRUCTION		$0.00	$0.00
14. CONVEYING SYSTEMS		$0.00	$0.00
15. MECHANICAL SYSTEMS	Y	$18209.00	$0.00
16. ELECTRICAL SYSTEMS		$0.00	$0.00
TOTALS		$116298.00	$96628.00

Figure 18.3   *Example Spreadsheet Template*

friendly). The purchaser usually does not have to be concerned about the error checks or data editing capabilities of the program. However, given the large number of available software vendors, the constructor needs to be able to evaluate alternative programs. Some of the concerns that should be taken into consideration when selecting a spreadsheet program are listed in Figure 18.5.

A second general application software program that has wide reaching applications for business and construction estimating is the database program. A database program enables the program user to accumulate, store, and sort vast pieces of information. For example, a business can store large mailing lists of vendors, suppliers, clients, subcontractors, etc. Once input into a database program, the program user can easily edit the lists, sort the lists (for example by zip codes), and search the list (for example,

---

POSSIBLE SPREADSHEET APPLICATIONS

1. Evaluate productivity under different weather conditions.

2. Tabulation of subcontractor bids.

3. Summation of job overhead costs.

4. Preparation of bid summaries.

5. Evaluate project feasibility estimate.

6. Evaluate equipment costs as a function of possible variable working conditions.

7. Establish costs as a function of possible project durations.

8. Determine labor costs as a function of changes in labor rates and wage negotiations.

9. Evaluate costs as a function of variable site conditions and soil conditions.

10. Evaluate percentage of completion under different cost to complete assumptions.

Figure 18.4   *Example Possible Spreadsheet Applications*

CONCERNS WHEN SELECTING A SPREADSHEET PROGRAM

1. Speed of programs for "at-risk" applications.

2. Number of rows and columns of data the program can handle.

3. Program commands should be such that a manual is not required to be referenced throughout use of program.

4. The recall of commands or menus should be efficient.

5. Some data storage should be available with the program.

6. The degree of mathematical rounding of numbers (significant figures) is of concern.

7. The program should be such that it has multiple template integration capabilities.

8. The amount of memory required to run the program efficiently is important.

9. The program should be such that column and row heading sizes can be easily changed.

10. The program should be such that it can be integrated with other programs.

Figure 18.5   *Spreadsheet Program Concerns*

---

searching for all names in a given state). Once the list has been input, tasks such as editing, sorting, searching, etc., take only a matter of seconds.

The storing and use of mailing lists is not the only application of database programs. Businesses use these programs for a diverse set of tasks, to include designing an accounts payable or accounts receivable system, storing of performance data, information retrieval, and attendance records.

One of the higher benefit applications of database programs for construction estimating entails the storage of past productivity information for work tasks or methods. Keeping data regarding past project performance for work tasks can enable the estimator to develop a database that can be reviewed when estimating a future project.

In computer terminology, each past sample of performance is referred to as a "record." The computer user can define or create the style of his or her records. For example, in regard to keeping past labor productivity data the estimator might format a record as follows:

Work Item	Quantity of work	Manhours Required	Productivity (/mh)	Superintendent	Weather	Location	Crew Size
Sch. 101	2,000	200	20/mh	Joe	Mild	Toledo	5
Sch. 102	5,000	400	12.5/mh	Steve	Cold	Cleveland	4

Hundreds of records of performance of the work task can be entered. Every sample or past project entered (School 101, 102, etc.) becomes a record that can be retrieved to aid in estimating a future project. In addition the database computer user can search the records for various defined conditions. For example, the estimator may want to examine only those records where Joe was the superintendent, and a crew size of five workers was utilized. In a matter of seconds a database program can locate these "specified" records and print them on the computer monitor or a hard copy printer.

Such an application of computers enables the estimator to "fit" the past data to the project being estimated. This can be referred to as correlation estimating. The process defeats the argument that every project is different, that past project data doesn't "fit" or represent future events. By searching the database for records that are similar if not the same as the event being estimated, the construction estimator should improve his labor productivity and labor costing accuracy.

Similar to spreadsheet programs, database programs are applicable to a diverse set of businesses and business applications. Database programs such as DBASE are widely sold. The software vendors are usually large firms and the programs are well written and tested. Given the relatively large number of available vendors, the purchaser has many available options. Considerations such as those listed in Figure 18.6 should be taken into consideration when selecting a specific database program.

In addition to spreadsheet and database programs there are other general application computer software programs that are useful to the construction firm and the construction estimating function. Word processing and desktop publishing programs are in wide usage by many firms. These programs can be utilized to improve the efficiency and accuracy of many information forms that are part of the construction process and estimating function.

```
┌─────────────────────────────────────────────────────┐
│                  DATABASE CONCERNS                    │
│   1. Memory required for data storage.                │
│   2. Mathematical roundoff of data stored.            │
│   3. Disk space required.                             │
│   4. Ability to integrate with other programs.        │
│   5. Speed of storing data.                           │
│   6. Speed of data retrieval.                         │
│   7. Efficiency of commands.                          │
│   8. Readability of manuals.                          │
│   9. Ability and speed of sorting data.               │
│  10. Ability to retrieve and use "partial" data.      │
└─────────────────────────────────────────────────────┘
```

Figure 18.6   *Database Program Concerns*

Because of the relatively large size and skills of the vendors that design and market general application programs, most general application programs are high quality programs. The construction estimator seldom has to worry about program defects or program support.

However, because by definition general application programs are written to serve a wide range of applications and businesses, the novice computer user will likely have to expend a fair amount of time learning the program commands and procedures. This can be done by studying program manuals or by attending training classes.

INDUSTRY PROGRAMS

Given the unique business and computer application needs of specific industries, software vendors have also designed computer software programs to meet the needs of a specific industry. Included in the targeted industries is the construction industry.

Numerous individuals and vendors have written software programs for construction firm and project applications. Included are many construction estimating programs.

An industry program might be a detailed and all-encompassing construction estimating program. For example, estimating programs sold by vendors such as TIMBERLINE, MC(2), and SOFTWARE SHOP are full-fledged estimating systems. These programs enable the construction estimator to use estimating formulas, work packages, and vendor supplied historical productivity and cost databases to prepare an entire detailed estimate. Because of the many options these programs provide the user, they may run in the thousands of dollars to purchase.

There are also many industry programs that serve more specific applications. Industry programs are available to perform such specific tasks as concrete mix design calculations, determining the cut and fill quantities for an estimate, and determining the contact area of concrete forming and needed forming hardware (walers, ties, wedge bolts, etc.). These types of programs are usually less costly in that their application and benefit are more limited.

Because industry programs are written for a specific business or industry, they are referred to as the vertical market, i.e., sold to one business type rather than to many (the horizontal market). Because of the smaller market for industry programs, the vendor designing the program can expect limited sales for any one program. The software vendor tends to be smaller than the firm that designs general application programs. The end result is that industry programs may be of a lesser quality than general application programs or may even contain slight flaws or defects. The purchaser of the program therefore must be more concerned about testing the program and evaluating the vendor. Some of these concerns are listed in Figure 18.7

While industry programs may not be of as high a quality as general application programs, they may in fact be more "user friendly." Because they are written to serve a more focused purpose, the vendor should be able to write the program in the industry language of the end user. For example, an industry program written for construction estimating can center around the needs of the estimator. Input and output commands should be easy for the user to understand. Even a novice at using a computer should be able to use such a program by merely following program driven instructions and commands. In summary, while an industry program may be less diverse in terms of possible applications, and may be more suspect in terms of quality, it should be more user friendly.

APPLICATIONS OF COMPUTERS TO ESTIMATING

Computers and accompanying software programs can significantly aid in improving the efficiency and accuracy of the construction estimating process. The speed, data storage, and accuracy of the computer can lessen the time it takes to prepare an estimate and also improve the accuracy of the estimate. There are essentially five applications of computers and accompanying software.

Computer Aided Design (CAD) applications
Math Conversion applications
"What If" or Sensitivity Analysis applications
Database applications
Simulation or Gaming applications

GENERAL CONCERNS WHEN SELECTING INDUSTRY SOFTWARE

1. **Price of software relative to application benefits**
   It is important to remember the following:
   a. The price of a vendor's software program likely has little to do with how much it cost the firm to develop the program or how "good" the program actually is. Instead, the price most reflects the vendor's anticipated number of sales. The larger the number of sales anticipated, the lower the price for which the vendor can sell the program. Too high a price probably means the vendor can't sell many and that there have been few installations.
   b. No matter how good a program is and no matter how many benefits it offers, if the buyer does not have a practical application for "their" firm, then the program offers no benefit to the firm.

2. **Does the program work. Does it have "bugs" in it?**
   Unfortunately the computer software business is a buyer beware business. The "best" way to test a program is to determine how many installations there are. The more installations there are, the more the program has been "field tested." Don't buy a program if there are only one or two field tested installations.
   Secondly, if possible, buy a program under an agreement that it can be returned within a determined number of days (say 30) if you do not like the program.

3. **Does the vendor offer software support?**
   The amount of support you need is dependent on how good the program is and how well it has been tested. A good program does not need support if it is written well and is well supported by a user manual.
   The issue of vendor support is often over-emphasized in that software programs written for the microcomputer are usually very user friendly and don't need support. They are sold like buying merchandise from a catalog.

Figure 18.7   *Concerns When Selecting Industry Programs*

Computer aided design (CAD) programs enable a user to generate construction drawings and in turn to establish the quantities of work to be performed. Using a CAD program, the computer drives a plotter/printer to prepare a complete set of construction drawings. To date, CAD programs have been primarily utilized by architects and engineers to prepare drawings. However, as the programs become universal in use, they are starting to be used to "read back" the drawings to establish the quantities of work that need to be performed. This computerized reading of drawings can significantly reduce estimating time.

Given the large market for CAD programs, there are many vendors. However one vendor, AUTOCAD, has captured a large percentage of the CAD market.

Given the significant estimating hours that the application of CAD can save, the use and popularity of these programs will continue to increase in the construction industry. It behooves the estimator to develop CAD skills.

A second application of computers and software to construction estimating is the performance of math calculations and conversions. The preparation of a construction estimate entails many math calculations and conversions. Calculations such as multiplying depths, widths, and lengths, conversions from volumes to weights, and additions, multiplications, and divisions of many numbers take a considerable amount of time if performed manually.

Many computer software programs to include general application programs and industry programs are available to speed up the performance of necessary mathematical calculations and conversions. While these programs are useful, it should be noted that they are mere extensions of what can be done using calculator programs.

The construction estimator should also be alerted to the fact that some mathematical calculation and conversion type programs speed up the estimating process at the expense of estimating accuracy. For example, a program may include formulas that automatically determine the amount of concrete forms required and reinforcing required as the cubic yards of concrete are input to the program. The formulas in effect assume certain ratios in determining the forming areas and rebar required as a function of the volume of concrete. While these ratios and formulas speed up the estimate, the resulting quantities may not be exact.

One of the higher benefit applications of computers to the construction estimating function is that of "what if" or sensitivity analysis applications. As was discussed in the previous section, "what if" or sensitivity applications center around subjecting a set of numbers or analyses to possible changes in one or more variables and studying the corresponding result. For example, an estimator, using a general application program or an industry program, might prepare an estimate to include identifying work item names, the quantity of work for each work item, a labor rate, productivity for each work item, and material and equipment costs. The software program can in turn be programmed to calculate the total cost of each work item and the total cost for the project.

The above process would yield one set of solution values for the entire project estimate. However changing any one variable that is input to the program, the estimator can evaluate other possible project costs. For example, the estimator might change one or more of the inputted labor productivities or rates. By changing these productivities or rates, the estimator can evaluate an entire range of possible project costs. This should enable the estimator to make better decisions regarding his willingness to bid the project, and enable him to determine a profit strategy.

The application of "what if" analysis or sensitivity analysis offers the construction estimator a very beneficial application of computers. This application by itself can justify the purchase and use of a computer. "What if" or sensitivity analysis applications are sometimes also referred to as financial modeling applications.

Along with "what if" or sensitivity analysis applications, perhaps the other application of software programs that offers the estimator the highest benefit is that of database management programs and applications. Database applications entail the storage of vast pieces of information that can be searched, sorted and printed according to the format defined by the computer user.

There are many applications of database management programs to include construction estimating. One of the best ways to estimate or predict the future is to analyze the past. Unfortunately, without a computer it is difficult if not impossible for the construction estimator to retrieve all the information he or she has seen or analyzed in the past. The end result is that the estimator in effect prepares each estimate without a complete database of information to draw upon. However, given the use of a computer with a database program, the estimator can access vast amounts of information almost instantaneously on the computer screen or via a hard copy printer. In effect, months and years of historical data can become available almost instantaneously. The storage and retrieval of past project information to include labor productivity data should enable the preparation of more accurate estimates.

Database management programs are available from both general application vendors and industry programs vendors. The advantages and disadvantages of each of these types of programs was discussed in the prior section.

A fifth application of computers and computer software to the construction estimating function is that of simulation or gaming. The majority of computers come equipped with a random generator number. Using this feature, a computer vendor can design a software program that enables the user to in effect "play the game" of building a construction project. The program can be programmed to generate uncertain events such as rain, equipment breakdowns, etc. The program user can then respond to a series of questions that require him or her to decide where to allocate resources, whether or not to repair a piece of equipment, etc. In effect, by playing such a game, the computer user can learn how to manage a project.

The same computer construction game can also be used to simulate the project cost. By running many simulations of a proposed project, the resulting cost responses can be studied as a basis of estimating the project cost.

Simulation or gaming applications to the managing of construction and the estimating of construction are relatively new. However, this application holds significant promise for the future.

While we have set out five separate types of applications of computers to the construction estimators, it should be noted that specific software programs may individually address more than a single application. For example, a single program may give the user the ability to perform mathematical calculations, "what if" or sensitivity analysis, and database management applications.

FUTURE APPLICATIONS OF COMPUTERS TO CONSTRUCTION

The construction firm and construction estimator can presently take advantage of many applications of computers to the management of the construction and the preparation of an estimate. The number of vendors and the number of estimating software programs that are available continue to increase.

Progress notwithstanding, the construction industry has been slow in adopting new technology to aid in the management and estimating functions. The fragmented nature of the industry is undoubtedly the reason for the slow implementation of new ideas and new technology.

The fact remains that the management and estimating tasks are such that they can take even more advantage of computers. Some of the applications of computers that are likely to gain in application are illustrated in Figure 18.8. While some of these applications are currently in use by large construction firms, smaller and middle size firms are still oriented toward manual procedures. This will change. The estimator needs to keep well abreast of new applications of software in order to remain competitive.

FUTURE SOFTWARE APPLICATIONS TO THE CONSTRUCTION INDUSTRY

1. Enable "correlation" estimating for the construction firm via the keeping of historical data and the factors unique to the data record that enable the estimator to "custom" design his estimate to the factors that are expected on the new project.

2. Enable the estimator to keep the risk of the expected productivity for a work item via the fitting of probability distributions to past data. This will also enable the estimator to calculate the "risk" of the project estimate as well as the expected cost of the project.

3. The determination of profit as a function of risk. This entails the development of a model that identifies and quantifies the risk elements in a project bid and adjusts the profit to be included in the bid to reflect the summation of the risk.

4. Through the collection of historical data, the determination of overhead formulae for applying overhead to projects as a function of the type of project, the duration of the project, the percentage of labor, material, equipment and subcontracting in a project, etc.

5. Integration of project management information systems to include the accounting function, estimating, scheduling, and cost control function.

6. Simulation games that enable the contractor to simulate a project before its occurrence with the objective of developing an action plan of performance and also the training of project managers, and superintendents.

7. Develop cost and productivity functions for individual work tasks that enable the constructor to determine the productivity and cost as a function of crew size, environmental factors, amount of work to be done, etc, for individual work items.

8. The use of CAD programs for the complete quantity take-off function, and the preparation of shop drawings.

9. The use of artificial intelligence programs that enable the on-site manager to determine "how to manage" a day's work to include how to allocate resources.

10. The use of computerized robotics to perform various office and on-site tasks.

11. The use of computer technology with bar code and video camera technology to automate the accounting function to include the monitoring and costing of on-site workers to cost codes, the preparation of the payroll, the preparation of cost reports, etc.

Figure 18.8   *Future Applications of Software Programs*

Appendix A

Interest Tables

2% COMPOUND INTEREST FACTORS

PERIOD	COMPOUND AMOUNT OF 1 $(1+i)^n$	PRESENT WORTH OF 1 $\dfrac{1}{(1+i)^n}$	COMPOUND AMOUNT OF 1 PER PERIOD $\dfrac{(1+i)^n-1}{i}$	UNIFORM SERIES THAT AMOUNTS TO 1 $\dfrac{i}{(1+i)^n-1}$	PRESENT WORTH OF 1 PER PERIOD $\dfrac{(1+i)^n-1}{i(1+i)^n}$	UNIFORM SERIES THAT 1 WILL BUY $\dfrac{i(1+i)^n}{(1+i)^n-1}$
1	1·0200	·98039	1·000	1·00000	·9803	1·02000
2	1·0403	·96116	2·019	·49505	1·9415	·51505
3	1·0612	·94232	3·060	·32675	2·8838	·34675
4	1·0824	·92384	4·121	·24262	3·8077	·26262
5	1·1040	·90573	5·204	·19215	4·7134	·21215
6	1·1261	·88797	6·308	·15852	5·6014	·17852
7	1·1486	·87056	7·434	·13451	6·4719	·15451
8	1·1716	·85349	8·582	·11650	7·3254	·13650
9	1·1950	·83675	9·754	·10251	8·1622	·12251
10	1·2189	·82034	10·949	·09132	8·9825	·11132
11	1·2433	·80426	12·168	·08217	9·7868	·10217
12	1·2682	·78849	13·412	·07455	10·5753	·09455
13	1·2936	·77303	14·680	·06811	11·3483	·08811
14	1·3194	·75787	15·973	·06260	12·1062	·08260
15	1·3458	·74301	17·293	·05782	12·8492	·07782
16	1·3727	·72844	18·639	·05365	13·5777	·07365
17	1·4002	·71416	20·012	·04996	14·2918	·06996
18	1·4282	·70015	21·412	·04670	14·9920	·06670
19	1·4568	·68643	22·840	·04378	15·6784	·06378
20	1·4859	·67297	24·297	·04115	16·3514	·06115
21	1·5156	·65977	25·783	·03878	17·0112	·05878
22	1·5459	·64683	27·298	·03663	17·6580	·05663
23	1·5768	·63415	28·844	·03460	18·2922	·05466
24	1·6084	·62172	30·421	·03287	18·9139	·05287
25	1·6406	·60953	32·030	·03122	19·5234	·05122
26	1·6734	·59757	33·670	·02969	20·1210	·04969
27	1·7068	·58586	35·344	·02829	20·7068	·04829
28	1·7410	·57437	37·051	·02698	21·2812	·04698
29	1·7758	·56311	38·792	·02577	21·8443	·04577
30	1·8113	·55207	40·568	·02464	22·3964	·04464
35	1·9998	·50002	49·994	·02000	24·9986	·04000
40	2·2080	·45289	60·401	·01655	27·3554	·03655
45	2·4378	·41019	71·892	·01390	29·4901	·03390
50	2·6915	·37152	84·579	·01182	31·4236	·03182
55	2·9717	·33650	98·586	·01014	33·1747	·03014
60	3·2810	·30478	114·051	·00876	34·7608	·02876
65	3·6225	·27605	131·126	·00762	36·1974	·02762
70	3·9995	·25002	149·977	·00666	37·4986	·02666
75	4·4158	·22645	170·791	·00585	38·6771	·02585
80	4·8754	·20510	193·771	·00516	39·7445	·02516
85	5·3828	·18577	219·143	·00456	40·7112	·02456
90	5·9431	·16826	247·156	·00404	41·5869	·02404
95	6·5616	·15239	278·084	·00359	42·3800	·02359
100	7·2446	·13803	312·232	·00320	43·0983	·02320

4% COMPOUND INTEREST FACTORS

PERIOD	COMPOUND AMOUNT OF 1	PRESENT WORTH OF 1	COMPOUND AMOUNT OF 1 PER PERIOD	UNIFORM SERIES THAT AMOUNTS TO 1	PRESENT WORTH OF 1 PER PERIOD	UNIFORM SERIES THAT 1 WILL BUY
	$(1+i)^n$	$\dfrac{1}{(1+i)^n}$	$\dfrac{(1+i)^n-1}{i}$	$\dfrac{i}{(1+i)^n-1}$	$\dfrac{(1+i)^n-1}{i(1+i)^n}$	$\dfrac{i(1+i)^n}{(1+i)^n-1}$
1	1·0400	·96153	1·000	1·00000	·9615	1·04000
2	1·0815	·92455	2·039	·49019	1·8860	·53019
3	1·1248	·88899	3·121	·32034	2·7750	·36034
4	1·1698	·85480	4·246	·23549	3·6298	·27549
5	1·2166	·82192	5·411	·18462	4·4518	·22462
6	1·2653	·79031	6·632	·15076	5·2421	·19076
7	1·3159	·75991	7·898	·12660	6·0020	·16660
8	1·3685	·73069	9·214	·10852	6·7327	·14852
9	1·4233	·70258	10·582	·09449	7·4353	·13449
10	1·4802	·67556	12·006	·08329	8·1108	·12329
11	1·5394	·64958	13·486	·07414	8·7604	·11414
12	1·6010	·62459	15·025	·06655	9·3850	·10655
13	1·6650	·60057	16·626	·06014	9·9856	·10014
14	1·7316	·57747	18·291	·05466	10·5631	·09466
15	1·8009	·55526	20·023	·04994	11·1183	·08994
16	1·8729	·53390	21·824	·04582	11·6522	·08582
17	1·9479	·51337	23·697	·04219	12·1656	·08219
18	2·0258	·49362	25·645	·03899	12·6592	·07899
19	2·1068	·47464	27·671	·03613	13·1339	·07613
20	2·1911	·45638	29·778	·03358	13·5903	·07358
21	2·2787	·43883	31·969	·03128	14·0291	·07128
22	2·3699	·42195	34·247	·02919	14·4511	·06919
23	2·4647	·40572	36·617	·02730	14·8568	·06730
24	2·5633	·39012	39·082	·02558	15·2469	·06558
25	2·6658	·37511	41·645	·02401	15·6220	·06401
26	2·7724	·36068	44·311	·02256	15·9827	·06256
27	2·8833	·34681	47·084	·02123	16·3295	·06123
28	2·9987	·33347	49·967	·02001	16·6630	·06001
29	3·1186	·32065	52·966	·01887	16·9837	·05887
30	3·2433	·30831	56·084	·01783	17·2920	·05783
35	3·9460	·25341	73·652	·01357	18·6646	·05357
40	4·8010	·20828	95·025	·01052	19·7927	·05052
45	5·8411	·17119	121·029	·00826	20·7200	·04826
50	7·1066	·14071	152·667	·00655	21·4821	·04655
55	8·6463	·11565	191·159	·00523	22·1086	·04532
60	10·5196	·09506	237·990	·00420	22·6234	·04420
65	12·7987	·07813	294·968	·00339	23·0466	·04339
70	15·5716	·06421	364·290	·00274	23·3945	·04274
75	18·9452	·05278	448·631	·00222	23·6804	·04222
80	23·0497	·04338	551·244	·00181	23·9153	·04181
85	28·0436	·03565	676·090	·00147	24·1085	·04147
90	34·1193	·02930	827·983	·00120	24·2672	·04120
95	41·5113	·02408	1012·784	·00098	24·3977	·04098
100	50·5049	·01980	1237·623	·00080	24·5049	·04080

6% COMPOUND INTEREST FACTORS

PERIOD	COMPOUND AMOUNT OF 1	PRESENT WORTH OF 1	COMPOUND AMOUNT OF 1 PER PERIOD	UNIFORM SERIES THAT AMOUNTS TO 1	PRESENT WORTH OF 1 PER PERIOD	UNIFORM SERIES THAT 1 WILL BUY
	$(1+i)^n$	$\dfrac{1}{(1+i)^n}$	$\dfrac{(1+i)^n-1}{i}$	$\dfrac{i}{(1+i)^n-1}$	$\dfrac{(1+i)^n-1}{i(1+i)^n}$	$\dfrac{i(1+i)^n}{(1+i)^n-1}$
1	1·0600	·94339	1·000	1·00000	·9433	1·06000
2	1·1235	·88999	2·059	·48543	1·8333	·54543
3	1·1910	·83961	3·183	·31410	2·6730	·37410
4	1·2624	·79209	4·374	·22859	3·4651	·28859
5	1·3382	·74725	5·637	·17739	4·2123	·23739
6	1·4185	·70496	6·975	·14336	4·9173	·20336
7	1·5036	·66505	8·393	·11913	5·5823	·17913
8	1·5938	·62741	9·897	·10103	6·2097	·16103
9	1·6894	·59189	11·491	·08702	6·8016	·14702
10	1·7908	·55839	13·180	·07586	7·3600	·13586
11	1·8982	·52678	14·971	·06679	7·8868	·12679
12	2·0121	·49696	16·869	·05927	8·3838	·11927
13	2·1329	·46883	18·882	·05296	8·8526	·11296
14	2·2609	·44230	21·015	·04748	9·2949	·10758
15	2·3965	·41726	23·275	·04296	9·7122	·10296
16	2·5403	·39364	25·672	·03895	10·1058	·09895
17	2·6927	·37136	28·212	·03544	10·4772	·09544
18	2·8543	·35034	30·905	·03235	10·8276	·09235
19	3·0255	·33051	33·759	·02962	11·1581	·08962
20	3·2071	·31180	36·785	·02718	11·4699	·08718
21	3·3995	·29415	39·992	·02500	11·7640	·08500
22	3·6035	·27750	43·392	·02304	12·0415	·08304
23	3·8197	·26179	46·995	·02127	12·3033	·08127
24	4·0489	·24697	50·815	·01967	12·5503	·07967
25	4·2918	·23299	54·864	·01822	12·7833	·07822
26	4·5493	·21981	59·156	·01690	13·0031	·07690
27	4·8223	·20736	63·705	·01569	13·2105	·07569
28	5·1116	·19563	68·528	·01459	13·4061	·07459
29	5·4183	·18455	73·639	·01357	13·5907	·07357
30	5·7434	·17411	79·058	·01264	13·7648	·07264
35	7·6860	·13010	111·434	·00897	14·4982	·06897
40	10·2857	·09722	154·761	·00646	15·0462	·06646
45	13·7646	·07265	212·743	·00470	15·4558	·06470
50	18·4201	·05428	290·335	·00344	15·7618	·06344
55	24·6503	·04056	394·172	·00253	15·9905	·06253
60	32·9876	·03031	533·128	·00187	16·1614	·06187
65	44·1449	·02265	719·082	·00139	16·2891	·06139
70	59·0759	·01692	967·932	·00103	16·3845	·06103
75	79·0569	·01264	1300·948	·00076	16·4558	·06076
80	105·7959	·00945	1746·599	·00057	16·5091	·06057
85	141·5788	·00706	2342·981	·00042	16·5489	·06042
90	189·4645	·00527	3141·075	·00031	16·5786	·06031
95	253·5462	·00394	4209·103	·00023	16·6009	·06023
100	339·3020	·00294	5638·367	·00017	16·6175	·06017

8% COMPOUND INTEREST FACTORS

PERIOD	COMPOUND AMOUNT OF 1	PRESENT WORTH OF 1	COMPOUND AMOUNT OF 1 PER PERIOD	UNIFORM SERIES THAT AMOUNTS TO 1	PRESENT WORTH OF 1 PER PERIOD	UNIFORM SERIES THAT 1 WILL BUY
	$(1+i)^n$	$\dfrac{1}{(1+i)^n}$	$\dfrac{(1+i)^n-1}{i}$	$\dfrac{i}{(1+i)^n-1}$	$\dfrac{(1+i)^n-1}{i(1+i)^n}$	$\dfrac{i(1+i)^n}{(1+i)^n-1}$
1	1·0800	·92592	1·000	1·00000	·9259	1·08000
2	1·1663	·85733	2·079	·48076	1·7832	·56076
3	1·2597	·79383	3·246	·30803	2·5770	·38803
4	1·3604	·73502	4·506	·22192	3·3121	·30192
5	1·4693	·68058	5·866	·17045	3·9927	·25045
6	1·5868	·63016	7·335	·13631	4·6228	·21631
7	1·7138	·58349	8·922	·11207	5·2063	·19207
8	1·8509	·54026	10·636	·09401	5·7466	·17401
9	1·9990	·50024	12·487	.08007	6·2468	·16007
10	2·1589	·46319	14·486	·06902	6·7100	·14902
11	2·3316	·42888	16·645	·06007	7·1389	·14007
12	2·5181	·39711	18·977	·05269	7·5360	·13269
13	2·7196	·36769	21·495	·04652	7·9037	·12652
14	2·9371	·34046	24·214	·04129	8·2442	·12129
15	3·1721	·31524	27·152	·03682	8·5594	·11682
16	3·4259	·29189	30·324	·03297	8·8513	·11297
17	3·7000	·27026	33·750	·02962	9·1216	·10962
18	3·9960	·25024	37·450	·02670	9·3718	·10670
19	4·3157	·23171	41·446	·02412	9·6035	·10412
20	4·6609	·21454	45·761	·02185	9·8181	·10185
21	5·0338	·19865	50·422	·01983	10·0168	·09983
22	5·4365	·18394	55·456	·01803	10·2007	·09303
23	5·8714	·17031	60·893	·01642	10·3710	·09642
24	6·3411	·15769	66·764	·01497	10·5287	·09497
25	6·8484	·14601	73·105	·01367	10·6747	·09367
26	7·3963	·13520	79·954	·01250	10·8099	·09250
27	7·9880	·12518	87·350	·01144	10·9351	·09144
28	8·6271	·11591	95·338	·01048	11·0510	·09048
29	9·3172	·10732	103·965	·00961	11·1584	·08961
30	10·0626	·09937	113·283	·00882	11·2577	·08882
35	14·7853	·06763	172·316	·00580	11·6545	·08580
40	21·7245	·04603	259·056	·00386	11·9246	·08386
45	31·9204	·03132	386·505	·00258	12·1084	·08258
50	46·9016	·02132	573·770	·00174	12·2334	·08174
55	68·9138	·01451	848·923	·00117	12·3186	·08117
60	101·2570	·00987	1253·213	·00079	12·3765	·08079
65	148·7798	·00672	1847·247	·00054	12·4159	·08054
70	218·6063	·00457	2720·079	·00036	12·4428	·08036
75	321·2045	·00311	4002·556	·00024	12·4610	·08024
80	471·9547	·00211	5886·934	·00016	12·4735	·08016
85	693·4564	·00144	8655·705	·00011	12·4819·	·08011
90	1018·9149	·00098	12723·936	·00007	12·4877	·08007
95	1497·1203	·00066	18701·503	·00005	12·4916	·08005
100	2199·7612	·00045	27484·515	·00003	12·4943	·08003

10% COMPOUND INTEREST FACTORS

PERIOD	COMPOUND AMOUNT OF 1	PRESENT WORTH OF 1	COMPOUND AMOUNT OF 1 PER PERIOD	UNIFORM SERIES THAT AMOUNTS TO 1	PRESENT WORTH OF 1 PER PERIOD	UNIFORM SERIES THAT 1 WILL BUY
	$(1+i)^n$	$\dfrac{1}{(1+i)^n}$	$\dfrac{(1+i)^n-1}{i}$	$\dfrac{i}{(1+i)^n-1}$	$\dfrac{(1+i)^n-1}{i(1+i)^n}$	$\dfrac{i(1+i)^n}{(1+i)^n-1}$
1	1·1000	·90909	1·000	1·00000	·9090	1·10000
2	1·2099	·82644	2·099	·47619	1·7355	·57619
3	1·3309	·75131	3·309	·30211	2·4868	·40211
4	1·4640	·68301	4·640	·21547	3·1698	·31547
5	1·6105	·62092	6·105	·16379	3·7907	·26379
6	1·7715	·56447	7·715	·12960	4·3552	·22960
7	1·9487	·51315	9·487	·10540	4·8684	·20540
8	2·1435	·46650	11·435	·08744	5·3349	·18744
9	2·3579	·42409	13·579	·07364	5·7590	·17364
10	2·5937	·38554	15·937	·06274	6·1445	·16274
11	2·8531	·35049	18·531	·05396	6·4950	·15396
12	3·1384	·31863	21·384	·04676	6·8136	·14676
13	3·4522	·28966	24·522	·04077	7·1033	·14077
14	3·7974	·26333	27·974	·03574	7·3666	·13574
15	4·1772	·23939	31·772	·03147	7·6060	·13147
16	4·5949	·21762	35·949	·02781	7·8237	·12781
17	5·0544	·19784	40.544	·02466	8·0215	·12466
18	5·5599	·17985	45·599	·02193	8·2014	·12193
19	6·1159	·16350	51·159	·01954	8·3649	·11954
20	6·7274	·14864	57·274	·01745	8·5135	·11745
21	7·4002	·13513	64·002	·01562	8·6486	·11562
22	8·1402	·12284	71·402	·01400	8·7715	·11400
23	8·9543	·11167	79·543	·01257	8·8832	·11257
24	9·8497	·10152	88·497	·01129	8·9847	·11129
25	10·8347	·09229	98·347	·01016	9·0770	·11016
26	11·9181	·08390	109·181	·00915	9·1609	·10915
27	13·1099	·07627	121·099	·00825	9·2372	·10825
28	14·4209	·06934	134·209	·00745	9·3065	·10745
29	15·8630	·06303	148·630	·00672	9·3696	·10672
30	17·4494	·05730	164·494	·00607	9·4269	·10607
35	28·1024	·03558	271·024	·00368	9·6441	·10368
40	45·2592	·02209	442·592	·00225	9·7790	·10225
45	72·8904	·01371	718·904	·00139	9·8628	·10139
50	117·3908	·00851	1163·908	·00085	9·9148	·10085
55	189·0591	·00528	1880·591	·00053	9·9471	·10053
60	304·4816	·00328	3034·816	·00032	9·9671	·10032
65	490·3706	·00203	4893·706	·00020	9·9796	·10020
70	789·7468	·00126	7887·468	·00012	9·9873	·10012
75	1271·8952	·00078	12708·952	·00007	9·9921	·10007
80	2048·4000	·00048	20474·000	·00004	9·9951	·10004
85	3298·9687	·00030	32979·687	·00003	9·9969	·10003
90	5313·0221	·00018	53120·221	·00001	9·9981	·10001
95	8556·6753	·00011	85556·753	·00001	9·9988	·10001
100	13780·6110	·00007	137796·110	·00000	9·9992	·10000

12% COMPOUND INTEREST FACTORS

PERIOD	COMPOUND AMOUNT OF 1	PRESENT WORTH OF 1	COMPOUND AMOUNT OF 1 PER PERIOD	UNIFORM SERIES THAT AMOUNTS TO 1	PRESENT WORTH OF 1 PER PERIOD	UNIFORM SERIES THAT 1 WILL BUY
	$(1+i)^n$	$\dfrac{1}{(1+i)^n}$	$\dfrac{(1+i)^n-1}{i}$	$\dfrac{i}{(1+i)^n-1}$	$\dfrac{(1+i)^n-1}{i(1+i)^n}$	$\dfrac{i(1+i)^n}{(1+i)^n-1}$
1	1·1200	·89285	1·000	1·00000	·8928	1·12000
2	1·2543	·79719	2·119	·47169	1·6900	·59169
3	1·4049	·71178	3·374	·29634	2·4018	·41634
4	1·5735	·63551	4·779	·20923	3·0373	·32923
5	1·7623	·56742	6·352	·15740	3·6047	·27740
6	1·9738	·50663	8·115	·12322	4·1114	·24322
7	2·2106	·45234	10·089	·09911	4·5637	·21911
8	2·4759	·40388	12·299	·08130	4·9676	·20130
9	2·7730	·36061	14·775	·06767	5·3282	·18767
10	3·1058	·32197	17·548	·05698	5·6502	·17698
11	3·4785	·28747	20·654	·04841	5·9376	·16841
12	3·8959	·25667	24·133	·04143	6·1943	·16143
13	4·3634	·22917	28·029	·03567	6·4235	·15567
14	4·8871	·20461	32·392	·03087	6·6281	·15087
15	5·4735	·18269	37·279	·02682	6·8108	·14682
16	6·1303	·16312	42·753	·02339	6·9739	·14339
17	6·8660	·14564	48·883	·02045	7·1196	·14045
18	7·6899	·13003	55·749	·01793	7·2496	·13793
19	8·6127	·11610	63·439	·01576	7·3657	·13576
20	9·6462	·10366	72·052	·01387	7·4694	·13387
21	10·8038	·09255	81·698	·01224	7·5620	·13224
22	12·1003	·08264	92·502	·01081	7·6446	·13081
23	13·5523	·07378	104·602	·00955	7·7184	·12955
24	15·1786	·06588	118·155	·00846	7·7843	·12846
25	17·0000	·05882	133·333	·00749	7·8431	·12749
26	19·0400	·05252	150·333	·00665	7·8956	·12665
27	21·3248	·04689	169·373	·00590	7·9425	·12590
28	23·8838	·04186	190·698	·00524	7·9844	·12524
29	26·7499	·03738	214·582	·00466	8·0218	·12466
30	29·9599	·03337	241·332	·00414	8·0551	·12414
35	52·7996	·01893	431·663	·00231	8·1755	·12231
40	93·0509	·01074	767·091	·00130	8·2437	·12130
45	163·9875	·00609	1358·229	·00073	8·2825	·12073
50	289·0021	·00346	2400·017	·00041	8·3044	·12041
55	509·3204	·00196	4236·003	·00023	8·3169	·12023
60	897·5966	·00111	7471·638	·00013	8·3240	·12013
65	1581·8719	·00063	13173·932	·00007	8·3280	·12007
70	2787·7987	·00035	23223·322	·00004	8·3303	·12004
75	4913·0538	·00020	40933·781	·00002	8·3316	·12002
80	8658·4794	·00011	72145·661	·00001	8·3323	·12001
85	15259·1980	·00006	127151·650	·00000	8·3327	·12000
90	26891·9150	·00003	224090·950	·00000	8·3330	·12000
95	47392·7240	·00002	394931·030	·00000	8·3331	·12000
100	83522·2210	·00001	696010·170	·00000	8·3332	·12000

14% COMPOUND INTEREST FACTORS

PERIOD	COMPOUND AMOUNT OF 1	PRESENT WORTH OF 1	COMPOUND AMOUNT OF 1 PER PERIOD	UNIFORM SERIES THAT AMOUNTS TO 1	PRESENT WORTH OF 1 PER PERIOD	UNIFORM SERIES THAT 1 WILL BUY
	$(1+i)^n$	$\dfrac{1}{(1+i)^n}$	$\dfrac{(1+i)^n-1}{i}$	$\dfrac{i}{(1+i)^n-1}$	$\dfrac{(1+i)^n-1}{i(1+i)^n}$	$\dfrac{i(1+i)^n}{(1+i)^n-1}$
1	1·1400	·87719	1·000	1·00000	·8771	1·14000
2	1·2995	·76946	2·139	·46728	1·6466	·60728
3	1·4815	·67497	3·439	·29073	2·3216	·43073
4	1·6889	·59208	4·921	·20320	2·9137	·34320
5	1·9254	·51936	6·610	·15128	3·4330	·29128
6	2·1949	·45558	8·535	·11715	3·8886	·25715
7	2·5022	·39963	10·730	·09319	4·2883	·23319
8	2·8525	·35055	13·232	·07557	4·6388	·21557
9	3·2519	·30750	16·085	·06216	4·9463	·20216
10	3·7072	·26974	19·337	·05171	5·2161	·19171
11	4·2262	·23661	23·044	·04339	5·4527	·18339
12	4·8179	·20755	27·270	·03666	5·6602	·17666
13	5·4924	·18206	32·088	·03116	5·8423	·17116
14	6·2613	·15971	37·581	·02660	6·0020	·16660
15	7·1379	·14009	43·842	·02280	6·1421	·16280
16	8·1372	·12289	50·980	·01961	6·2650	·15961
17	9·2764	·10779	59·117	·01691	6·3728	·15691
18	10·5751	·09456	68·394	·01462	6·4674	·15462
19	12·0556	·08294	78·969	·01266	6·5503	·15266
20	13·7434	·07276	91·024	·01098	6·6231	·15098
21	15·6675	·06382	104·768	·00954	6·6869	·14954
22	17·8610	·05598	120·435	·00830	6·7429	·14830
23	20·3615	·04911	138·297	·00723	6·7920	·14723
24	23·2122	·04308	158·658	·00630	6·8351	·14630
25	26·4619	·03779	181·870	·00549	6·8729	·14549
26	30·1665	·03314	208·332	·00480	6·9060	·14480
27	34·3899	·02907	238·499	·00419	6·9351	·14419
28	39·2044	·02550	272·889	·00366	6·9606	·14366
29	44·6931	·02237	312·093	·00320	6·9830	·14320
30	50·9501	·01962	356·786	·00280	7·0026	·14280
35	98·1001	·01019	693·572	·00144	7·0700	·14144
40	188·8834	·00529	1342·024	·00074	7·1050	·14074
45	363·6790	·00274	2590·564	·00038	7·1232	·14038
50	700·2329	·00142	4994·520	·00020	7·1326	·14020

16% COMPOUND INTEREST FACTORS

PERIOD	COMPOUND AMOUNT OF 1 $(1+i)^n$	PRESENT WORTH OF 1 $\dfrac{1}{(1+i)^n}$	COMPOUND AMOUNT OF 1 PER PERIOD $\dfrac{(1+i)^n-1}{i}$	UNIFORM SERIES THAT AMOUNTS TO 1 $\dfrac{i}{(1+i)^n-1}$	PRESENT WORTH OF 1 PER PERIOD $\dfrac{(1+i)^n-1}{i(1+i)^n}$	UNIFORM SERIES THAT 1 WILL BUY $\dfrac{i(1+i)^n}{(1+i)^n-1}$
1	1·1600	·86206	1·000	1·00000	·8620	1·16000
2	1·3455	·74316	2·159	·46296	1·6052	·62296
3	1·5608	·64065	3·505	·28525	2·2458	·44525
4	1·8106	·55229	5·066	·19737	2·7981	·35737
5	2·1003	·47611	6·877	·14540	3·2742	·30540
6	2·4363	·41044	8·977	·11138	3·6847	·27138
7	2·8262	·35382	11·413	·08761	4·0385	·24761
8	3·2784	·30502	14·240	·07022	4·3435	·23022
9	3·8029	·26295	17·518	·05708	4·6065	·21708
10	4·4114	·22668	21·321	·04690	4·8332	·20690
11	5·1172	·19541	25·732	·03886	5·0286	·19886
12	5·9360	·16846	30·850	·03241	5·1971	·19241
13	6·8857	·14522	36·786	·02718	5·3423	·18718
14	7·9875	·12519	43·671	·02289	5·4675	·18289
15	9·2655	·10792	51·659	·01935	5·5754	·17935
16	10·7480	·09304	60·925	·01641	5·6684	·17641
17	12·4676	·08020	71·673	·01395	5·7487	·17395
18	14·4625	·06914	84·140	·01188	5·8178	·17188
19	16·7765	·05960	98·603	·01014	5·8774	·17014
20	19·4607	·05138	115·379	·00866	5·9288	·16866
21	22·5744	·04429	134·840	·00741	5·9731	·16741
22	26·1863	·03818	157·414	·00635	6·0113	·16635
23	30·3762	·03292	183·601	·00544	6·0442	·16544
24	35·2364	·02837	213·977	·00467	6·0726	·16467
25	40·8742	·02446	249·213	·00401	6·0970	·16401
26	47·4141	·02109	290·088	·00344	6·1181	·16344
27	55·0003	·01818	337·502	·00296	6·1363	·16296
28	63·8004	·01567	392·502	·00254	6·1520	·16254
29	74·0085	·01351	456·303	·00219	6·1655	·16219
30	85·8498	·01164	530·311	·00188	6·1771	·16188
35	180·3140	·00554	1120·712	·00089	6·2153	·16089
40	378·7210	·00264	2360·756	·00042	6·2334	·16042
45	795·4436	·00125	4965·272	·00020	6·2421	·16020
50	1670·7033	·00059	10435·645	·00009	6·2462	·16009

18% COMPOUND INTEREST FACTORS

PERIOD	COMPOUND AMOUNT OF 1	PRESENT WORTH OF 1	COMPOUND AMOUNT OF 1 PER PERIOD	UNIFORM SERIES THAT AMOUNTS TO 1	PRESENT WORTH OF 1 PER PERIOD	UNIFORM SERIES THAT 1 WILL BUY
	$(1+i)^n$	$\dfrac{1}{(1+i)^n}$	$\dfrac{(1+i)^n-1}{i}$	$\dfrac{i}{(1+i)^n-1}$	$\dfrac{(1+i)^n-1}{i(1+i)^n}$	$\dfrac{i(1+i)^n}{(1+i)^n-1}$
1	1·1800	·84745	1·000	1·00000	·8474	1·18000
2	1·3923	·71818	2·179	·45871	1·5656	·63871
3	1·6430	·60863	3·572	·27992	2·1742	·45992
4	1·9387	·51578	5·215	·19173	2·6900	·37173
5	2·2877	·43710	7·154	·13977	3·1271	·31977
6	2·6995	·37043	9·441	·10591	3·4976	·28591
7	3·1854	·31392	12·141	·08236	3·8115	·26236
8	3·7588	·26603	15·326	·06524	4·0775	·24524
9	4·4354	·22545	19·085	·05239	4·3030	·23239
10	5·2338	·19106	23·521	·04251	4·4940	·22251
11	6·1759	·16191	28·755	·03477	4·6560	·21477
12	7·2875	·13721	34·931	·02862	4·7932	·20862
13	8·5993	·11628	42·218	·02368	4·9095	·20368
14	10·1472	·09854	50·818	·01967	5·0080	·19967
15	11·9737	·08351	60·965	·01640	5·0915	·19640
16	14·1290	·07077	72·938	·01371	5·1623	·19371
17	16·6722	·05997	87·068	·01148	5·2223	·19148
18	19·6732	·05083	103·740	·00963	5·2731	·18963
19	23·2144	·04307	123·413	·00810	5·3162	·18810
20	27·3930	·03650	146·627	·00681	5·3527	·18681
21	32·3237	·03093	174·020	·00574	5·3836	·18574
22	38·1420	·02621	206·344	·00484	5·4099	·18484
23	45·0076	·02221	244·486	·00409	5·4321	·18409
24	53·1089	·01882	289·494	·00345	5·4509	·18345
25	62·6686	·01595	342·603	·00291	5·4669	·18291
26	73·9489	·01352	405·271	·00246	5·4804	·18246
27	87·2597	·01146	479·220	·00208	5·4918	·18208
28	102·9665	·00971	566·480	·00176	5·5016	·18176
29	121·5005	·00823	669·447	·00149	5·5098	·18149
30	143·3706	·00697	790·947	·00126	5·5168	·18126
35	327·9971	·00304	1816·650	·00055	5·5386	·18055
40	750·3780	·00133	4163·211	·00024	5·5481	·18024
45	1716·6831	·00058	9531·572	·00010	5·5523	·18010
50	3927·3551	·00025	21813·083	·00004	5·5541	·18004

20% COMPOUND INTEREST FACTORS

PERIOD	COMPOUND AMOUNT OF 1	PRESENT WORTH OF 1	COMPOUND AMOUNT OF 1 PER PERIOD	UNIFORM SERIES THAT AMOUNTS TO 1	PRESENT WORTH OF 1 PER PERIOD	UNIFORM SERIES THAT 1 WILL BUY
	$(1+i)^n$	$\dfrac{1}{(1+i)^n}$	$\dfrac{(1+i)^n-1}{i}$	$\dfrac{i}{(1+i)^n-1}$	$\dfrac{(1+i)^n-1}{i(1+i)^n}$	$\dfrac{i(1+i)^n}{(1+i)^n-1}$
1	1·2000	·83333	1·000	1·00000	·8333	1·20000
2	1·4399	·69444	2·199	·45454	1·5277	·65454
3	1·7279	·57870	3·639	·27472	2·1064	·47472
4	2·0735	·48225	5·367	·18628	2·5887	·38628
5	2·4883	·40187	7·441	·13437	2·9906	·33437
6	2·9859	·33489	9·929	·10070	3·3255	·30070
7	3·5831	·27908	12·915	·07742	3·6045	·27742
8	4·2998	·23256	16·499	·06060	3·8371	·26060
9	5·1597	·19380	20·798	·04807	4·0309	·24807
10	6·1917	·16150	25·958	·03852	4·1924	·23852
11	7·4300	·13458	32·150	·03110	4·3270	·23110
12	8·9160	·11215	39·580	·02526	4·4392	·22526
13	10·6993	·09346	48·496	·02062	4·5326	·22062
14	12·8391	·07788	59·195	·01689	4·6105	·21689
15	15·4070	·06490	72·035	·01388	4·6754	·21388
16	18·4884	·05408	87·442	·01143	4·7295	·21143
17	22·1861	·04507	105·930	·00944	4·7746	·20944
18	26·6233	·03756	128·116	·00780	4·8121	·20780
19	31·9479	·03130	154·739	·00646	4·8434	·20646
20	38·3375	·02608	186·687	·00535	4·8695	·20535
21	46·0051	·02173	225·025	·00444	4·8913	·20444
22	55·2061	·01811	271·030	·00368	4·9094	·20368
23	66·2473	·01509	326·236	·00306	4·9245	·20306
24	79·4968	·01257	392·484	·00254	4·9371	·20254
25	95·3961	·01048	471·980	·00211	4·9475	·20211
26	114·4754	·00873	567·377	·00176	4·9563	·20176
27	137·3705	·00727	681·852	·00146	4·9636	·20146
28	164·8446	·00606	819·223	·00122	4·9696	·20122
29	197·8135	·00505	984·067	·00101	4·9747	·20101
30	237·3762	·00421	1181·881	·00084	4·9789	·20084
35	590·6680	·00169	2948·340	·00033	4·9915	·20033
40	1469·7711	·00068	7343·855	·00013	4·9965	·20013
45	3657·2606	·00027	18281·303	·00005	4·9986	·20005
50	9100·4350	·00010	45497·175	·00002	4·9994	·20002

Appendix B

GENERAL REQUIREMENTS

01020. Allowances
01100. Alternatives
01200. Project Meetings
01300. Submittals
01400. Quality Control
01500. Temporary Facilities and Controls
01600. Material and Equipment
01701. Project Closeout

SITE WORK

02000. Alternatives
02010. Subsurface Exploration
02100. Clearing
02110. Demolition
02200. Earthwork
02250. Soil Treatment
02300. Pile Foundations
02350. Caissons
02400. Shoring
02500. Site Drainage
02550. Site Utilities
02600. Paving & Surfacing
02700. Site Improvements
02800. Landscaping
02850. Railroad Work
02900. Marine Work
02950. Tunneling

CONCRETE

03000. Alternatives
03100. Concrete Formwork
03150. Expansion & Contraction Joints
03200. Concrete Reinforcement
03300. Cast-in-Place Concrete
03350. Specially Finished Concrete
03360. Specially Placed Concrete
03400. Precast Concrete
03500. Cementitious Decks

MASONRY

04000. Alternatives
04100. Mortar
04150. Masonry Accessories
04200. Unit Masonry
04400. Stone
04500. Masonry Restoration & Cleaning
04550. Refractories

METALS

05000. Alternatives
05100. Structural Metal Framing
05200. Metal Joists
05300. Metal Decking
05400. Lightgage Metal Framing
05500. Metal Fabrications
05700. Ornamental Metal

WOOD AND PLASTICS

06000.   Alternatives
06100.   Rough Carpentry
06130.   Heavy Timber Construction
06150.   Trestles
06170.   Prefabricated Structural Wood
06200.   Finish Carpentry
06300.   Wood Treatment
06400.   Architectural Woodwork
06500.   Prefabricated Structural Plastics
06600.   Plastic Fabrications

THERMAL & MOISTURE PROTECTION

07000.   Alternatives
07100.   Waterproofing
07150.   Dampproofing
07200.   Insulation
07300.   Shingles & Roofing Tiles
07400.   Preformed Roofing & Siding
07500.   Membrane Roofing
07570.   Traffic Topping
07600.   Flashing & Sheet Metal
07800.   Roof Accessories
07900.   Sealants

DOORS & WINDOWS

08000.   Alternatives
08100.   Metal Doors & Frames
08200.   Wood & Plastic Doors
08300.   Special Doors
08400.   Entrances & Storefronts
08500.   Metal Windows
08600.   Wood & Plastic Windows
08550.   Special Windows
08700.   Hardware & Specialities

08800.   Glazing
08900.   Window Walls/Curtain Walls

FINISHES

09000.   Alternatives
09100.   Lath & Plaster
09250.   Gypsum Wallboard
09400.   Terrazzo
09500.   Acoustical Treatment
09540.   Suspension Systems
09550.   Wood Flooring
09650.   Resilient Flooring
09680.   Carpeting
09700.   Special Flooring
09760.   Floor Treatment
09800.   Special Coatings
09900.   Painting
09950.   Wall Covering

SPECIALTIES

10000.   Alternatives
10100.   Chalkboards and Tackboards
10150.   Compartments and Cubicles
10200.   Louvers and Vents
10240.   Grilles and Screens
10260.   Wall and Corner Guards
10270.   Access Flooring
10280.   Specialty Modules
10290.   Pest Control
10300.   Fireplaces
10350.   Flagpoles
10400.   Identifying Devices
10450.   Pedestrian Control Devices
10500.   Lockers
10530.   Protective Covers
10550.   Postal Specialties
10600.   Partitions

10650. Scales
10670. Storage Shelving
10700. Sun Control Devices (Exterior)
10750. Telephone Enclosures
10800. Toilet & Bath Accessories
10900. Wardrobe Specialties

EQUIPMENT

11000. Alternatives
11050. Built-In Maintenance Equipment
11100. Bank & Vault Equipment
11150. Commercial Equipment
11170. Checkroom Equipment
11180. Darkroom Equipment
11200. Ecclesiastical Equipment
11300. Educational Equipment
11400. Food Service Equipment
11480. Vending Equipment
11500. Athletic Equipment
11550. Industrial Equipment
11600. Laboratory Equipment
11630. Laundry Equipment
11650. Library Equipment
11700. Medical Equipment
11800. Mortuary Equipment
11830. Musical Equipment
11850. Parking Equipment
11860. Waste Handling Equipment
11870. Loading Dock Equipment
11880. Detention Equipment
11900. Residential Equipment
11970. Theater & Stage Equipment
11990. Registration Equipment

FURNISHINGS

12000. Alternatives
12100. Artwork

12300. Cabinets and Storage
12500. Window Treatment
12550. Fabrics
12600. Furniture
12670. Rugs & Mats
12700. Seating
12800. Furnishing Accessories

SPECIAL CONSTRUCTION

13000. Alternatives
13010. Air-Supported Structures
13050. Integrated Assemblies
13100. Audiometric Room
13250. Clean Room
13350. Hyperbaric Room
13400. Incinerators
13440. Instrumentation
13450. Insulated Room
13500. Integrated Ceilings
13540. Nuclear Reactors
13550. Observatory
13600. Prefabricated Buildings
13700. Special Purpose Rooms & Buildings
13750. Radiation Protection
13770. Sound & Vibration Control
13800. Vaults
13850. Swimming Pools

CONVEYING SYSTEMS

14000. Alternatives
14100. Dumbwaiters
14200. Elevators
14300. Hoists & Cranes
14400. Lifts
14500. Material Handling Systems
14550. Conveyors & Chutes

14570.	Turntables
14600.	Moving Stairs & Walks
14700.	Pneumatic Tube Systems
14800.	Powered Scaffolding

15700.	Liquid Heat Transfer
15800.	Air Distribution
15900.	Controls & Instrumentation

MECHANICAL

15000.	Alternatives
15010.	General Provisions
15050.	Basic Materials and Methods
15180.	Insulation
15200.	Water Supply & Treatment
15300.	Waste Water Disposal & Treatment
15400.	Plumbing
15500.	Fire Protection
15600.	Power or Heat Generation
15650.	Refrigation

ELECTRICAL

16000.	Alternatives
16010.	General Provisions
16100.	Basic Materials and Methods
16200.	Power Generation
16300.	Power Transmission
16400.	Service & Distribution
16500.	Lighting
16600.	Special Systems
16700.	Communications
16850.	Heating & Cooling
16900.	Controls & Instrumentation

Index